Studies in the Structure, Physiology and Ecology of Molluscs

FRONTISPIECE: Above, *Leptopoma*; below, *Papuina*. (Photographs by courtesy of John F. Peake).

SYMPOSIA OF THE ZOOLOGICAL SOCIETY OF LONDON
AND
THE MALACOLOGICAL SOCIETY OF LONDON
NUMBER 22

Studies in the Structure, Physiology and Ecology of Molluscs

*(The Proceedings of a Symposium held at The Zoological
Society of London on 8 and 9 March, 1967)*

Edited by

VERA FRETTER

*Department of Zoology,
The University, Reading, England*

Published for

THE ZOOLOGICAL SOCIETY OF LONDON

BY

ACADEMIC PRESS

1968

ACADEMIC PRESS INC. (LONDON) LTD

Berkeley Square House
Berkeley Square
London, W.1

U.S. Edition published by

ACADEMIC PRESS INC.

111 Fifth Avenue

New York, New York 10003

Library of Congress Catalog Card Number: 68–17680

PRINTED IN GREAT BRITAIN BY
J. W. ARROWSMITH LTD., BRISTOL

CONTRIBUTORS

ABOLINŠ-KROGIS, ANNA, *Institute of Zoophysiology, University of Uppsala, Sweden* (p. 75)

BOER, H. H., *Department of Zoology, Free University, Amsterdam, The Netherlands* (p. 237)

BOER, MARIA H., *Department of Zoology, Free University, Amsterdam, The Netherlands* (p. 213)

CORNELISSE, C. J., *Department of Zoology, Free University, Amsterdam, The Netherlands* (p. 213)

DE VLIEGER, T. A., *Department of Zoology, Free University, Amsterdam, The Netherlands* (p. 257)

DOUMA, ELISABETH, *Zoological Department, Free University, Amsterdam, The Netherlands* (p. 237)

DIGBY, PETER S. B., *Department of Zoology, McGill University, Montreal, Canada* (p. 93)

EVANS, J. G., *Institute of Archaeology, London, England* (p. 293)

HEDGES, ANNEMARIE, *Department of Physiology and Biochemistry, The University, Southampton, England* (p. 33)

HURST, ANNE, *Zoology Department, University of Reading, England* (p. 151)

JOOSSE, J., *Department of Zoology, Free University, Amsterdam, The Netherlands* (p. 213)

KAY, ALISON, *General Science Department, University of Hawaii, Hawaii, U.S.A.* (p. 109)

KERKUT, G. S., *Department of Physiology and Biochemistry, Southampton University, England* (p. 1), (p. 19)

KERNEY, M. P., *Department of Geology, Imperial College, London, England* (p. 273)

KOKSMA, JENNEKE M. A., *Zoological Department, Free University, Amsterdam, The Netherlands* (p. 237)

LEVER, J., *Department of Zoology, Free University, Amsterdam, The Netherlands* (p. 259)

NEWMAN, G., *Department of Physiology and Biochemistry, Southampton University, England* (p. 1)

NISBET, R. H., *Department of Physiology, Royal Veterinary College, University of London, England* (p. 193)

PEAKE, JOHN F., *British Museum, London, England* (p. 319)

PLUMMER, JENIFER M., *Department of Physiology, Royal Veterinary College, University of London, England* (p. 193)

POTTS, W. T. W., *Department of Biology, University of Lancaster, England* (p. 187)

SEDDEN, CHRISTINE B., *West Lodge, Edmonstone, Gilmerton, Edinburgh, Scotland* (p. 19)

SWEDMARK, BERTIL, *Kristineberg Zoological Station, Fiskebäckskil, Sweden* (p. 135)

THIJSSEN, R., *Department of Zoology, Free University, Amsterdam, The Netherlands* (p. 259)

TRUEMAN, E. R., *Zoology Department, The University, Hull, England,* (p. 167)

WALKER, R. J., *Department of Physiology and Biochemistry, Southampton University, England* (p. 1), (p. 19), (p. 33)

WOODRUFF, G. N., *Department of Physiology and Biochemistry, Southampton University, England* (p. 33)

ORGANIZER AND CHAIRMEN

ORGANIZER
VERA FRETTER, *on behalf of The Zoological Society of London*

CHAIRMEN OF SESSIONS
A. J. CAIN, *Department of Zoology, University of Manchester, England*

VERA FRETTER, *Department of Zoology, The University, Reading, England*

G. M. HUGHES, *Department of Zoology, University of Bristol, England*

J. LEVER, *Department of Zoology, Free University, Amsterdam, The Netherlands*

R. D. PURCHON, *Chelsea College of Science and Technology, University of London, England*

FOREWORD

Man has been attracted to molluscs since earliest times. They lived around his home and provided easily accessible food. Their shells were heaped on his kitchen midden to await the curiosity of the archaeologist and palaeontologist who used these remains to reconstruct facts about predator and prey.

The shell, secreted and moulded by the soft parts to their particular requirements, leaves in the fossil record a manifestation of the success of molluscan organization. This organization has proved its adaptability in a vast number of ecological niches, giving us over 80,000 different species, with some in more obscure habitats which have remained unknown until very recent years. Recently discovered species have revolutionized our ideas of molluscan evolution: the hope of finding others is a challenge to the naturalist. Variety in structure is a hall-mark of the phylum, and provides the anatomist with virtually inexhaustible material. Abundance in numbers, easy availability and endurance of laboratory conditions are attractions to the physiologist who regularly employs land, freshwater and marine snails as objects for the most advanced scientific techniques.

This ubiquitous appeal of molluscs to the scientist is illustrated in the list of chairmen and speakers who took part in this symposium jointly organized by the Zoological and Malacological Societies of London. It brought together people of widely divergent interests in the phylum, although only gastropods and to a lesser extent lamellibranchs were under consideration. The occasion permitted us to discuss malacology with scientists from other countries and we are grateful to those who came from Europe, the United States and elsewhere to participate so freely in formal and informal talks.

Some will read the papers included in this volume to learn of advances in fine structure and physiology, others to appreciate the functional anatomy and ecology of unfamiliar species, or to concentrate on facts which can be deduced from the molluscan shell, living, fossilized or artificial. Whatever topic is selected the reader will see that the mollusc maintains its position as a challenge to human interest. We look forward to another occasion when we may survey the results of this challenge.

May 1968   VERA FRETTER

CONTENTS

The Structure of the Brain of *Helix aspersa*. Electron
Microscope Localization of Cholinesterase and Amines

G. NEWMAN, G. A. KERKUT, and R. J. WALKER

The Localization of Dopamine and 5-hydroxytryptamine in
Neurones of *Helix aspersa*

C. B. SEDDEN, R. J. WALKER and G. A. KERKUT

The Pharmacology of the Neurones of *Helix aspersa*

R. J. WALKER, A. HEDGES, and G. N. WOODRUFF

Shell Regeneration in *Helix pomatia* with Special Reference to the Elementary Calcifying Particles

ANNA ABOLIŅŠ-KROGIS

The Mechanism of Calcification in the Molluscan Shell

PETER S. B. DIGBY

A Review of the Bivalved Gastropods and a Discussion of Evolution within the Sacoglossa

E. ALISON KAY

The Biology of Interstitial Mollusca

BERTIL SWEDMARK

The Feeding Mechanism and Behaviour of the Opisthobranch
Melibe leonina
ANNE HURST

The Burrowing Activities of Bivalves
E. R. TRUEMAN

Aspects of Excretion in the Molluscs
W. T. W. POTTS

The Fine Structure of Cardiac and Other Molluscan Muscle
R. H. NISBET and JENIFER M. PLUMMER

Gametogenesis and Oviposition in *Lymnaea stagnalis* as
Influenced by γ-Irradiation and Hunger

J. JOOSSE, MARIA H. BOER and C. J. CORNELISSE

Electron Microscopy Study of Neurosecretory Cell and
Neurohaemal Organs in the Pond Snail *Lymnaea stagnalis*

H. H. BOER, ELISABETH DOUMA and JENNEKE M. A. KOKSMA

Spontaneous Activity and Tactile Pathways in the Central
Nervous System of *Lymnaea stagnalis*

T. A. DE VLIEGER

Sorting Phenomena During the Transport of Shell Valves on
Sandy Beaches; Studies with the Use of Artificial Valves

J. LEVER and R. THIJSSEN

Britain's Fauna of Land Mollusca and its Relation to the
Post-Glacial Thermal Optimum

M. P. KERNEY

Changes in the Composition of Land Molluscan Populations in
North Wiltshire During the Last 5000 Years

J. G. EVANS

Habitat Distribution of Solomon Island Land Mollusca
JOHN F. PEAKE

Symp. zool. Soc. Lond. (1968) No. 22, 1–17.

THE STRUCTURE OF THE BRAIN OF *HELIX ASPERSA* ELECTRON MICROSCOPE LOCALIZATION OF CHOLINESTERASE AND AMINES

G. NEWMAN, G. A. KERKUT and R. J. WALKER

Department of Physiology and Biochemistry,
Southampton University, Southampton, England

SYNOPSIS

The general structure of the brain, nerve trunk and nerve muscle junction of *Helix aspersa* is described from the viewpoint of both light microscopy and electron microscopy (e.m.). The location of the large neurones in the ganglia is given.

Esterases are found at only a few synaptic junctions in the neuropil. The presynaptic vesicles at these locations contain electron-dense vesicles of 800–1000 Å or clear vesicles of 250–300 Å.

The chromaffin reaction allows location of amines in e.m. sections. It appears that there is not a good correlation between the size of a vesicle and the nature of the chemical that it probably contains.

INTRODUCTION

The large nerve cells present in the central nervous system of gastropods have been the object of study by cytologists (Thomas, 1947; Chou, 1957a,b; Lane, 1964; Sumner, 1965; Fernandez, 1966) and by electron microscopists (Schlote, 1957; Gerschenfeld, 1963; Rosenbluth, 1963; Baxter and Nisbet, 1963; Amoroso, Baxter, Chiquoine and Nisbet, 1964; Simpson, Bern and Nishioka, 1966). The present study on the nervous system of the common British snail *Helix aspersa* was undertaken to provide a morphological framework for the neurophysiological studies that are being carried out in the Department of Physiology and Biochemistry, Southampton University (Kerkut and Walker, 1961, 1962; Kerkut and Thomas, 1964, 1965; Kerkut and Meech, 1966; Kerkut and Gardner, 1967; Kerkut, 1967), and also to see if some information can be obtained as to the nature of the chemical transmission systems within the central nervous system. In particular, we have applied histochemical methods with reference to the localization of choline esterase and to amines to see what light this throws on the nature of the granules and vesicles seen in electron microscope section of the snail brain.

1

METHODS

Standard techniques were applied to the study of the sections of snail brain; in all cases both light and electron microscopy were applied together to the same series of sections so as to allow a more ready interpretation of the results. The e.m. tissues were all fixed in glutaraldehyde and usually treated with buffered osmium tetroxide. The tissues were embedded in Epon and the only modifications were in the histochemical methods.

Esterase localization. 2·5 ml of 0·5 M thiolacetic acid was adjusted to pH 7·2 with NaOH and diluted to 200 ml with cacodylate buffer pH 7·2; 20 ml of this solution was mixed with 5 ml of 0·01 M lead nitrate in cacodylate buffer pH 7·2. This solution is double the strength of the solution used by Smith and Treherne (1965) and we found it gave the best result for our tissues.

Amines. The tissue was placed for 3–5 h in 2·5% glutaraldehyde buffered in cacodylate at pH 7·2 at 1°C. It was washed in cacodylate buffer pH 7·2 for 12 h at 5°C and then incubated at either 5°C or room temperature for 24 h in the following solution: 10 ml 2·5% potassium dichromate solution + 1% sodium hydroxide in 0·2 M acetate buffer at pH 4·1. The tissue was then washed in buffer at 7·2, dehydrated in alcohol and embedded in Epon. It was not osmic treated. The control animals were injected with soluble reserpine (1%) for 24 h prior to sacrifice, and then the brains were rapidly removed and treated as above.

RESULTS

The snail brain consists of a series of ganglia surrounding the oesophagus (Fig. 1). There are two cerebral ganglia lying dorsal to the oesophagus and a series of seven suboesophageal ganglia. Between the two pedal ganglia and the five other ganglia lies the aorta. The right parietal and the right pallial ganglia are larger than the corresponding left ganglia.

The disposition of some cells larger than 100 μ is shown in Fig. 2. These cells are particularly well represented in the visceral and in the right parietal ganglia. The cerebral ganglia each contain one large nerve cell.

Figure 3 shows diagrammatically a transverse section through the seven suboesophageal ganglia showing the disposition of the ganglion cells on the periphery of the ganglia and the more dense neuropil in the centre of each ganglionic mass.

The ganglia are covered by a layer of connective tissue. This tissue is well developed and thick in the older (3-year-old) snails but relatively thin in young 1-year-old snails. In electrophysiological studies it is usual to remove this connective tissue to get to the nerve cells and often the ganglia swell slightly after removal, perhaps indicative of a restraining pressure exerted by the connective tissue. Electron micrographs of the connective tissues shows it to contain several interesting types of cells whose structure is indicative of something more than just

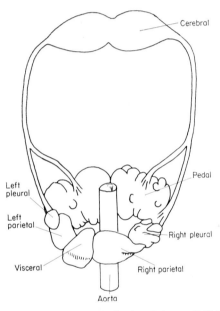

Fig. 1. Diagram of the structure of the brain of the snail *Helix aspersa*. The brain consists of two cerebral ganglia joined by circumoesophageal commissures to the seven suboesophageal ganglia. The aorta runs between the pedal ganglia and the other five ganglia.

a supporting role. The outermost layer of the connective tissue is rather loose and contains collagen fibres, electron-dense vesicles (lipids?) and fibroblasts. There is a more dense inner region of the connective tissue where the collagen fibres are orientated together and packed more closely. There are many cells which contain large vesicles some of which contain electron-dense material. One is indicated in Fig. 4, another is shown at higher magnification in Fig. 5. Note the large vesicles containing electron-dense material and also the smaller vesicles at the periphery of the cell and the opening of the vesicle to the outside.

Figure 6 shows an enlargement of the periphery of one of these cells
and the vesicular openings are now clearly seen together with thickenings at the pores. It would be interesting to know the functions of these
cells and what the vesicles contain. One possibility is that they could

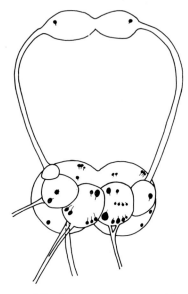

FIG. 2. Positions of some of the large neurones present in the snail brain. The large
cells are shown in black. Only those larger than 100 μ are shown.

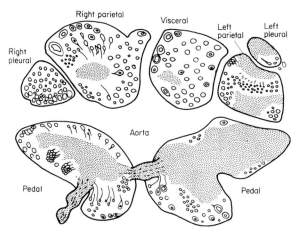

FIG. 3. Diagram of a transverse section through the suboesophageal ganglia showing
the positions of the cells in the periphery of the seven ganglia.

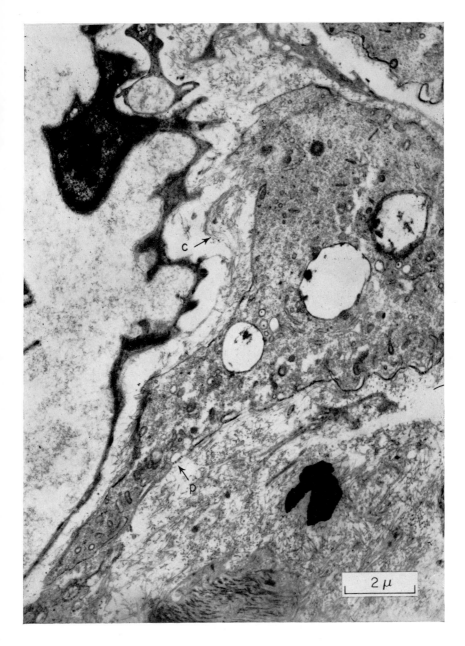

Fig. 4. Electron micrograph through the middle region of the connective tissue around the brain showing fibroblast with vesicles. Note the pores (*p*) at the periphery of the cell and the secreted collagen (*c*).

be secreting collagen. Another possibility is that the cells are taking material up from the outside and hence the dense large vesicles in the inner regions of the cell.

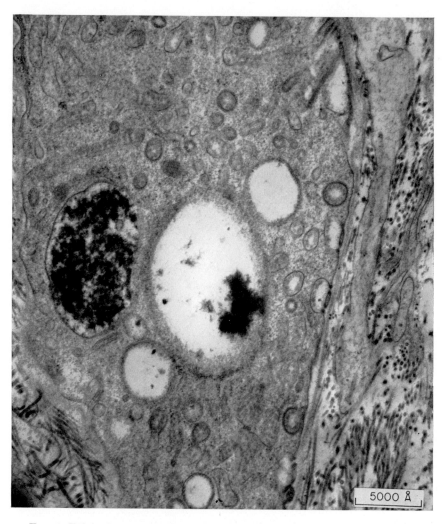

FIG. 5. Enlarged region of connective tissue cell showing large vesicles and the periphery vesicles. Calibration 5000 Å.

The neurones within the connective tissue are surrounded by glial cells and glial processes. Their structure has well been described by other authors (Baxter and Nisbet, 1963; Amoroso et al., 1964). Within the cytoplasm of the cells are many inclusions, Golgi systems, mito-chondria, endoplasmic reticulum, ribosomes and vesicles. The vesicles are of particular interest and can be classified according to their approximate size range and the relative electron density

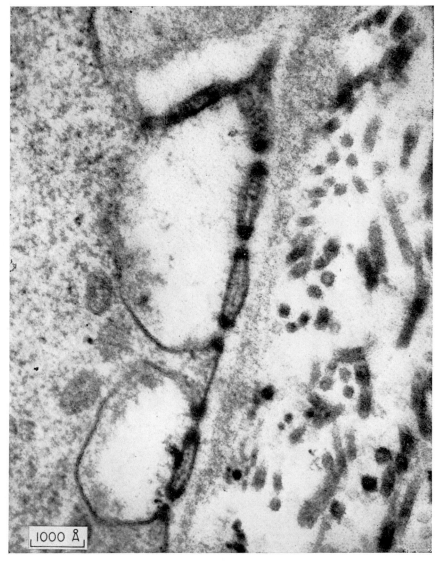

FIG. 6. Enlarged section of connective tissue cell showing the pores at the peripheral region and the thickening of the pore entrances. Calibration line 1000 Å.

Gerschenfeld, 1963; Simpson *et al.*, 1966). Some of the larger more dense vesicles arise in close proximity to the Golgi systems.

Localization of choline esterase

Figure 7 shows a section through the neuropil from a brain that has been incubated in thiolacetic acid (Barrnett, 1962; Smith and Treherne, 1965). There is a marked deposit at a specific site in the neuropil at the junction between the two axons. In one axon (presynaptic?) there are many electron-dense bodies 800–1000 Å in size. Figure 8 shows a group of nerve axons terminating in the region of another axon and a layer of esterase at the region of the junction (the figure is a little difficult to interpret because the section has been cut at an angle to the junction so that the full contact area is probably out of the plane of the section). It is possible that this could be a case where many presynaptic impulses are necessary to elicit an action potential in the post synaptic fibre (low synaptic security). Figure 9 shows a case where two synapses are localized onto one axon. The area of contact is about 2000 Å in each case. It is of interest that the synaptic vesicles (shown in more detail in Fig. 10) are here of 250–300 Å in size and are clear as compared with the 800–1000 Å dense vesicles seen in the presynaptic region of Fig. 7. We know that some of the nerve cells in the *Helix* brain are cholinergic (Kerkut and Walker, 1962; Kerkut and Thomas, 1965; Kerkut and Meech, 1966) and we know that acetylcholine is present in the snail brain (Kerkut and Cottrell, 1963). The electron micrographs show the localization of choline esterase at some of the synapses, and it would seem a reasonable supposition that the vesicles in the pre-synaptic junction contain acetylcholine. This then raises a problem. Can it be that acetylcholine is present in two different types of vesicles, dense 800–1000 Å ones and clear 250–300 Å ones (Fig. 11)?

Vesicles are also found in the axons in nerve trunks running to the muscles and in the nerve muscle junction (Kerkut, Woodhouse and Newman, 1966). The vesicles at the nerve muscle junction of the snail are 250 Å in size and electron clear. We have not been able to localize any esterase at the nerve muscle junction nor has the material given a positive response for amines. There is some experimental evidence to suggest that the material could be glutamate (Kerkut, Leake, Shapira, Cowan and Walker, 1965).

Amines

The chromaffin reaction has been applied to electron microscopy by Coupland and Hopwood (1966) and G. A. Cottrell (personal communication). In sections of snail brain so treated a positive reaction is found for vesicles of 800–1000 Å in size present in nerve cells and axons. The vesicles take up the stain and appear electron dense. In brains taken from snails treated for 24 h with reserpine, most of these dense vesicles have disappeared and only clear vesicles are seen (Fig. 12). Reserpine

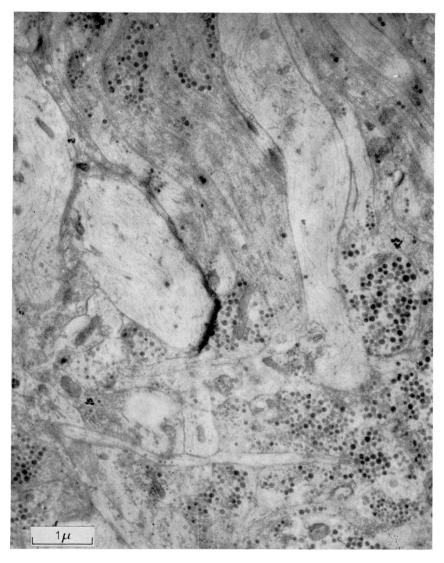

Fig. 7. Localization of choline esterase at the junction between two axons. The vesicles in the presynaptic axon are 1000 Å. Calibration 1 μ.

tends to deplete the nerve cells of catecholamine content and this is
further evidence that the chromaffin reaction is staining catecholamines
or indolealkylamines. We know that the snail brain contains both
3-hydroxytyramine (dopamine) and also 5-hydroxytryptamine (5HT)
(Kerkut and Cottrell, 1963; Kerkut, Sedden and Walker, 1966;

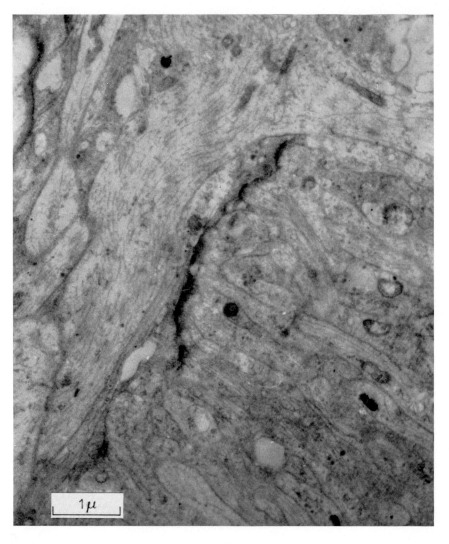

1 μ

FIG. 8. Localization of choline esterase at the junctions between axons. Note that
many axons synapse onto one axon. Calibration 1 μ.

pp. 19–32 below). At present our studies on the localization of amines are only preliminary. It would be most useful to check with e.m. sections on the structure of nerve cells that have been shown by the fluorescent technique to contain either dopa of 5HT (pp. 19–32).

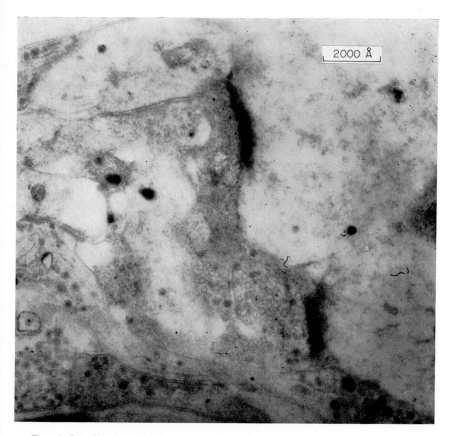

2000 Å

FIG. 9. Localization of choline esterase. Two synapses end on one axon. The full length of contact is 2000 Å.

DISCUSSION

A major problem facing electron microscopists today is to provide a functional interpretation of the structures seen in the micrographs. The neurophysiologist's dilemma is clearly seen when he comes to consider the vesicles and dense bodies found in nerve cells and axons. These can be classified as synaptic vesicles, presynaptic vesicles, post-synaptic vesicles, dense bodies, granules, pinocytotic vesicles, etc. and in most cases the names are more descriptions than diagnostic identifications.

The problem has been clearly presented by Bern and his colleagues (Simpson *et al.*, 1966) in their examination of the evidence for neurosecretion in the nervous system of the pulmonate *Heliosoma tenue*.

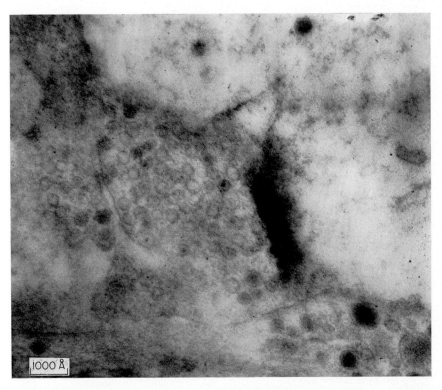

FIG. 10. Localization of choline esterase. Note that the presynaptic vesicles are clear and about 250–300 Å in size.

They correlated the cells that were positive to paraldehyde fuschin (fuschinophilic) under light microscopy with those that showed neurosecretory granules under electron microscopy. They found that some cells were positively fuschinophilic but did not contain neurosecretory granules; instead they contained glycogen as revealed by the electron microscope. Other fuschinophilic cells did contain neurosecretory granules of which there were two sizes, 750 Å and 1500 Å. The smaller granules were the only ones found in the neurosecretory neuropil whilst both sizes were found in the cell body. This work shows the problem of correlating e.m. studies with light microscopical studies. An account by Brady (1967) shows the further difficulty of correlating the light microscopical studies of neurosecretory cells with the physiological studies of activity.

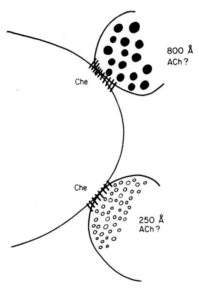

Fɪɢ. 11. Suggested synaptic endings onto axons in the snail neuropil. The endings are cholinergic, as shown by the localized presence of esterase (Che). Could acetylcholine (ACh) be contained in vesicles of different sizes, i.e. dense 800–1000 Å vesicles and clear 250 Å vesicles?

In a strictly electron microscopical study Gerschenfeld (1963) classified the synapses in the nervous system of the slug *Vagunula solea* and the snail *Cryptomphallus aspersa* into three main types.

Type 1. Synaptic endings containing only clear vesicles of 600–800 Å.

Type 2. Synaptic endings containing mainly dense synaptic vesicles of 800–1100 Å but also some clear vesicles of 600–800 Å.

Type 3. Synaptic endings containing neurosecretory vesicles 1200–1400 Å and clear vesicles of two sizes, 800–900 Å and 400–600 Å. Gerschenfeld suggested that type 1 was cholinergic, type 2 contained 5HT or catecholamines, and type 3 was neurosecretory.

The present study shows that it is possible to go a little further in determining the function of a given set of vesicles. If the postsynaptic membrane gives a positive reaction for esterase and if the response is a discrete one, then there is a reasonable chance that the presynaptic vesicles immediately next to this site contain acetylcholine.

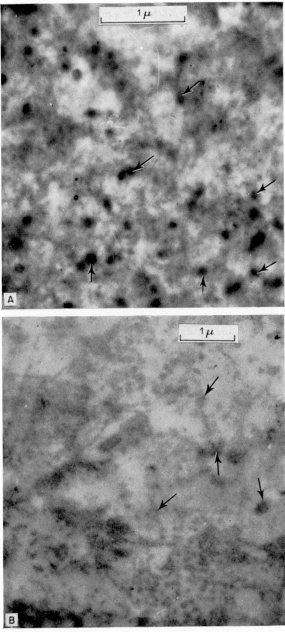

FIG. 12. Localization of amines in electron microscopic sections, showing dense granules of 800–1000 Å in the normal brain (A), and clear granules of the same size in the brain that had been pretreated with reserpine (B). Reserpine is known to reduce the concentration of amines in the snail brain.

The method of esterase staining used gave fairly discrete localization of the synaptic site. Not all synapses stained up (for many are non-cholinergic) and the reaction was inhibited in the presence of 10^{-5}g/ml of physostigmine. If higher concentrations of substrate were used, then a more diffuse reaction was seen and the mitochondria and certain cell inclusions also showed a deposit of lead sulphide. Another region where the esterase showed a regional localization was around the cell bodies immediately beneath the connective tissue sheath. These regions gave a good reaction with acetylthiocholine or indoxyacetate. Esterases have been found present in the snail nervous system by Korne (1964) and Nagy and Salánki (1965).

One problem that this study has shown is that at least two different types of vesicles can be found in the presumptive presynaptic cholinergic ending. It could be that one of the endings was adrenergic in which there was also an implication of acetylcholine (Burn and Rand, 1959; Burn, 1963) but then we would expect to find two types of vesicles in the one ending.

The amine localization technique should prove to be especially valuable if the fixation of the cytoplasm can be improved and if finer detail can be seen. It should also be possible to check on the localization of the amines in snail neurones known to contain 5HT, DOPA or both as indicated by the fluorescent technique (Sedden *et al.*, pp. 19–32).

SUMMARY

1. Esterases can be localized at certain synaptic junctions in the neuropil of the snail brain. The reaction is inhibited by physostigmine.

2. The presynaptic endings at these cholinergic junctions contain either electron-dense vesicles of 800–1000 Å or else clear vesicles of 250–300 Å.

3. The chromaffin reaction indicates that certain vesicles contain amines. These vesicles loose their material if the brain is pretreated with reserpine.

4. The problem of interpreting the function of the various types of vesicles is discussed.

ACKNOWLEDGEMENT

We are indebted to the Science Research Council for financial support towards this work.

B

References

Amoroso, E. C., Baxter, M. I., Chiquoine, A. O. and Nisbet, R. H. (1964). The fine structure of neurones and other elements in the nervous system of the giant African land snail *Archachatina marginata*. *Proc. R. Soc.* (B) **160**: 167–180.

Barrnett, R. J. (1962). The fine structure localization of acetylcholinesterase at the myoneural junction. *J. Cell Biol.* **12**: 247–262.

Baxter, M. I. and Nisbet, R. H. (1963). Features of the nervous system and heart of *Archachatina* revealed by the electron microscope and by electro-physiological recording. *Proc. malac. Soc. Lond.* **35**: 167–177.

Brady, J. (1967). Histological observations on circadian changes in the neuro-secretory cells of the cockroach suboesophageal ganglia. *J. Insect Physiol.* **13**: 210–214.

Burn, J. H. (1963). The release of norepinephrine from the postsympathetic postganglionic fibre. *Bull. Johns Hopkins Hosp.* **112**: 167–182.

Burn, J. H. and Rand, M. J. (1959). Sympathetic postganglionic mechanisms. *Nature, Lond.* **184**: 163–165.

Chou, J. T. Y. (1957a). The cytoplasmic inclusions in the neurones of the nervous system of *Helix aspersa* and *Limnaea stagnalis*. *Q. Jl microsc. Sci.* **98**: 47–58.

Chou, J. T. Y. (1957b). The chemical composition of the lipid globules in the neurones of *Helix aspersa*. *Q. Jl microsc. Sc.* **98**: 59–64.

Coupland, R. E. and Hopwood, D. (1966). Mechanism of a histochemical reaction differentiating between Adrenaline- and Noradrenaline-storing cells in the electron microscope. *Nature Lond.* **209**: 590–591.

Fernandez, J. (1966). Nervous system of *Helix aspersa*. Structure and histo-chemistry of ganglionic sheath and neuroglia. *J. comp. Neurol.* **127**: 157–182.

Gerschenfeld, H. M. (1963). Observations on the ultrastructure of synapses in some pulmonate molluscs. *Z. Zellforsch. mikrosk. anat.* **60**: 258–275.

Kerkut, G. A. (1967). Biochemical aspects of invertebrate nerve cells. In *Invertebrate nervous systems*: 5–37. Giersma, C.A.G. (ed.). Chicago: U.P.

Kerkut, G. A. and Cottrell, G. A. (1963). Acetylcholine and 5HT in the snail brain. *Comp. Biochem. Physiol.* **8**: 53–63.

Kerkut, G. A. and Gardner, D. R. (1967). The role of calcium ions in the action potentials of *Helix aspersa* neurones. *Comp. Biochem. Physiol.* **20**: 147–162.

Kerkut, G. A., Leake, L. D., Shapira, A., Cowan, S. and Walker, R. J. (1965). The presence of glutamate in the nerve muscle perfusates of *Helix, Carcinus* and *Periplaneta*. *Comp. Biochem. Physiol.* **15**: 485–502.

Kerkut, G. A. and Meech, R. W. (1966). The internal chloride concentration of H and D cells in the snail brain. *Comp. Biochem. Physiol.* **19**: 819–832.

Kerkut, G. A., Sedden, C. B. and Walker, R. J. (1966). The effect of DOPA, methyldopa and reserpine on the dopamine content of the brain of the snail *Helix aspersa*. *Comp. Biochem. Physiol.* **18**: 921–930.

Kerkut, G. A. and Thomas, R. C. (1964). The effect of anion injection and changes in the external potassium and chloride concentration on the reversal potentials of the IPSP and acetylcholine. *Comp. Biochem. Physiol.* **11**: 199–213.

Kerkut, G. A. and Thomas, R. C. (1965). An electrogenic sodium pump in snail nerve cells. *Comp. Biochem. Physiol.* **14**: 167–183.

Kerkut, G. A. and Walker, R. J. (1961). The effect of drugs on the neurones of the snail. *Comp. Biochem. Physiol.* **3**: 143–160.

Kerkut, G. A. and Walker, R. J. (1962). The specific chemical sensitivity of *Helix* nerve cells. *Comp. Biochem. Physiol.* **7**: 277–288.

Kerkut, G. A., Woodhouse, M. and Newman, G. (1966). Nerve muscle junction in the snail *Helix aspersa*. *Comp. Biochem. Physiol.* **19**: 309–311.

Korne, M. E. (1964). *Some problems of neuromuscular mediation in the higher invertebrates.* D. Phil. Thesis. University of Oxford.

Lane, N. J. (1964). Elementary neurosecretory granules in the neurones of the snail *Helix aspersa*. *Q. Jl microsc. Sci.* **105**: 31–34.

Nagy, I. Zs.- and Salánki, J. (1965). Histochemical investigation of the choline esterase in different molluscs with reference to functional conditions. *Nature, Lond.* **206**: 842–843.

Rosenbluth, J. (1963). The visceral ganglion of *Aplysia californica*. *Z. Zellforsch. mikrosk. Anat.* **60**: 213–236.

Schlote, F. W. (1957). Submikroscopische Morphologie von Gastropodnerven *Z. Zellforsch. mikrosk. Anat.* **45**: 543–568.

Simpson, L., Bern, H. A. and Nishioka, R. S. (1966). Examination of the evidence for neurosecretion in the nervous system of *Heliosoma tenue* (Gastropoda Pulmonata). *Gen. Comp. Endocr.* **7**: 525–548.

Smith, D. S. and Treherne, J. E. (1965). Electron microscope localization of acetylthiocholinesterase activity in the central nervous system of an insect (*Periplaneta americana*). *J. Cell Biol.* **26**: 445–465.

Sumner, A. T. (1965). The cytology and histochemistry of the digestive gland cells of *Helix*. *Q. Jl microsc. Sci.* **196**: 173–192.

Thomas, O. L. (1947). The cytology of the neurones of *Helix aspersa*. *Q. Jl microsc. Sci.* **88**: 445–461.

Symp. zool. Soc. Lond. (1968) No. 22, 19–32.

THE LOCALIZATION OF DOPAMINE AND 5-HYDROXYTRYPTAMINE IN NEURONES OF *HELIX ASPERSA*

C. B. SEDDEN, R. J. WALKER and G. A. KERKUT

*Department of Physiology and Biochemistry,
Southampton University, Southampton, England*

SYNOPSIS

The presence of 3-hydroxytyramine (dopamine) in the brain of *Helix aspersa* has been demonstrated and the concentration estimated at $5 \cdot 5$ $\mu g/g$ wet wt. No 3,4-dihydroxy-phenylalanine (dopa), the precursor of dopamine, is present in the brain but $0 \cdot 18$ μg dopa per ml is present in the blood of the snail. Increasing this concentration by injection of dopa leads to the presence of dopa in the brain and raises the dopamine content by 44%.

Fluorescence microscopy of the snail brain shows the presence of a large 5-hydroxy-tryptamine (5HT) containing cell in the cerebral ganglion with a group of closely associated small green neurones. The visceral and right parietal ganglia contain many fluorescent cells whose colour suggests that both dopamine and 5HT are present. The neuropil of all the ganglia is fluorescent and contains many varicosities. The neuropil of the pedal ganglia is intensely fluorescent and has several small dopamine-containing neurones closely associated with it. Situated ventrally in the pedal ganglia is a large group of 5HT-containing neurones.

Injection of dopa into the snail leads to a green fluorescence in the cells of the visceral and right parietal ganglia and in the neuropil of the pedal ganglia. Similarly injection of 5-hydroxytryptophan (5HTP) leads to a yellow fluorescence in the cells of the visceral and right parietal ganglia and in the neuropil of the pedal ganglia. These experiments would suggest that both dopamine and 5HT are present in the cells at the junction of the visceral and right parietal ganglia, and that dopamine and 5HT vari-cosities are present in the neuropil of the pedal ganglia.

INTRODUCTION

Many workers have shown the presence of 3-hydroxytyramine (dopa-mine) in invertebrate nervous systems. In a survey of the Mollusca, Sweeney (1963) found that dopamine occurred in many species. He particularly investigated gastropods and pelecypods, where he found the maximum concentration of 261 μg dopamine per g tissue from the cerebral, pedal and visceral ganglia of *Mercenaria mercenaria*.

Cardot (1963) working on *Helix pomatia* estimated colorimetrically that 2–4 μg dopamine per g nervous tissue were present. This confirmed the observation of Dahl, Falck, Lindquist and von Mecklenberg (1962) that $7 \cdot 25$ $\mu g/g$ were contained in the cerebral ganglia of this animal. They also estimated the 5-hydroxytryptamine (5HT) content of the

19

cerebral ganglia using a spectrophotofluorimeter and found $3\cdot77\ \mu g/g$ were present. No other catecholamines were found.

The 5HT content of the circumoesophageal ganglia of *Helix aspersa* has been calculated by Kerkut and Cottrell (1963) to be $0\cdot5\text{–}4\ \mu g/g$.

Using fluorescence microscopy Dahl, Falck, von Mecklenburg, Myhrberg and Rosengren (1966) have studied the cellular distribution of monoamines in *Helix pomatia*. Many neurones have been shown to contain either 5HT or a primary catecholamine. They have also shown numerous strongly fluorescent varicosities on the nerve fibres in the neuropil. If these varicosities represent the presynaptic endings of the nerve fibre as suggested by Elfvin (1963), the monoamines are located in high concentrations in the region of the synapse.

An investigation into the physiological action of dopamine was carried out by Kerkut and Walker (1961) using an isolated preparation of the *Helix aspersa* brain. They found that, overall, dopamine had an inhibitory action on the neurones, causing a decrease in the frequency of spontaneous action potentials and hyperpolarizing the membrane. The usual threshold for inhibition was 10^{-9} g/ml.

A similar action for dopamine has been shown by Gerschenfeld (1964) on the brain of *Cryptomphallus aspersa* (Argentine land snail). He reported some cells which on stimulation of the visceral nerve and sometimes the right pallial nerve show inhibitory postsynaptic potentials. Dopamine has a strong inhibitory action on these cells at 10^{-9} M, but α-adrenergic blockers do not alter the inhibitory response to stimulation.

METHODS

Chromatography

Crude nervous tissue extracts were prepared by finely chopping fifty brains of *Helix aspersa* in acidified ethanol at $0°C$. After centrifugation the supernatant was spotted on No. 1 Whatman paper and developed in various solvents by the descending method. Any spots present were located by spraying with ethylenediamine (Weil-Malherbe and Bone, 1957).

Spectrophotofluorimetry

Extraction of the catecholamines was performed according to the method of Brownlee and Spriggs (1965). The estimation of 3,4-dihydroxyphenylalanine (dopa) and dopamine was carried out using the methods of Bertler, Carlsson and Rosengren (1958) and Carlsson and Waldeck (1958). The solutions of the fluorophores produced with

0·25% $K_3Fe(CN)_6$ were read at 350/465 mμ (uncorrected values) and those produced with 0·02 N iodine solution were read at 320/382 mμ.

Fluorescence microscopy

The method described by Falck and Owman (1965) was followed. The tissues were quenched by coating them in talc and immersing them in liquid nitrogen. They were freeze-dried at $-40°C$. This was followed by exposure to formaldehyde vapour at 80°C for 1–3 h. The tissues were embedded in paraffin wax and sectioned at 10 μ. Controls were treated similarly, but were not exposed to the para-formaldehyde vapour. The sections were examined in a Zeiss fluorescence microscope using a Schott BG 12, a Zeiss 50 and a Zeiss 44 filter.

The first step in the development of the fluorescence is a condensation between the formaldehyde and the catecholamine to produce a 6,7-dihydroxy-1,2,3,4-tetrahydroisoquinoline. The next step is a protein catalysed dehydrogenation giving a 6,7-dihydroxy-3,4-dihydro-isoquinoline. 5HT is similarly converted to the fluorescent 3,4-dihydro-β-carboline. The two fluorescent products, although they have the same activation peak, have different fluorescent peaks, that is 480 mμ for the catecholamines and 530 mμ for the 5HT. This enables the two compounds to be distinguished, as the former fluoresces green and the latter yellow.

The specificity of the fluorescence was checked by subjecting some of the slides to the sodium borohydride reduction test (Corrodi, Hillarp and Jonsson, 1964).

Injection of drugs

Soluble reserpine was made up in distilled water. Dopa was dissolved in a little hydrochloric acid, made up with distilled water and the pH adjusted to 6 with sodium hydroxide. 0·1 ml of each solution was injected into the foot of the snail; the controls were injected with an equal volume of distilled water.

RESULTS

Chromatography and spectrophotofluorimetry

The chromatographic investigation of crude extracts of snail brain produced a fluorescent spot on spraying with ethylene diamine, which had the same R_F as authentic dopamine in five different solvents (Table I). There did not appear to be any other catecholamines present. Using the spectrophotofluorimeter the presence of dopamine was established at a concentration of 5·5 μg/g wet wt and the absence of adrenaline, noradrenaline and dopa was confirmed (Kerkut, Sedden and Walker, 1966).

Injection of reserpine at a concentration of 5 mg per animal for 48 h depleted the brain dopamine content so that no fluorescent spot was produced after chromatography. Spectrophotofluorimetric estimations showed that after 5 h the dopamine level had fallen to 3·6 μg/g and after 48 h it was less than 0·5 μg/g.

TABLE I

Solvent systems employed to demonstrate the presence of dopamine in extracts of the Helix *brain*

Solvent		$R_F \times 100$ Dopamine	Extract
Butanol/acetic acid/water		35	35
Butanol/ethanol/water		50	49
Chloroform/acetic acid/water		85	84
Methanol/butanol/benzene		58	58
Butanol/N HCl	R_D	100	100

An investigation of snail blood for a precursor of dopamine produced a fluorescent spot chromatographically with the same R_F as authentic dopa in butanol saturated with N HCl. This was confirmed on the spectrophotofluorimeter, the concentration of dopa being 0·18 μg/ml.

An increase in blood dopa level by injection of 0·1 mg per animal increased both the dopamine and dopa levels in the brain. Initially the brain dopa concentration was zero, but this rose rapidly to 7·1 μg/g after 15 min (Fig. 1). A peak value of 10·1 μg/g was obtained 75 min after injection, the dopa level then falling to 2·6 μg/g after 6 h. The dopamine content rose to a maximum of 9·8 μg/g after 3 h and then fell back to the resting level but more slowly than the dopa.

Fluorescence microscopy

For the fluorescence microscopy the whole ganglionic ring was dissected out and the ganglia studied individually.

Cerebral ganglia

On the outer edge of the cerebral ganglion there is a large yellow fluorescent cell. It is 130 μ in diameter and has a non-fluorescent nucleus. Around one edge of this cell is a group of intensely fluorescent small green cells (Fig. 2). These are always present in this position. Each cell is only 10–15 μ in size and the nucleus is again non-fluorescent, although it may appear green due to the thinness of the cells. The axons of these small neurones are smooth and only moderately fluorescent. They run through the centre of the cluster of cells towards the neuropil of the cerebral ganglion, which is located centrally. The fibres in the neuropil are varicose in nature and exhibit a bright green fluorescence. On the edge of this fibrous mass is a small group of yellow cells, but these are not arranged in a cluster. Where the commissures leave the cerebral ganglia there are a few medium sized green cells.

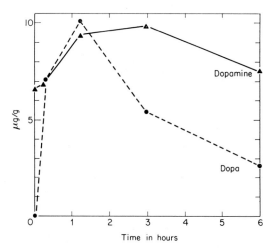

Fig. 1. The effect of dopa injection on the dopa and dopamine levels in the *Helix* brain.

The commissure running to the right pedal ganglion is surrounded by a tube of yellow fluorescent cells. The other commissure does not appear to possess any similar cells.

Visceral and right parietal ganglia

These ganglia contain many fluorescent cells, which tend to be located at their adjacent sides (Fig. 3). Although the position of these cells varies from snail to snail, one is usually found at the posterior end of the visceral ganglion next to the right parietal. The fluorescent cells range in size from 40 to 150 μ and are green in colour with yellow areas. It would appear, therefore, that they contain both dopamine and 5HT.

FIGS. 2–7. Fluorescence micrographs of *Helix aspersa* brain. The magnification is the same in each, see scale on Fig. 2.

FIG. 2. Cerebral ganglion showing a bunch of green cells. These are always adjacent to a large yellow cell, which is below this section. Note the smooth axons running through the cluster of cells and the varicosities in the neuropil (arrow).

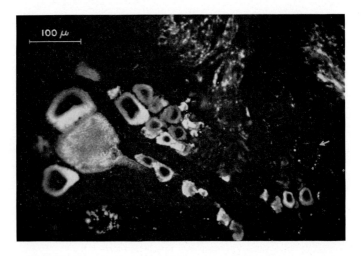

FIG. 3. The junction of the visceral and right parietal ganglia. The right parietal (lower half of the picture) contains three large fluorescent cells. The visceral shows only one, but it is usually located in this position. Numerous small fluorescent cells are present. A few non-fluorescent cells can be seen containing orange autofluorescent granules (arrow).

However, on exposure to u.v. light for long periods, the fluorescence fades to that of the background, which is indicative of 5HT. The larger fluorescent cells possess weaker but still distinctly fluorescent axons (Fig. 3). In the region of the axon hillock these are 20 μ in diameter (Fig. 4) but they then taper as they run towards the central neuropil as

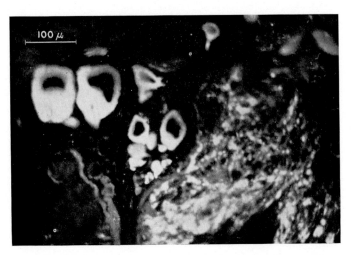

FIG. 4. The first of two serial sections showing two large green and yellow fluorescent cells (see text for explanation.) The axon hillock is 20 μ in diameter.

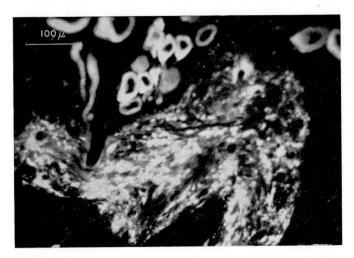

FIG. 5. The next section showing the axons running from the two cells, which although they are only weakly fluorescent can still be seen 200 μ from the perikaryon.

can be seen in the next section (Fig. 5). The axons can be followed for varying distances, sometimes up to 200 μ from the perikaryon. In these ganglia approximately one-tenth of the cells is fluorescent. The remainder are completely non-fluorescent except for a few bright orange autofluorescent granules. The central neuropil, as in the cerebral

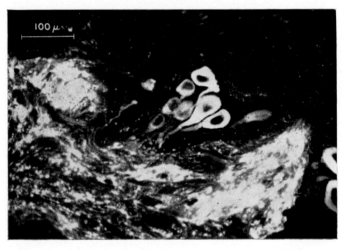

Fig. 6. The neuropil of the visceral and right parietal ganglia, showing the numerous varicosities. Also several fluorescent cells can be seen near a large non-fluorescent neurone.

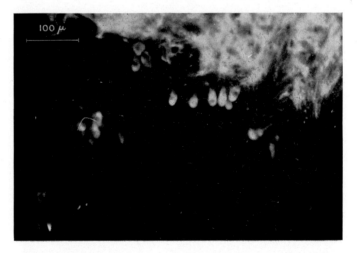

Fig. 7. The pedal ganglion showing a group of small green cells closely associated with the neuropil. They have non-fluorescent nuclei and are unipolar.

ganglia, shows a green fluorescence. However, in certain areas where the fibres are closely packed together and there is a concentration of the fluorescence, the colour produced is yellow. Without a micro-spectrophotometer it is difficult to determine whether the fibres here are green or yellow or a mixture of both. In the less intensely fluorescent areas (Fig. 6), where the morphology of the fibres can be seen, numerous varicosities are present varying from $1 \cdot 8$ to $3 \cdot 8 \mu$ in diameter. The number per 100μ length of fibre ranges from 15 to 35. The connecting fibre is very weakly fluorescent and can only occasionally be seen. The varicosities appear to be arranged at fairly regular intervals along the fibres.

Left parietal and pleural ganglia

There are usually no fluorescent cells found in these ganglia. Orange autofluorescent granules are present in many of the neurones. The neuropil which connects all the suboesophageal ganglia is again fluorescent in areas, but there are many more non-fluorescent fibres than in the visceral and right parietal.

Pedal ganglia

The neuropil of the pedal ganglia is intensely fluorescent. Again in the densely fibrous areas the colour is bright yellow, but at the peri-phery of the neuropil a green fluorescence is seen. Varicosities are present but as the pedal nerves emerge the number of varicosities decreases and the fibres become smooth. Immediately dorsal to and in the same plane as the pedal nerves numerous small intensely fluorescent green cells are present (Fig. 7). These are $20-25 \mu$ in length, have non-fluorescent nuclei and are closely associated with the neuropil. Although they are very similar in size and shape to the small green cells found in the cerebral ganglia, they are not grouped together in a cluster but are arranged in a line.

Also present near the junction of the two pedal ganglia are small bunches of yellow cells. Each neurone is slightly smaller than the green cells, being only $15-20 \mu$ in diameter. As in the bunch of cells in the cerebral ganglia the axons run from the centre of the cluster and are smooth in nature. Ventrally to the neuropil in the pedal ganglia there is a large group of yellow fluorescent cells. The axons of these cells fluoresce with less intensity than the cell bodies, but they can be followed as they run towards the central neuropil. The size of these neurones is $25-45 \mu$ and although they are all yellow some fluoresce more intensely than others. The position of all the fluorescent cells is summarized in Fig. 8 and 9.

Effect of injection of drugs on the fluorescence

After injection of dopa into the snail the cells of the visceral and right parietal ganglia, which before were a mixture of green and

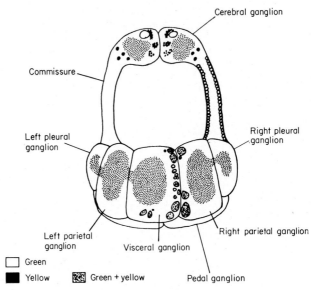

Fig. 8. A summary diagram showing the position of the fluorescent cells. Stippled area represents neuropil.

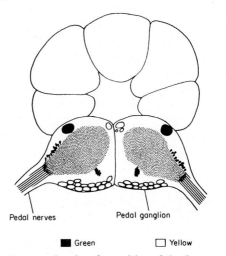

Fig. 9. A summary diagram showing the position of the fluorescent cells in the pedal ganglia. The shaded circle on the upper edge of each ganglion is the otocyst. Stippled area represents neuropil.

yellow, become a distinct green. The neuropil of the pedal ganglia similarly shows a more dominant green colour.

Pretreatment of the snails with NSD 1024, a dopa decarboxylase inhibitor, prevents this green fluorescence of the cells in the visceral and right parietal ganglia after dopa injection.

Injection of 5-hydroxytryptophan (5HTP), the precursor of 5HT, increases the yellow fluorescence of these cells in the visceral and right parietal ganglia. It also causes the neuropil of the pedal ganglia to fluoresce more brightly yellow.

DISCUSSION

The presence of dopamine has been shown in the brain of *Helix aspersa*, which agrees with the observations of Dahl *et al.* (1962) for *H. pomatia*. The value of $7·25\,\mu g$ dopamine per g cerebral ganglia tissue obtained by these authors for *H. pomatia* is in agreement with $5·5\,\mu g/g$ for the whole circumoesophageal mass of *H. aspersa*. No connective tissue was dissected away when making these determinations, so the value for actual nervous tissue is slightly higher than indicated here.

Reserpine has been shown by a number of authors to deplete the brain level of not only catecholamines but also 5HT. Thus Bertaccini (1961) found that reserpine depleted the 5HT and catecholamine content of the optic ganglia of *Eledone moschata*. After 36 h the value for catecholamines had fallen from $1·12$ to $0·12\,\mu g/g$. *Helix aspersa* has been found to behave similarly, the dopamine value falling to $0·5\,\mu g/g$ from $5·5\,\mu g/g$ after 48 h. The 5HT content also falls, as after fluorescence microscopy no fluorescent cells or axons are seen.

The presence of dopa in the blood would be expected if it were the precursor for dopamine synthesis in the brain and indeed a value of $0·18\,\mu g/ml$ was found. No dopa was present in the brain of the snail, suggesting a fairly high turnover rate. Injection of dopa into the animal produced a rise in brain dopamine from $6·6\,\mu g/g$ to $9·5\,\mu g/g$ in 75 min. A similar increase after dopa injection was also found in the catecholamine content of the optica ganglia of *Eledone moschata* by Bertaccini (1961) where the value rose from $1·2$ to $3\,\mu g/g$ in 1 h.

Fluorescence microscopy has shown that the dominant amine in the cerebral ganglia is dopamine. It is found in the neuropil, in many small cells of $20–25\,\mu$ and in a few larger cells of $40–50\,\mu$. 5HT is present in a large cell, $150\,\mu$ in diameter on the outer edge of the cerebral ganglia. The visceral and right parietal again show the presence of dopamine in the neuropil but the content of the fluorescent cells is not clearly established. Normally these cells are green with yellow areas. However, after treatment with the precursors dopa and 5HTP they appear completely green and yellow respectively. This would suggest that they contain both dopamine and 5HT. The colour of the neuropil in the pedal ganglia changes in a similar manner and it would appear that both 5HT and dopamine fibres are present. Around the neuropil and closely associated with it are numerous small dopamine-containing neurones. These cells are only found in the pedal ganglia and are located near the pedal nerves. It has been suggested that in the earthworm small unipolar local neurones are the interneurones and these cells in the snail would appear to fit this description. Also present at the ventral end of the pedal ganglia is a large group of 5HT-containing neurones 25–45 μ in diameter. It has been suggested by Myrhberg (1966) that yellow fluorescence is connected with motor functioning.

SUMMARY

1. Using descending paper chromatography and five different solvent systems, dopamine has been shown to be the only catecholamine present in the circumoesophageal ganglionic mass of *Helix aspersa*.

2. Spectrophotofluorimetric analysis shows that the dopamine concentration is $5 \cdot 5$ μg/g wet wt.

3. Dopa at a concentration of $0 \cdot 18$ μg/ml is present in the blood of *Helix*.

4. Reserpine was found to deplete the brain dopamine content, the value dropping to $0 \cdot 5$ μg/g wet wt 48 h after treatment.

5. Injection of dopa leads to the presence of dopa in the brain and to a raised dopamine level by 44%.

6. Fluorescence microscopy of the cerebral ganglia has shown the presence of a large 5HT-containing cell around one side of which a group of dopamine cells are found. The neuropil contains numerous varicosities.

7. The visceral and right parietal ganglia contain cells in which both dopamine and 5HT appear to be present.

8. The neuropil of the pedal ganglia is intensely fluorescent containing varicosities of both dopamine and 5HT. Small dopamine-containing neurones are closely associated with this neuropil. Situated ventrally in the pedal ganglia is a large group of 5HT-containing neurones.

9. Injection of dopa leads to a green fluorescence in the cells of the visceral and right parietal ganglia and in the neuropil of the pedal ganglia.

10. Injection of 5HTP leads to a yellow fluorescence in the cells of the visceral and right parietal ganglia and in the neuropil of the pedal ganglia.

ACKNOWLEDGEMENTS

Gifts of Serpasil from Ciba Ltd are gratefully acknowledged. We are indebted to the Medical Research Council for a training grant to one of us (C.B.S.), and to the Royal Society for a Government Grant for the purchase of the spectrophotofluorimeter and the fluorescence microscope.

REFERENCES

Bertaccini, G. (1961). A discussion. In *Regional neurochemistry*: 305–306. Kety, S. S. and Elkes, J. (ed.). Oxford: Pergamon Press.

Bertler, A., Carlsson, A. and Rosengren, E. (1958). A method for the fluorimetric determination of adrenaline and noradrenaline in tissues. *Acta physiol. scand.* **44**: 273–292.

Brownlee, G. and Spriggs, T. L. B. (1965). Estimation of dopamine, noradrenaline, adrenaline and 5-hydroxytryptamine from single rat brains. *J. Pharm. Pharmacol.* **17**: 429–433.

Cardot, J. (1963). Sur la présence de dopamine dans le système nerveux et ses relations avec la décarboxylation de la dioxyphénylalanine chez le Mollusque *Helix pomatia. C.r. hebd. Séanc. Acad. Sci., Paris* **257**: 1346–1366.

Carlsson, A. and Waldeck, B. (1958). A fluorimetric method for the determination of dopamine. *Acta physiol. scand.* **44**: 293–298.

Corrodi, H., Hillarp, N-Å. and Jonsson, G. (1964). Fluorescence methods for the histochemical demonstration of monoamines. 3. Sodium borohydride reduction of the fluorescent compounds as a specifity test. *J. Histochem. Cytochem.* **12**: 582–586.

Dahl, E., Falck, B., Lindquist, M. and von Mecklenberg, C. (1962). Monoamines in mollusc neurones. *K. fysiogr. Sällsk. Lund Forh.* **32**: 89–92.

Dahl, E., Falck, B., von Mecklenburg, C., Myhrberg, H. and Rosengren, E. (1966). Neuronal localization of dopamine and 5-hydroxytryptamine in some Mollusca. *Z. Zellforsch. mikrosk. Anat.* **71**: 489–498.

Elfvin, L-G. (1963). The ultrastructure of the superior cervical sympathetic ganglion of the cat. II. The structure of the preganglionic end fibres and the synapses as studied by serial sections. *J. Ultrastruct. Res.* **8**: 441–476.

Falck, B. and Owman, C. (1965). A detailed methodological description of the fluorescence method for the cellular demonstration of biogenic monoamines. *Acta Univ. Lund.* Section II **1965** (7): 1–23.

Gerschenfeld, H. M. (1964). A non-cholinergic synaptic inhibition in the central nervous system of a mollusc. *Nature, Lond.* **203**: 415–416.

Kerkut, G. A. and Cottrell, G. A. (1963). Acetylcholine and 6-hydroxytryptamine in the snail brain. *Comp. Biochem. Physiol.* **8**: 53–63.

Kerkut, G. A., Sedden, C. B. and Walker, R. J. (1966). The effect of dopa, α-methyldopa and reserpine on the dopamine content of the brain of the snail, *Helix aspersa*. *Comp. Biochem. Physiol.* **18**: 921–930.

Kerkut, G. A. and Walker, R. J. (1961). The effect of drugs on the neurones of the snail *Helix aspersa*. *Comp. Biochem. Physiol.* **3**: 143–160.

Myhrberg, H. (1966). Personal communication in Clark, M. E. (1966). Histochemical localization of monoamines in the nervous system of the polychaete *Nephtys*. *Proc. R. Soc.* (B) **165**: 308–325.

Sweeney, D. (1963). Dopamine: its occurrence in molluscan ganglia. *Science, N.Y.* **139**: 1051.

Weil-Malherbe, H. and Bone, A. D. (1957). The fluorimetric estimation of adrenaline and noradrenaline in plasma. *Biochem. J.* **67**: 65–72.

Symp. zool. Soc. Lond. (1968) No. 22, 33–74.

THE PHARMACOLOGY OF THE NEURONES OF *HELIX ASPERSA*

R. J. WALKER, A. HEDGES and G. N. WOODRUFF

Department of Physiology and Biochemistry,
Southampton University, Southampton, England

SYNOPSIS

Acetylcholine inhibits the spontaneous activity of certain neurones of the snail *Helix aspersa*, H cells, and excites the spontaneous activity of other neurones, D cells. This acetylcholine response can be antagonized by both muscarinic blocking agents, for example atropine and scopolamine, and by nicotinic blocking agents, such as tubocurarine and gallamine. One microgram of these antagonists will block the response to 10 μg acetylcholine. Hexamethonium, tetraethylammonium, decamethonium, succinylcholine and benzoquinonium fail to block the acetylcholine response when they are applied at concentrations equal to or more dilute than the dose of acetylcholine. High concentrations of hexamethonium, tetraethylammonium, decamethonium, succinylcholine and benzoquinonium will antagonize the response to acetylcholine in D cells. Decamethonium, succinylcholine and benzoquinonium will also antagonize the acetylcholine response in H cells. Atropine will completely antagonize the response to mecholine while tubocurarine will completely antagonize the response to nicotine. Neurones respond to a number of muscarinic agonists, for example muscarine, muscarone, pilocarpine, arecoline and oxotremorine. Neurones will also respond to certain nicotinic agonists, for example dimethylphenylpiperazinium and tetramethylammonium. Certain neurones are more sensitive to muscarinic agonists than they are to nicotinic agonists while other neurones are more sensitive to nicotinic agonists than to muscarinic agonists. Other neurones are equally sensitive to both nicotinic and muscarinic agonists. Pretreatment with eserine increases the duration of the acetylcholine response which then resembles the carbachol response. The responses of mecholine, propionylcholine and butyrylcholine have a longer duration than equipotent concentrations of acetylcholine. Histamine hyperpolarizes and inhibits the activity of certain *Helix* neurones and depolarizes and excites other neurones. Both effects are antagonized reversibly by mepyramine. Dopamine hyperpolarizes and inhibits the activity of certain neurones. This effect is antagonized reversibly by dibenyline.

INTRODUCTION

Our present knowledge of the pharmacology of compounds which mimic and antagonize possible chemical transmitter compounds between nerve and nerve depends largely on the pharmacology of the autonomic nervous system. The postganglionic fibres of the autonomic nervous system and the motoneurone endings onto striated muscle have afforded further preparations for the study of the pharmacology of neurotransmitters.

From such investigations it has been possible to postulate two types of receptor at vertebrate cholinergic synapses. At the cholinergic synapse onto the autonomic ganglion and at the nerve striated muscle junction, nicotine at low concentrations mimics the actions of acetylcholine. The cholinergic receptor at these two sites has been termed the nicotinic receptor. The action of both acetylcholine and nicotine at the former site is antagonized by hexamethonium and at the latter site by decamethonium. The action of acetylcholine at the parasympathetic nerve smooth muscle junction is mimicked by muscarine and the action of both compounds is antagonized by atropine. This receptor is termed the muscarinic site for acetylcholine release. These points are summarized in Fig. 1. A considerable number of compounds have

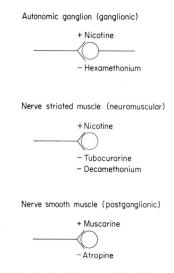

FIG. 1. Diagram to illustrate the three peripheral vertebrate cholinergic receptor sites; + indicates an agonist; − indicates an antagonist.

been tested and classified as to their agonistic or antagonistic effect when compared to acetylcholine at these three sites. Those compounds which will be investigated in the present study are listed in Table I.

A number of cholinoceptive sites have been postulated in the vertebrate central nervous system. Some of these sites are muscarinic, some nicotinic, while others would appear to contain both types of receptor. For reviews of this subject the reader is referred to Curtis (1965), Krnjevic (1965), Bradley and Wolstencroft (1965) and Phillis (1965). For the general pharmacology of acetylcholine and other possible transmitter agents, the reader is referred to Barlow (1964), Goodman and Gilman (1965) and Triggle (1965).

TABLE I

Summary of cholinergic pharmacology from the vertebrate autonomic nervous system and the nerve striated muscle junction

Compound	Receptor site	Type of action	Mode of action
Nicotine	Ganglionic	Agonist	Depolarize and excite
	Neuromuscular	Agonist	Depolarize and excite
Tetramethylammonium	Ganglionic	Agonist	Depolarize and excite
	Neuromuscular	Agonist	Depolarize and excite
Dimethylphenyl-piperazinium	Ganglionic	Agonist	Depolarize and excite
Benzoylcholine	Ganglionic	Agonist	Depolarize and excite
Butyrylcholine	Ganglionic	Agonist	Depolarize and excite
	Neuromuscular	Agonist	Depolarize and excite
Propionylcholine	Ganglionic	Agonist	Depolarize and excite
	Postganglionic	Agonist	Depolarize and excite
	Neuromuscular	Agonist	Depolarize and excite
Carbamoylcholine (Carbachol)	Ganglionic	Agonist	Depolarize and excite
	Postganglionic	Agonist	Depolarize and excite
	Neuromuscular	Agonist	Depolarize and excite
Tetraethylammonium	Ganglionic	Antagonist	Competitive, no effect on r.p.*
Hexamethonium	Ganglionic	Antagonist	Competitive, no effect on r.p.
Muscarine	Postganglionic	Agonist	Depolarize and excite
Muscarone	Postganglionic	Agonist	Depolarize and excite
Acetyl-β-methylcholine (Mecholine)	Postganglionic	Agonist	Depolarize and excite
Pilocarpine	Postganglionic	Agonist	Depolarize and excite
Arecoline	Postganglionic	Agonist	Depolarize and excite
Oxotremorine	Postganglionic	Agonist	Depolarize and excite
Atropine	Postganglionic	Antagonist	Competitive, no effect on r.p.
Scopolamine	Postganglionic	Antagonist	Competitive, no effect on r.p.
Tubocurarine	Neuromuscular	Antagonist	Competitive, no effect on r.p.
	Ganglionic	Antagonist	Competitive, no effect on r.p.
β-Erythroidine	Neuromuscular	Antagonist	Competitive, no effect on r.p.
	Ganglionic	Antagonist	Competitive, no effect on r.p.
Gallamine	Neuromuscular	Antagonist	Competitive, no effect on r.p.

TABLE I—*continued*

Decamethonium	Neuromuscular	Antagonist	Competitive, depolarizes r.p.
Succinylcholine	Neuromuscular	Antagonist	Competitive, depolarizes r.p.
Benzoquinonium	Neuromuscular	Antagonist	Competitive, depolarizes r.p.

* r.p., Resting potential.

As yet little is known regarding the structure of the vertebrate cholinergic receptor at the muscarinic and nicotinic sites. Present interpretation of the receptors is largely based on agonist–antagonist studies. Vast series of compounds have been synthesized in which one part or other of the acetylcholine molecule has been systematically altered, for example using simple onium salts, altering the onium group, altering the acyl group, altering the choline part of the molecule. The activity of a series of analogues can then be compared with acetylcholine for agonist activity in terms of affinity and efficacy or intrinsic activity. In some cases changing the molecule transforms an agonist into an antagonist. From such structure-activity relationships an idea of the parameters of the receptor can be obtained and a possible shape postulated. Until other methods of investigation can be developed, such as biochemical studies on the isolated receptor or electron microscope studies of the synaptic membrane, this is the most profitable method for research. Reviews of receptor structure have been written by Waser (1961) for the muscarinic receptor, Gill (1959) for the nicotinic ganglion receptor, and Khromov-Borisov and Michelson (1966) for the nicotinic nerve striated muscle receptor. Barlow (1964) has reviewed current concepts for all three receptor sites.

A cholinergic inhibitory synapse has been postulated to be present in the central nervous system of the snail *Helix* (Tauc and Gerschenfeld, 1962; Kerkut and Thomas, 1963, 1964). At these synapses, acetylcholine inhibits the spontaneous activity of the neurones and may hyperpolarize the membrane. Cells which respond to acetylcholine in this way are termed H cells. Other cells are depolarized and excited by acetylcholine; these cells are termed D cells. An example of an H cell is shown in Fig. 3, and an example of a D cell is shown in Fig. 2. It has been shown by Kerkut and Meech (1966a,b) that cells which are H to acetylcholine have a low chloride concentration, while those which are D have a high chloride concentration. It had already been shown by Kerkut and Thomas (1964) that the H effect to acetylcholine, at least in certain cells, is due mainly to a selective increase in the permeability of the membrane to chloride ions. The present paper presents information concerning certain aspects of the pharmacology of the H and D response to acetylcholine in these neurones (Walker and Hedges, (1967; 1968)).

In addition to the pharmacology of the response of *Helix* neurones to acetylcholine, some preliminary results will be described in which it will be shown that the histamine response of these cells can be antagonized by mepyramine and the dopamine response can be antagonized by dibenyline.

METHODS

All experiments in this study were performed on the isolated sub-oesophageal ganglionic mass of the garden snail, *Helix aspersa*, which were obtained locally. The outer connective tissue was removed from over the ganglionic mass prior to penetration with the intracellular recording electrodes. The electrodes were pulled from Pyrex glass tubing, using a vertical electrode puller, and filled with molar solution of potassium acetate. The electrode tips ranged in diameter from $0 \cdot 1$ to $0 \cdot 5 \, \mu$, and had a resistance of between 5 and 30 MΩ. The spontaneous activity of the neurones was recorded by means of a conventional cathode follower device and displayed on a 502 Tetronix oscilloscope. The activity was then filmed or recorded on an Ediswan pen oscillograph. The Ringer solution used had the following formula: 80 mM sodium chloride, 4 mM potassium chloride, 7 mM calcium chloride, 5 mM magnesium chloride, and 5 mM Tris HCl. The pH of this Ringer is 8. The compounds to be tested were either added to the bath or iontophoretically injected from an electrode containing the test compound.

RESULTS

Acetylcholine antagonists

The action of acetylcholine on both H and D neurones can readily be blocked by atropine, scopolamine, tubocurarine and gallamine. Examples of these blocking effects can be seen in Figs. 2–4. It is clear from Fig. 2 that a neurone contains both nicotinic and muscarinic receptor sites. All four antagonists can produce complete block at a concentration ten times less than the dose of acetylcholine applied. Other cholinergic antagonists were also able to block or reduce the effect of acetylcholine, but they required much higher doses. Figures 5–7 show the effect on an acetylcholine D response of pretreatment with hexamethonium, tetraethylammonium, decamethonium, succinylcholine, and benzoquinonium. In most cases with these antagonists doses between 100 and 1000 times the concentration of acetylcholine

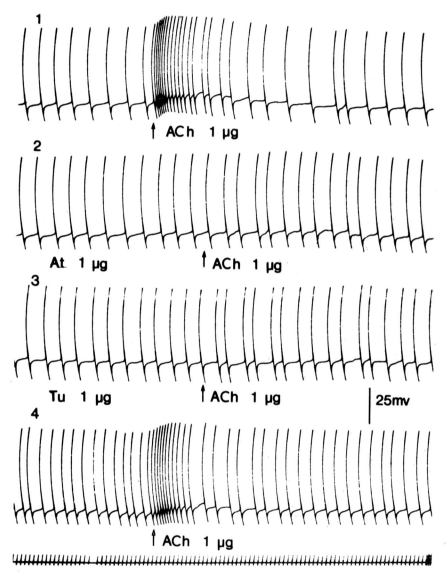

FIG. 2. The effect of pretreatment with 1 μg atropine and 1 μg tubocurarine on the acetylcholine response of a D cell. Time trace in this and subsequent figures is in intervals of 1 sec.

were required to reduce the acetylcholine response. At these high concentrations the antagonists often have depressant effects on the spontaneous activity of the cell. This effect is clearly shown in the case of decamethonium (Fig. 6, trace 2) and benzoquinonium (Fig. 7, trace 2).

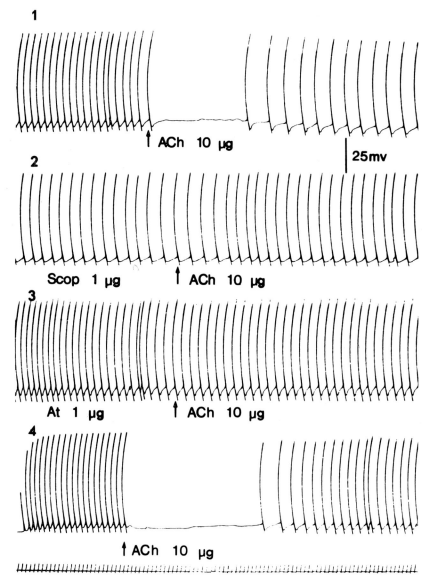

FIG. 3. The effect of pretreatment with 1 μg atropine and 1 μg scopolamine on the acetylcholine response of an H cell.

However, on some occasions the benzoquinonium had no effect on the spontaneous activity. In a few D cells decamethonium excited the cell and in these cases the acetylcholine response was not blocked. The response of these antagonists on H cells is more difficult to interpret

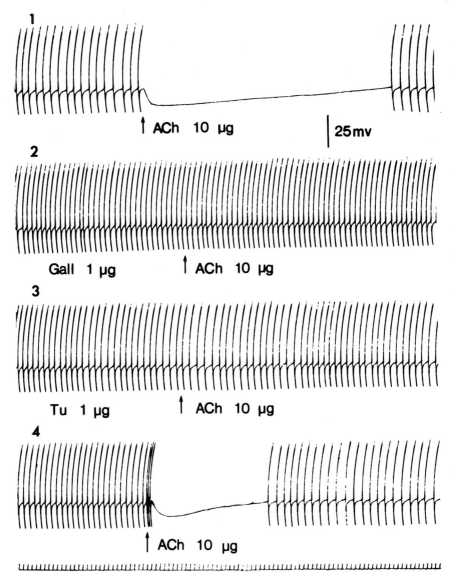

Fig. 4. The effect of pretreatment with 1 μg gallamine and 1 μg tubocurarine on the acetylcholine response of an H cell.

since at high concentrations they often block the spontaneous activity of the cell. Under these conditions it is difficult to determine whether or not the H response to acetylcholine has been blocked. However,

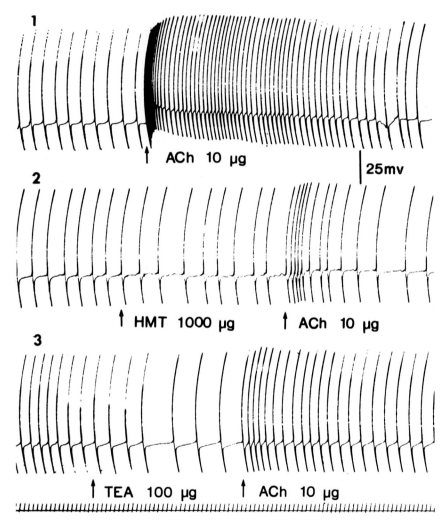

FIG. 5. The effect of pretreatment with 1000 μg hexamethonium and 100 μg tetraethylammonium on the acetylcholine response of a D cell.

benzoquinonium clearly does antagonize the acetylcholine response in H cells when present at a concentration 100 times greater than the dose of acetylcholine applied.

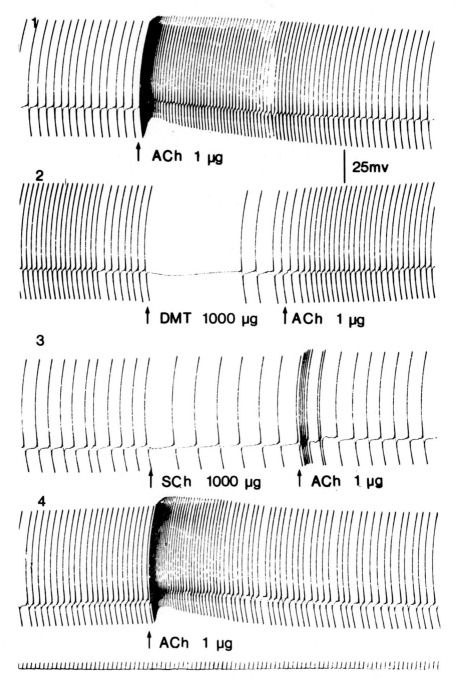

FIG. 6. The effect of pretreatment with 1000 μg decamethonium and 1000 μg succinylcholine on the acetylcholine response of a D cell.

At least in certain cells, tetraethylammonium at a concentration of 10 000 μg fails to block the response to 10 μg acetylcholine. At high concentrations, tetraethylammonium slows the firing rate of the cell,

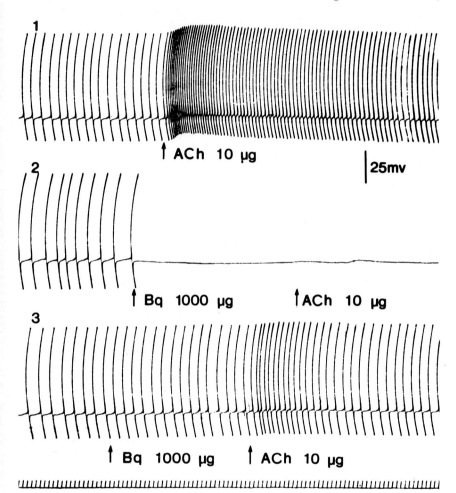

Fig. 7. The effect of pretreatment with 1000 μg benzoquinonium on the acetylcholine response of a D cell. In (2) the benzoquinonium also blocked the spontaneous activity of the cell.

but never completely blocks the acetylcholine response. Hexamethonium does not block the response to acetylcholine in H cells, even at a concentration 1000 times greater than the dose of acetylcholine. However the response to acetylcholine is reduced. Hexamethonium always slows the spontaneous activity of the cell but rarely causes hyperpolarization. Decamethonium does block the H response to acetylcholine, although a concentration between 100 and 1000 μg is required. In an H cell decamethonium always inhibits the activity. On certain occasions decamethonium at low concentrations, 10 and 100 μg, potentiates the acetylcholine response but then blocks the response to acetylcholine at 1 000 μg. Similarly succinylcholine blocks the spontaneous activity of the cell.

Mecholine antagonists

Addition to the bath of 1 μg atropine completely antagonizes the response to 50 μg mecholine (Fig. 8, trace 2). Pretreatment with 1 μg tubocurarine has a slight effect on the mecholine response (Fig. 8, trace 3). The antagonism is completely reversible, as is seen in Fig. 8, trace 4. Atropine at a concentration of 1 μg does not alter the rate of the spontaneous activity of the cell. Atropine antagonizes the mecholine response in both H and D neurones.

Nicotine antagonists

Both the D and H responses to nicotine are antagonized by pre-treatment with tubocurarine (Figs 9 and 10 respectively). The response of the cell to nicotine very quickly exhibits tachyphylaxis. This can clearly be seen in Fig. 9, traces 1 and 4). This effect is less marked in Fig. 10, traces 1 and 4). Atropine has a slight effect on the nicotine response, but this is difficult to distinguish from the tachyphylaxis effect.

Muscarinic agonists

The effects of a number of muscarinic agonists were tested for their relative potency on both H and D cells. The effect of acetylcholine, muscarine, muscarone and mecholine can be seen in Fig. 11. In this preparation the responses to acetylcholine, muscarine and muscarone were of a similar intensity, although muscarine inhibited the activity for a slightly longer period. However, the degree of hyperpolarization is greater with muscarine than with either muscarone or acetylcholine. Mecholine is less active and caused no hyperpolarization. For this cell the order of potency would be: muscarine > muscarone $\approx$ acetylcholine > mecholine. In another H cell, the potency of acetylcholine was compared with pilocarpine, arecoline and oxotremorine (Fig. 12). In this case the order of potency was: acetylcholine > pilocarpine > arecoline $\approx$ oxotremorine. A similar pattern in the order of potency can be seen in D cells (Figs 13 and 14). The order of potency for

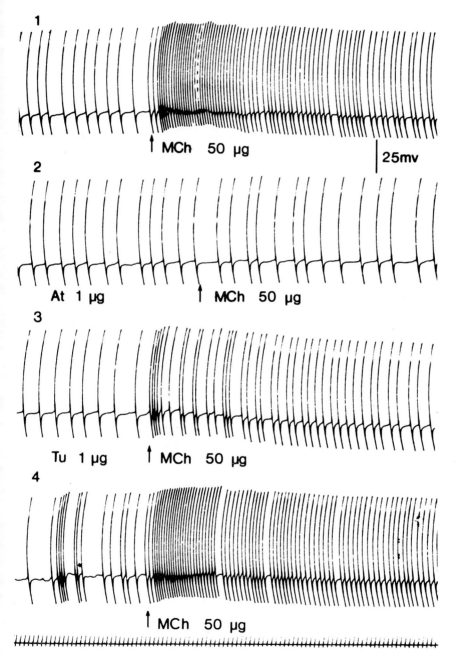

Fig. 8. The effect of pretreatment with 1 μg atropine and 1 μg tubocurarine on the mecholine response of a D cell.

acetylcholine, muscarine, muscarone and mecholine was: acetylcholine > muscarine > muscarone > mecholine. Mecholine was the least active compound on this cell, 1 μg having no effect on the activity.

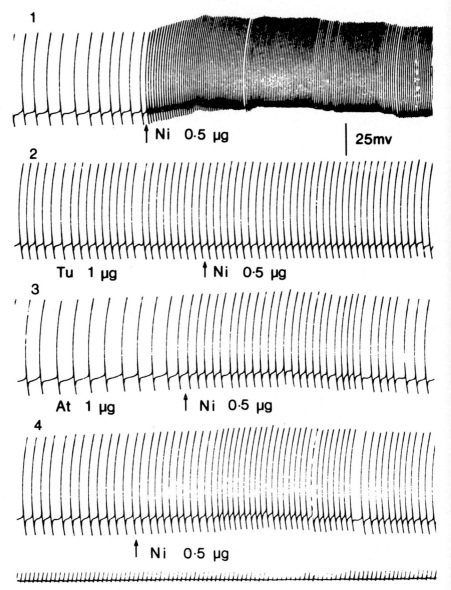

FIG. 9. The effect of pretreatment with 1 μg tubocurarine and 1 μg atropine on the nicotine response of a D cell.

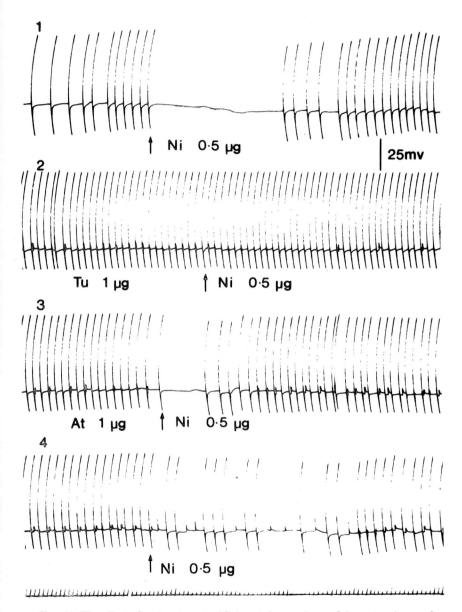

Fɪɢ. 10. The effect of pretreatment with 1 μg tubocurarine and 1 μg atropine on the nicotine response of an H cell.

c

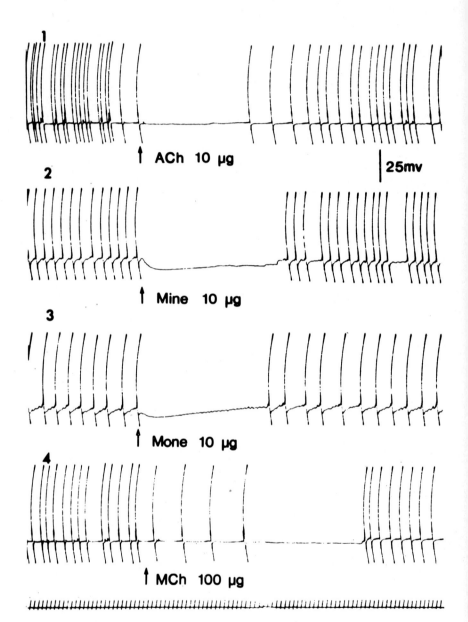

FIG. 11. The effect of 10 μg acetylcholine, 10 μg muscarine, 10 μg muscarone, and 100 μg mecholine on the spontaneous activity of an H cell.

Similarly in a D cell with acetylcholine, pilocarpine, oxotremorine and arecoline, oxotremorine and arecoline are the least active (Fig. 14). In this cell the order of potency was: acetylcholine ≫ pilocarpine ≫ arecoline > oxotremorine.

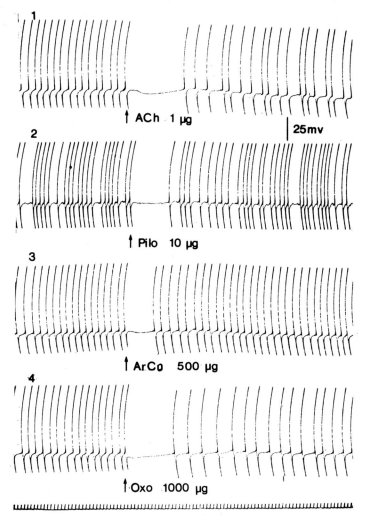

FIG. 12. The effect of 1 μg acetylcholine, 10 μg pilocarpine, 500 μg arecoline, and 1000 μg oxotremorine on the spontaneous activity of an H cell.

Nicotine agonists

The responses of both D and H cells to nicotine and acetylcholine were compared to those obtained with dimethylphenylpiperazinium

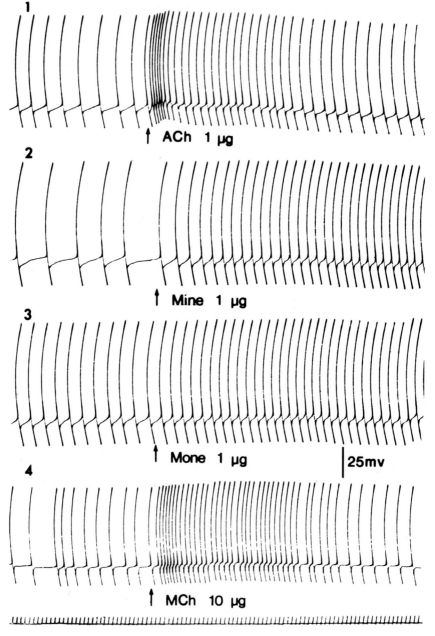

FIG. 13. The effect of 1 μg acetylcholine, 1 μg muscarine, 1 μg muscarone, and 10 μg mecholine on the spontaneous activity of a D cell.

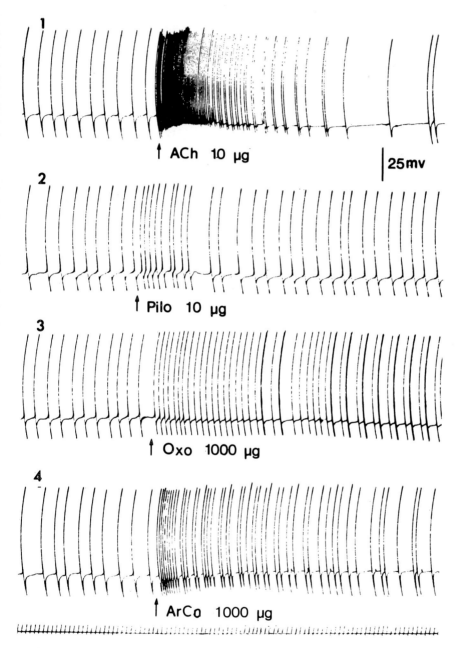

FIG. 14. The effect of 10 μg acetylcholine, 10 μg pilocarpine, 1000 μg oxotremorine, and 1000 μg arecoline on the spontaneous activity of a D cell.

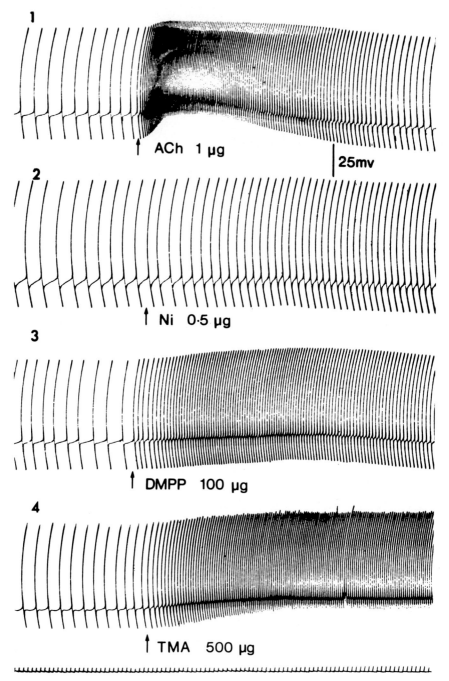

FIG. 15. The effect of 1 μg acetylcholine, 0·5 μg nicotine, 100 μg dimethylphenyl-piperazinium, and 500 μg tetramethylammonium on the spontaneous activity of a D cell.

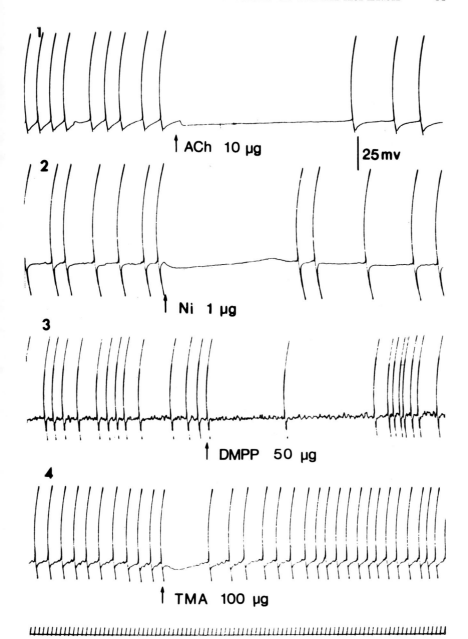

FIG. 16. The effect of 10 μg acetylcholine, 1 μg nicotine, 10 μg dimethylphenyl-piperazinium, and 100 μg tetramethylammonium on the spontaneous activity of an H cell.

and tetramethylammonium. The results are shown in Figs 15 and 16 respectively. In both cases the cells were much more sensitive to acetylcholine and nicotine than to dimethylphenylpiperazinium and tetramethylammonium. The order of potency in the D cell was: acetylcholine > nicotine ≫ dimethylphenylpiperazinium > tetramethylammonium. The H cell was more sensitive to nicotine than to acetylcholine; the order of potency was: nicotine > acetylcholine > dimethylphenylpiperazinium > tetramethylammonium.

Muscarinic and nicotinic cells

During this study it became clear that while certain cells were sensitive to muscarine agonists, such as muscarine, muscarone and mecholine (Fig. 17), other cells were comparatively insensitive (Fig. 18). In some cases the cells sensitive to muscarine agonists were relatively insensitive to nicotine (Fig. 17, trace 5). Conversely certain cells which responded to the addition of low concentrations of nicotine (Fig. 18, trace 2) were relatively insensitive to muscarinic agonists (Fig. 18, traces 3, 4 and 5). However, even in cells sensitive to muscarinic agonists, mecholine was less potent than either muscarine or muscarone. The order of potency in the muscarinic cell was: acetylcholine ≈ muscarine ≈ muscarone > mecholine ≫ nicotine. While the order of potency in the nicotinic cell was: nicotine > acetylcholine ≫ muscarine > mecholine > muscarone. Other cells were approximately equisensitive to both nicotine and muscarine. Figure 19 shows a cell which was sensitive to 1 μg nicotine and 1 μg muscarine. This cell was much less sensitive to the nicotinic agonist tetramethylammonium, although the duration of the tetramethylammonium response was approximately equal to the duration of the muscarine response. The duration of the nicotine response was much longer. The order of potency in this cell was: acetylcholine > nicotine ≈ muscarine ≫ tetramethylammonium.

Anticholinesterases

Pretreatment of the cell with eserine (physostigmine) increases both the initial depolarization of the acetylcholine response and the duration of the response. The acetylcholine response in the presence of eserine resembles the normal carbachol response. Carbachol is not hydrolysed by cholinesterase. The potentiation of acetylcholine occurs in both D and H cells (Figs 20 and 21 respectively) and suggests the

FIG. 17. The effect of 1 μg acetylcholine, 0·5 μg muscarine, 1 μg muscarone, 10 μg mecholine, and 10 μg nicotine on the spontaneous activity of a D cell.

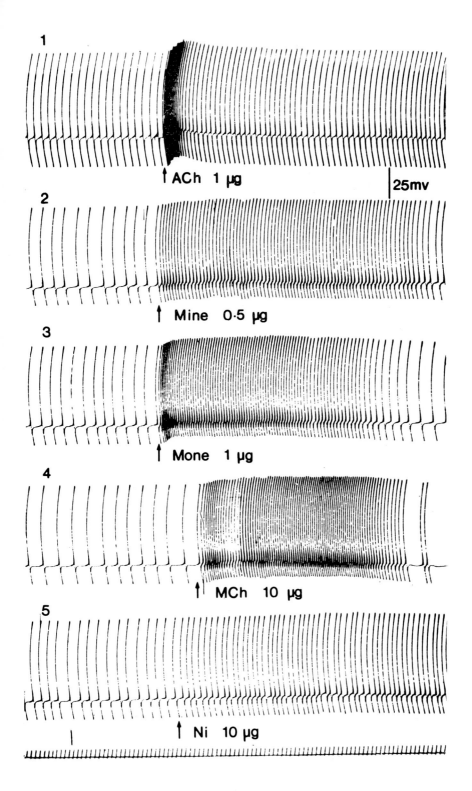

1

↑ ACh 1 µg

25mv

2

↑ Mine 0·5 µg

3

↑ Mone 1 µg

4

↑ MCh 10 µg

5

↑ Ni 10 µg

presence of an esterase in the *Helix* brain. The increased duration of the acetylcholine response following pretreatment with eserine can be seen in Fig. 19.

Cholines

The effect on cell activity of different choline compounds has also been investigated. Butyrylcholine and propionylcholine both have more pronounced effects than acetylcholine on both H and D cells (Figs 22 and 23). However, these compounds are hydrolysed more slowly by cholinesterase than is acetylcholine and this could in part account for their greater effect. This may not be the complete answer

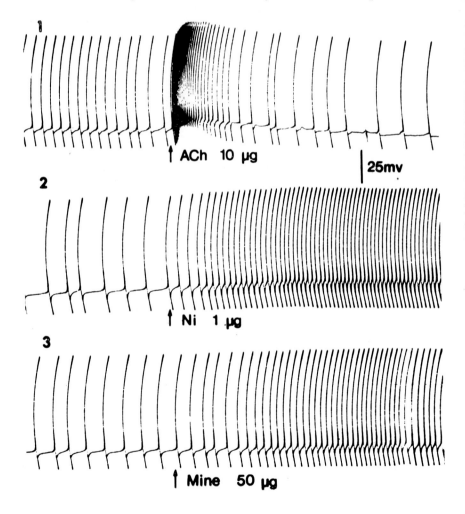

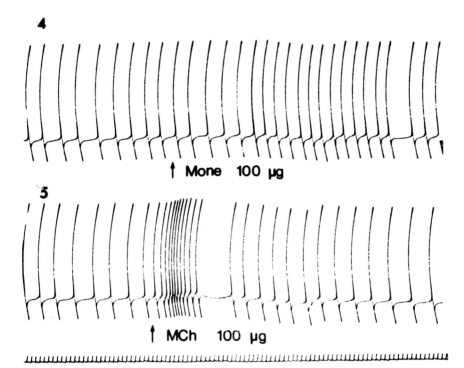

FIG. 18. The effect of 10 µg acetylcholine, 1 µg nicotine, 50 µg muscarine, 100 µg muscarone, and 100 µg mecholine on the spontaneous activity of a D cell.

since mecholine, which is also hydrolysed at a slower rate than acetylcholine, is less potent than acetylcholine (Figs 22, trace 4, and 23, trace 4). The traces in Fig. 23 are displayed as a graph on Fig. 24, which clearly shows the prolonged effect of all three cholines over the acetylcholine response. From Fig. 23 the relative potencies are: butyrylcholine > propionylcholine > acetylcholine > mecholine. In Fig. 25 can be seen the comparative potencies of acetylcholine, carbachol, choline and benzoylcholine: acetylcholine > carbachol > choline > benzoylcholine. The addition of a benzene ring, in the case of benzoylcholine, would appear to reduce the potency of the choline compound. A surprising observation in this cell was the comparatively potent effect of choline (Fig. 25, trace 3) which in vertebrate preparations is generally considered to be at least 1000 times less active than acetylcholine. In this case choline is approximately ten times less active than acetylcholine. In an H cell (Fig. 26) choline is approximately fifty times less active than acetylcholine. In this cell (Fig. 26, trace 3) benzoylcholine was almost equipotent with acetylcholine.

Removal of the onium group of a choline, as in dimethylaminoethanol, would appear to convert the compound into an inhibitory
agent, since dimethylaminoethanol inhibits the spontaneous activity
in both H and D cells (Fig. 27). In this figure, traces 1 and 2 are from
a D cell, and traces 3 and 4 are from an H cell. The sensitivity of
H and D cells to dimethylaminoethanol would appear to be similar.

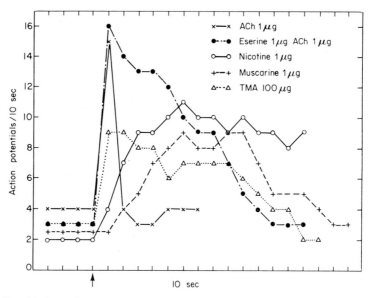

Fig. 19. A graph to show the response of a D cell to 1 μg acetylcholine, 1 μg eserine
followed by 1 μg acetylcholine, 1 μg nicotine, 1 μg muscarine, and 100 μg tetramethylammonium.

Histamine antagonists

Certain neurones respond to histamine by being depolarized and
excited (Fig. 28, trace 1), while histamine inhibits the spontaneous
activity of other neurones (Fig. 28, trace 2). This response to histamine
can be blocked by pretreating the preparation with mepyramine
which specifically antagonizes the response to histamine in vertebrate
tissues. Approximately two-thirds of the neurones tested responded to
histamine, in doses up to 100 μg.

Dopamine antagonists

The spontaneous activity of certain neurones in the *Helix* brain is
inhibited following the addition of dopamine to the preparation

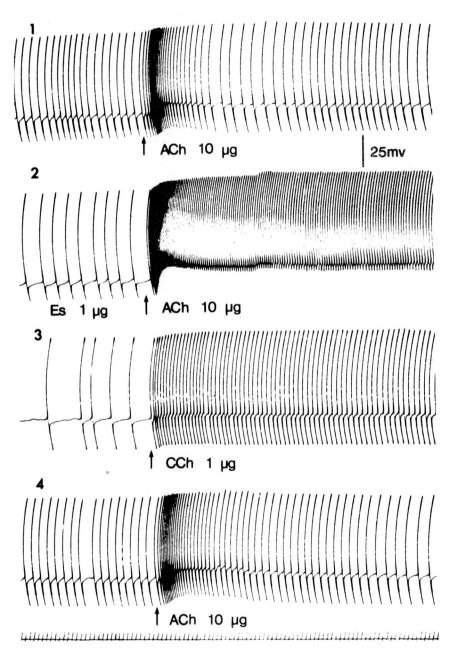

Fig. 20. The effect of 10 μg acetylcholine, 1 μg eserine followed by 10 μg acetylcholine, 1 μg carbachol, and 10 μg acetylcholine on the spontaneous activity of a D cell.

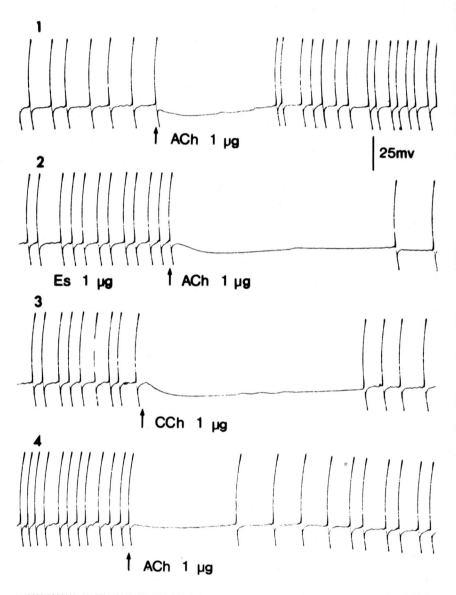

FIG. 21. The effect of 1 μg acetylcholine, 1 μg eserine followed by 1 μg acetylcholine, 1 μg carbachol, and 1 μg acetylcholine on the spontaneous activity of an H cell.

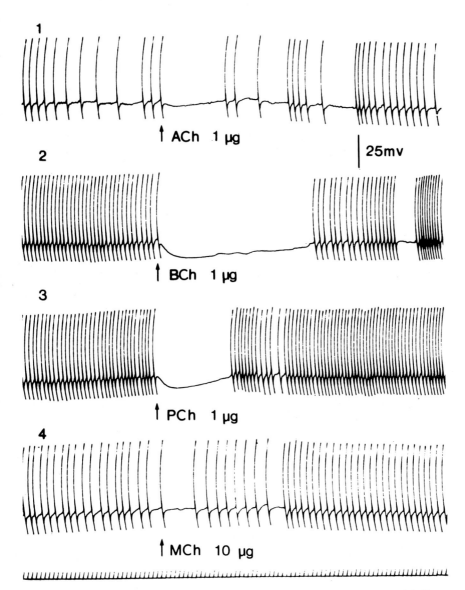

Fig. 22. The effect of 1 µg acetylcholine, 1 µg butyrylcholine, 1 µg propionylcholine, and 10 µg mecholine on the spontaneous activity of an H cell.

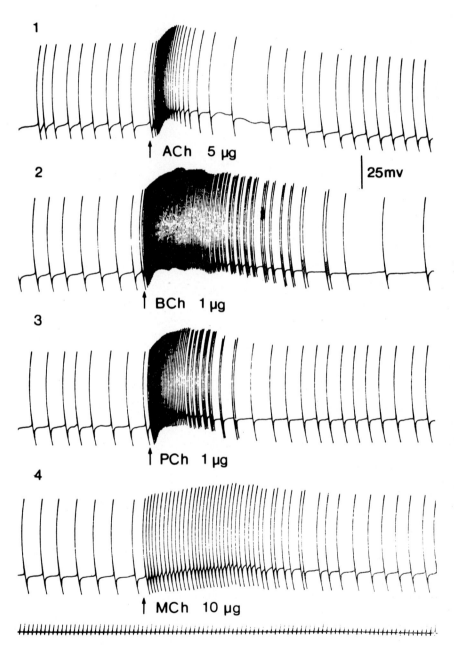

FIG. 23. The effect of 5 μg acetylcholine, 1 μg butyrylcholine, 1 μg propionylcholine, and 10 μg mecholine on the spontaneous activity of a D cell.

(Fig. 29); this inhibitory effect is often accompanied by the hyper-polarization of the cell membrane. This inhibitory response is antagon-ized by the vertebrate catecholamine α-receptor blocking agent, dibenyline (Fig. 29, trace 2). This antagonistic effect is reversible, as can be seen in the third trace of Fig. 29. The time between the addition of the α blocker and the second application of dopamine, as shown in the third trace was 6 min. The response had not quite fully re-covered.

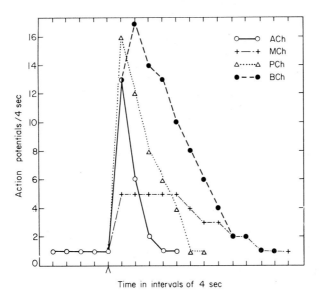

FIG. 24. A graph to show the response of a D cell to 5 μg acetylcholine, 1 μg butyryl-choline, 1 μg propionylcholine, and 10 μg mecholine. The traces from which this graph was drawn are shown in Fig. 23.

DISCUSSION

The neurones of *Helix aspersa* contain cholinoceptive receptors, and these appear to be both muscarinic and nicotinic, since the action of acetylcholine is antagonized equally well by both atropine and tubo-curarine. However, it is possible that the proportion of each type of receptor varies from one cell to another. It is clear that while certain cells are equally sensitive to both nicotine and muscarine, other cells are much more sensitive to one compound than the other. It would be interesting to see if in, for example, a nicotine cell, tubocurarine has a greater ability to antagonize acetylcholine than atropine.

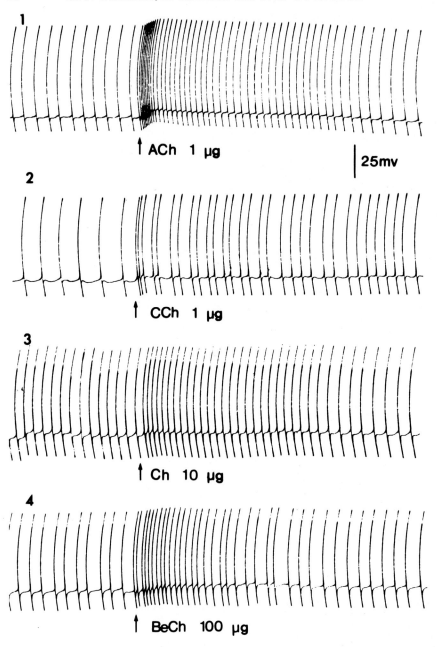

FIG. 25. The effect of 1 μg acetylcholine, 1 μg carbachol, 10 μg choline, and 100 μg benzoylcholine on the spontaneous activity of a D cell.

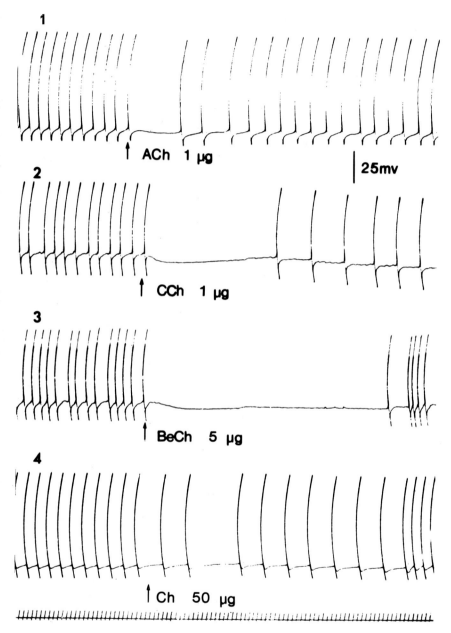

FIG. 26. The effect of 1 μg acetylcholine, 1 μg carbachol, 5 μg benzoylcholine, and 50 μg choline on the spontaneous activity of an H cell.

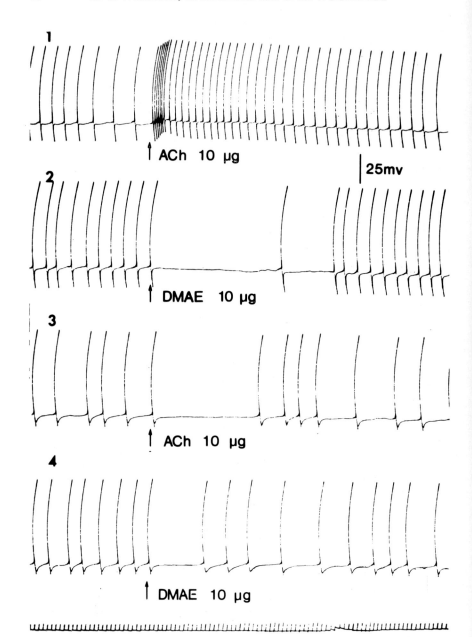

FIG. 27. The effect of 10 µg acetylcholine and 10 µg dimethylaminoethanol on the spontaneous activity of a D cell; the effect of 10 µg acetylcholine and 10 µg dimethylaminoethanol on the spontaneous activity of an H cell.

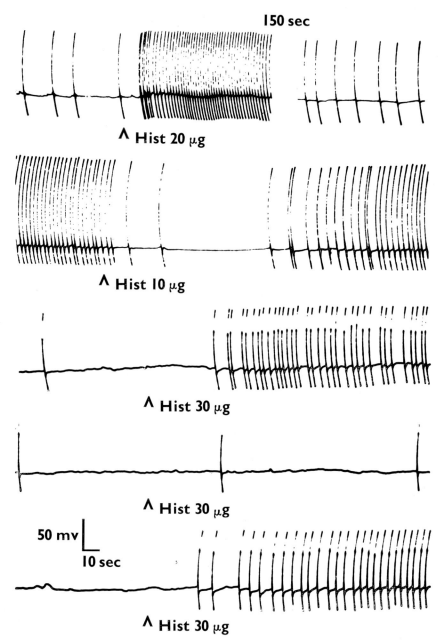

FIG. 28. The effect of histamine on a D cell to histamine, and on an H cell to histamine. Trace 3, the response of a D cell to histamine; trace 4, 1 μg mepyramine followed by 30 μg histamine; trace 5, 30 μg histamine.

There is a distinct division in the response of the *Helix* cholino-ceptive system to the compounds which antagonize the action of acetylcholine at vertebrate cholinergic sites. Thus while atropine, scopolamine, tubocurarine and gallamine antagonize acetylcholine at a concentration ten times less than the dose of acetylcholine applied, other antagonists require much higher concentrations for blockage.

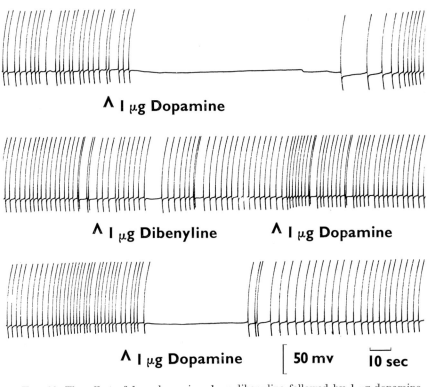

Λ 1 μg Dopamine

Λ 1 μg Dibenyline   Λ 1 μg Dopamine

Λ 1 μg Dopamine   [50 mv   10 sec

FIG. 29. The effect of 1 μg dopamine, 1 μg dibenyline followed by 1 μg dopamine, and 1 μg dopamine on the spontaneous activity of a D cell to acetylcholine.

Atropine and scopolamine are both competitive antagonists at verte-brate smooth muscle junctions while tubocurarine and gallamine are competitive antagonists at the nerve striated muscle junction. None of these antagonists affect the membrane potential, either in vertebrates or in *Helix*. Hexamethonium which is a competitive antagonist at autonomic ganglia will only antagonize the acetylcholine response in *Helix* at concentrations at least 100 times more concentrated than the dose of acetylcholine applied. This antagonism clearly acts on D cells to acetylcholine; however, high doses of hexamethonium often block the spontaneous activity and so mask the H effect to acetylcholine. In cells where acetylcholine clearly hyperpolarizes the membrane potential, the hyperpolarization is not antagonized by hexamethonium when present at a concentration in the bath of 1000 μg. This lack of antagonism of hexamethonium to the H response of acetylcholine agrees with the observation of Tauc and Gerschenfeld (1962) on *Aplysia* neurones. Another cholinergic ganglionic blocker, tetraethyl-ammonium, is also without effect on the acetylcholine response in H and D cells at low concentrations. Since neither ganglionic blocker is effective at low concentrations, it is possible that the nicotinic component of the receptor resembles that at vertebrate nerve striated muscle junctions rather than the receptor at the ganglion. Cholinergic antagonists at the vertebrate nerve striated muscle junction which depolarize and block decamethonium, succinylcholine and benzo-quinonium likewise require very high concentrations to antagonize both H and D acetylcholine responses. It is tempting to correlate their lack of antagonism at low concentrations with their inability to depolarize the *Helix* neurone membrane. At high concentrations these antagonists sometimes depressed the spontaneous activity but did not depolarize the membrane in D cells. However, decamethonium some-times hyperpolarized the membrane potential of H cells, especially when acetylcholine also had a hyperpolarizing effect. Thus, in H cells, decamethonium could be mimicking the vertebrate nerve striated muscle response, but in the snail the effect would be hyperpolarize and block rather than depolarize and block. Benzoquinonium at vertebrate striated muscle behaves both as a competitive antagonist and also as a depolarizing and blocking agent. Benzoquinonium in *Helix* has two effects: in some cells it blocks action potentials in both H and D cells, possibly acting like decamethonium and succinylcholine, while in other cells it has no effect on the spontaneous activity, possibly acting like tubocurarine. In either case, at high concentrations, benzoquinonium blocks the response to acetylcholine in both H and D cells.

Certain central vertebrate cholinoceptive neurones, for example those in the brain stem, contain both nicotine and muscarinic receptors (Bradley and Wolstencroft, 1965). These neurones are excited by acetylcholine. This excitatory effect of acetylcholine can be antagonized by atropine, hexamethonium and gallamine. Thus from vertebrate studies it would appear that receptors on a single neurone can be antagonized by antagonists at all three peripheral cholinergic sites. At the cholinoceptive sites in the vertebrate cerebral cortex the receptor would appear to be muscarinic from a study of agonists.

However, this muscarinic site is antagonized by atropine and gallamine. This resembles the *Helix* receptor in that it is very sensitive to atropine and gallamine.

It is interesting that mecholine is consistently less active than muscarine or muscarone on both H and D cells. Mecholine is approximately equipotent with acetylcholine at nerve smooth muscle junctions. However, in the snail, neurones are usually at least ten times less sensitive to mecholine than to acetylcholine and the initial depolarization and fast firing rate seen in a D cell following the application of acetylcholine is not seen with mecholine. Again at the cholinoceptive muscarinic sites in the cerebral cortex mecholine is equipotent with acetylcholine (Krnjevic, 1965). Other muscarinic agonists, such as pilocarpine, arecoline and oxotremorine, are even less potent than mecholine in both H and D cells. All three compounds are tertiary bases. Of these three compounds, pilocarpine is generally the most active, being about ten times less active than acetylcholine on certain neurones. Oxotremorine and arecoline are generally at least 100–1000 times less active than acetylcholine. At vertebrate muscarinic sites both arecoline and pilocarpine are about 100 times less active than acetylcholine. Oxotremorine is approximately equipotent with acetylcholine at vertebrate muscarine sites (Cho, Haslett and Jenden, 1962). It would be of interest to investigate the action of acetylcholine antagonists on these muscarinic agonists. The only muscarinic agonist so far tested with antagonists is mecholine. This compound is antagonized by atropine but not by tubocurarine.

Nicotinic agonists such as tetramethylammonium and dimethylphenylpiperazinium are also far less active than acetylcholine or nicotine. These two compounds often require at least a concentration 100–1000 times greater than either acetylcholine or nicotine for a threshold response. Tetramethylammonium is about 100 times less active than acetylcholine at the vertebrate striated muscle junction. However, at the vertebrate autonomic ganglion tetramethylammonium is about equipotent with nicotine which is about ten times more active than acetylcholine. The action of tetramethylammonium in the snail would tend to suggest that the nicotinic site resembles that at the nerve striated muscle junction rather than the autonomic ganglion. This finding agrees with the results obtained from nicotinic antagonists where those at the autonomic ganglion are far less effective than the competitive antagonists at the nerve striated muscle junction. Tetramethylammonium is a very feeble agonist at vertebrate muscarinic sites, requiring about 1 000 times the concentration of acetylcholine for an equivalent response. Dimethylphenylpiperazinium is considered to be a potent agonist of nicotine at autonomic ganglia, being about ten times more potent than nicotine itself. However, in *Helix* this compound is relatively inactive, although in some cases it may be ten times more active than tetramethylammonium. Recalling that tetramethylammonium is equipotent with nicotine, then dimethylphenylpiperazinium is also about ten times more active at autonomic ganglia. Nicotine is antagonized by tubocurarine at both H and D cells in *Helix*, while atropine fails to antagonize the response to nicotine. However, the response to nicotine often decreases markedly on succeeding additions. This effect interferes with the response in the presence of antagonists.

Removal of a methyl group from the terminal nitrogen of acetylcholine, as in the case of dimethylaminoethanol, results in inhibition of the spontaneous activity in both H and D cells to acetylcholine. This compound at concentrations above the threshold also hyperpolarizes the membrane. The effect of cholinergic antagonists has not been investigated on the dimethylaminoethanol response.

The relative potencies of different cholinergic agonists are complicated by the presence of active esterases which rapidly hydrolyse acetylcholine and other cholines. Korne (1964), using starch gel electrophoresis, found four esterase bands in homogenates of *Helix* brains. Two of these bands are probably aliesterases while a third is a non-specific cholinesterase. To elimate this problem, it would be necessary to examine the relative potencies of the different agonists in the presence of eserine or prostigmine. The presence of eserine greatly increases the duration of the acetylcholine response so that it resembles the carbachol response. Butyrylcholine would appear to be hydrolysed more slowly than propionylcholine. However, one interesting observation that has emerged from the studies on different choline compounds is the relatively high sensitivity of certain neurones to choline. Thus 10 μg choline will excite D cells and inhibit H cells. These concentrations of choline compare in their response with 1 μg acetylcholine. Choline is about 1000 times less potent than acetylcholine at the vertebrate nerve striated muscle junction. The surprising sensitivity of certain neurones to choline makes it inadvisable to use choline as a substitute for sodium in sodium-free Ringer solutions in this preparation. Pretreatment of the cell with either 1 μg atropine or 1 μg tubocurarine blocks the D response to 100 μg choline. The action of choline clearly demonstrates the importance of the onium group in the action of acetylcholine on the cell.

It is interesting that in addition to acetylcholine, other compounds which are active on *Helix* neurones, for example histamine, dopamine

and 5-hydroxytryptamine are also antagonized by the specific vertebrate antagonists. Thus histamine is antagonized by mepyramine,
dopamine by dibenyline, 5-hydroxytryptamine by LSD 25 (Gerschenfeld and Stefani, 1966). It would appear that the receptors for these
agents on *Helix* neurones resemble the vertebrate receptors.

SUMMARY

1. The acetylcholine response in both H and D cells can be blocked
by muscarinic blocking agents, for example atropine and scopolamine,
and by nicotinic blocking agents, for example tubocurarine and
gallamine.

2. One microgram atropine, scopolamine, tubocurarine or gallamine
will antagonize the response to 10 μg acetylcholine.

3. Hexamethonium, tetraethylammonium, decamethonium, succinylcholine, and benzoquinonium fail to block the acetylcholine
response when they are applied at concentrations equal to or more
dilute than the dose of acetylcholine.

4. Large doses of hexamethonium, tetraethylammonium, decamethonium, succinylcholine and benzoquinonium will block or reduce
the response to acetylcholine in D cells. Decamethonium, succinylcholine and benzoquinonium will also antagonize the acetylcholine
response in H cells. Hexamethonium and tetraethylammonium do not
antagonize completely the response to acetylcholine in H cells.

5. Atropine will completely antagonize the response to mecholine
while tubocurarine only partially antagonizes the response.

6. Tubocurarine will completely antagonize the response to
nicotine while atropine only partially blocks the response.

7. Neurones respond to a variety of muscarinic agonists, muscarine,
muscarone, pilocarpine, arecoline, and oxotremorine. Cells are relatively
insensitive to arecoline and oxotremorine.

8. Neurones respond to nicotinic agonists, dimethylphenylpiperazinium and tetramethylammonium. However, cells are much more
sensitive to acetylcholine and nicotine than they are to either dimethylphenylpiperazinium or tetramethylammonium.

9. Certain cells are much more sensitive to muscarinic agonists than
they are to nicotinic agonists.

10. Certain cells are much more sensitive to nicotinic agonists than
they are to muscarinic agonists.

11. Other cells are equally sensitive to both nicotinic and muscarinic
agonists.

12. Pretreatment with eserine increases the duration of the

acetylcholine response which then resembles that of carbachol. This indicates the presence of an esterase in the brain for the hydrolysis of acetylcholine.

13. The responses to mecholine, propionylcholine and butyrylcholine have a longer duration than equipotent concentrations of acetylcholine.

14. Choline is only about ten times less active than acetylcholine on certain neurones.

15. Histamine hyperpolarizes and inhibits the activity of certain cells and depolarizes and excites other cells. Both effects are antagonized reversibly by mepyramine.

16. Dopamine hyperpolarizes and inhibits the activity of certain cells. This effect is antagonized by dibenyline and is reversible.

ACKNOWLEDGEMENTS

Gifts of D,L-muscarine iodide and D,L-muscarone iodide from Messrs Geigy Pharmaceutical Co. Ltd and of benzoquinonium chloride from the Bayer Products Co. are gratefully acknowledged.

We are grateful to the Science Research Council for a training grant to one of us (A.H.).

REFERENCES

Barlow, R. B. (1964). *Introduction to chemical pharmacology.* 2nd edit. London: Methuen.

Bradley, P. B. and Wolstencroft, J. H. (1965). Actions of drugs on single neurones in the brain-stem. *Br. med. Bull.* **21**: 15–18.

Cho, A. K., Haslett, W. L. and Jenden, D. J. (1962). The peripheral action of oxotremorine, a metabolite of tremorine. *J. Pharmac. exp. Ther.* **138**: 249–257.

Curtis, D. R. (1965). Actions of drugs on single neurones in the spinal cord and thalamus. *Br. med. Bull.* **21**: 5–9.

Gerschenfeld, H. M. and Stefani, E. (1966). An electrophysiological study of 5-hydroxytryptamine receptors of neurones in the molluscan nervous system. *J. Physiol.* **185**: 684–700.

Gill, E. W. (1959). Interquaternary distance and ganglionic blocking activity in bis-quaternary compounds. *Proc. R. Soc.* (B) **150**: 381–402.

Goodman, L. S. and Gilman, A. (1965). *The pharmacological basis of therapeutics.* 3rd edit. New York: Macmillan.

Kerkut, G. A. and Meech, R. W. (1966a). Microelectrode determination of the intracellular chloride concentration in nerve cells. *Life Sci.* **5**: 453–456.

Kerkut, G. A. and Meech, R. W. (1966b). The internal chloride concentration of H and D cells in the snail brain. *Comp. Biochem. Physiol.* **19**: 819–832.

Kerkut, G. A. and Thomas, R. C. (1963). Acetylcholine and the spontaneous inhibitory postsynaptic potentials in the snail neurone. *Comp. Biochem. Physiol.* **8**: 39-45.

Kerkut, G. A. and Thomas, R. C. (1964). The effect of anion injection and changes in the external potassium and chloride concentration on the reversal potentials of the ipsp and acetylcholine. *Comp. Biochem. Physiol.* **11**: 199—213.

Khromov-Borisov, N. V. and Michelson, M. J. (1966). The mutual disposition of cholinoreceptors of locomotor muscles, and the changes in their disposition in the course of evolution. *Pharmac. Rev.* **18**: 1051–1090.

Korne, M. E. (1964). *Some problems of neuromuscular mediation in the higher invertebrates.* D.Phil. Thesis. University of Oxford.

Krnjevic, K. (1965). Actions of drugs on single neurones in cerebral cortex. *Br. med. Bull.* **21**: 10–14.

Phillis, J. W. (1965). Cholinergic mechanisms in the cerebellum. *Br. med. Bull.* **21**: 26–29.

Tauc, L. and Gerschenfeld, H. M. (1962). A cholinergic mechanism of inhibitory synaptic transmission in a molluscan nervous system. *J. Neurophysiol.* **25**: 236–262.

Triggle, D. J. (1965). *Chemical aspects of the autonomic nervous system.* London: Academic Press.

Walker, R. J. and Hedges, A. (1967). The effect of cholinergic antagonists on the response to acetylcholine, acetyl-β-methylcholine and nicotine of neurones of *Helix aspersa. Comp. Biochem. Physiol.* **23**: 977–987.

Walker, R. J. and Hedges, A. (1968). The effect of cholinergic agonists on the spontaneous activity of neurones of *Helix aspersa. Comp. Biochem. Physiol.* **24**: 355–376.

Waser, P. G. (1961). Chemistry and pharmacology of muscarine, miscarone and some related compounds. *Pharmac. Rev.* **13**: 465–515.

Symp. zool. Soc. Lond. (1968) No. 22, 75–92.

SHELL REGENERATION IN *HELIX POMATIA* WITH SPECIAL REFERENCE TO THE ELEMENTARY CALCIFYING PARTICLES

ANNA ABOLIŅŠ-KROGIS

Institute of Zoophysiology, University of Uppsala, Sweden

SYNOPSIS

The process of mineralization of shell-regenerating membranes was studied in *Helix pomatia*.

In the first phase of mineralization different types of *organic* crystalline bodies are formed within the organic matrix. They show positive reactions for the presence of proteins, acid mucopolysaccharides and lipids. They were completely destroyed after the successive action of trypsin and testicular hyaluronidase.

During the second phase of mineralization the organic crystalline bodies in the shell-regenerating matrix are transformed into the inorganic calcium carbonate crystals. Crystal types of calcite and aragonite, and probably also of vaterite, are present in the matrix simultaneously.

Every organic crystalline body contains a set of strongly refractive particles, the b-granules, *ca* 0·2–0·4 μ in diameter, which are arranged in a pattern characteristic of the peculiar type of the body. Moreover, free particles are randomly dispersed in the matrix. A working hypothesis was advanced by the author according to which the b-granules represent the elementary calcifying particles in the mineralization of the shell. They act as kind of "calcification centres".

The damage to the shell activates the ordinary cell functions of the hepatopancreas and the mantle. A suddenly increased disintegration and output of various accumulated cell inclusions and substances occur. Both the released structural elements, the refractive b-granules, and the required substances are transported to the area of shell repair by the migratory cells.

The content of b-granules constitutes a chemical system capable of hardening the structural proteins and binding the calcium ions.

At least part of the substances integrated in b-granules originate in the Golgi region of the cytoplasm of hepatopancreatic cells. Small Golgi vesicles with dense osmiophilic material fuse together and form membrane-bounded granules. These membrane-bounded granules are successively transformed into large concrements by incorporation of new materials from the cytoplasm. It is supposed that within the concrements the substances of b-granules or their precursors are formed.

INTRODUCTION

The shells of molluscs are good material for studying the basic processes of tissue mineralization. The mineralizing substratum is easily accessible, the gradual process of the mineralization is readily seen and is accomplished in a relatively short time.

Generally the shells of molluscs consist of three parts: (1) the outer cuticular part, the periostracum; (2) the outer crystalline part, the prismatic layer, and (3) the inner crystalline part, the nacreous layer or nacre. In *Helix pomatia* the prismatic constituent is five or six layers. The calcium carbonate crystals in each alternate layer are arranged in palisades or in lattices and only traces of organic material appear between them. The nacreous layer is relatively reduced. It is formed of a few mineralized sheets and the calcium carbonate crystals are mostly of the spherulite type; the organic component consists of extremely thin sheets, which alternate with the crystalline lamellae (Grégoire, 1957).

The essential processes in normal growth and in repair of the damaged shell seem to be comparable. When the damaged area is not in contact with the mantle border and the mantle groove, which are responsible for the normal shell growth, the periostracum is not produced in the regenerated shell (Levetzow, 1932; Kessel, 1933). The structure of the outer mineralized part of the shell regenerate also differs from the usual one and resembles the structure of the nacre.

The growth and regeneration of the shells of molluscs begin with the formation of an organic matrix and are followed by subsequent mineralization of this matrix. The mineralization of the matrix of *H. pomatia* is quite a complicated process and has at least two phases.

RESULTS AND DISCUSSIONS

The organic crystalline structures of the
shell-regenerating matrix

The present paper is concerned with the first phase of mineralization of the organic matrix during shell repair in *Helix*. It has already been shown that in early shell-regenerating matrices particular crystalline structures occur (Abolinš-Krogis, 1958): their origin and development are connected with the beginning of the mineralization. In this early phase minute round disks, *ca* 0·6–5·0 μ in diameter, boat-shaped disks, *ca* 2·0–5·5 μ in length, and spherulites, *ca* 4·5–5·5 μ in diameter, appear in the matrix (Figs. 1–3). These crystalline bodies undergo growth and morphological transformations typical of the crystals of all natural and synthetic organic polymers formed *in vitro* as described by Bernauer (1929) and Rånby (1957). At least some of them appear to be single crystals of constant and well-defined form. The round disks and the spherulites apparently grow by the equal peripheral addition of new substances, but in the boat-shaped disks the opposite ends of the longitudinal axes grow faster than the central parts. This results in the formation of double-fan-shaped bodies. Minute round and boat-shaped disks often join with the larger solitary crystalline bodies deforming the pattern of their growth. Through the aggregation of the bodies of various kinds complex organic polycrystalline structures of more irregular shape arise.

These crystalline structures show positive histochemical reactions for proteins of the collagen type, acid mucopolysaccharides of chondroitin type and lipids (Aboliņš-Krogis, 1958). The reaction with pyronin-Y was also positive after treatment of the preparations with RNase. The results indicate that in the crystalline structures of the early shell-regenerating membrane of *Helix* a complex substance is present that is similar to the osteoid, the calcifiable ground substance (matrix, collagen fibres, mucopolysaccharides) in the mineralizing tissues of vertebrates (McLean and Budy, 1959; Sobel, Laurence and Burger, 1960). Therefore the crystalline bodies and complex structures should be called organic bodies. Even the minute bodies show most frequently a characteristic birefringence which changes its pattern of growth. It is not yet possible to decide how much of the pattern is due to the intrinsic birefringence of the substances and how much to their arrangement in the crystals and to the presence of the calcium.

The incorporation of calcium into the organic crystalline bodies probably starts early in their development, but at the beginning the addition of the calcium is relatively slow. The gradual mineralization of the organic bodies seems to be accompanied by the rearrangement of their inner structure, the originally rounded outer surfaces of the organic crystals becoming polyhedral (Aboliņš-Krogis, 1958).

Interesting details of the inner organization of the organic crystalline bodies were revealed when they were stained with toluidine blue in citrate buffer at pH 4·5 or in distilled water. The crystals were partially destroyed, but lost their form only a little. The incorporated calcium was lost, probably due to the disruption of the chemical bonds between the calcium ions and the reactive groups of the mucopolysaccharide sulphate esters. The remnants of the organic crystalline structures, the proteins and the mucopolysaccharides, were stained with toluidine blue (Aboliņš-Krogis, 1958). Usually the deformed "organic lattice" of the crystals showed a fibrous or granular structure (compare Figs 9 and 10). The organic crystalline structures were completely destroyed only after the successive action of trypsin and testicular hyaluronidase. This kind of enzymatic destruction indicated once more that the main constituents of the crystalline structures are proteins and acid mucopolysaccharides, probably in a complex form.

The early stages of calcification in shell regenerates of marine molluscs have been described by Bévelander and Benzer (1948). They observed in the matrix typical growing crystals which came to be enclosed in a thin layer of organic matrix and assumed a polyhedral shape. It is possible that the observed crystals correspond to the above-described organic crystalline bodies in the early shell-regenerating matrix of *Helix pomatia*. At the same time, when the existence of the organic crystalline bodies was established (Abolinš-Krogis, 1958), a paper containing results obtained by electron microscope studies on the growth of calcite crystals of the nacreous layer in the shell of the oyster was published by Watabe, Sharp and Wilbur (1958). It appears that the round and elongate particles observed by these authors in the

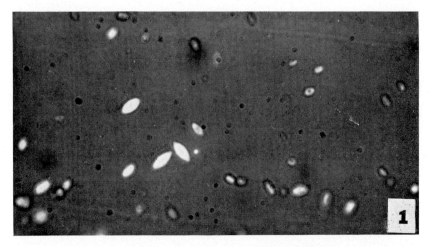

FIG. 1. Photomicrograph of early stages of organic crystalline bodies in the shell-regenerating membrane, showing several birefringent boat-shaped disks of different sizes. At the bottom (left) and in the middle boat-shaped disks are in positions of extinction (black) with a pair of bright b-granules within them. In the middle is a bright birefringent b-granule, and on the left several b-granules in the position of extinction (black). There are some minute round disks. Polarizing microscope. × 1100.

conchiolin matrix and called by them crystal seeds are similar to the early stages of the organic crystals—minute, round and boat-shaped disks, as well as the b-granules to be referred to later. The matter will be discussed in another paper dealing with the results of electron microscope studies on the formation of the organic crystalline bodies (Abolinš-Krogis, in preparation). Wada (1961, 1964b) noticed in the mineralized matrix of marine molluscs the presence of round-shaped metachromatic areas when the glass coverslip preparations were stained with toluidine blue. These areas probably correspond to the remnants of the organic crystalline bodies which after the staining with toluidine blue were partially destroyed. The crystalline (hand-drum-shaped) deposits of pearl oyster described by Wada (1964a) are probably identical with the different types of organic crystals (boat-shaped disks, double-fan-shaped bodies and spherulites) within the regenerating matrix of *H. pomatia*.

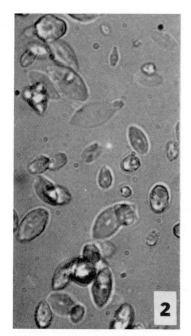 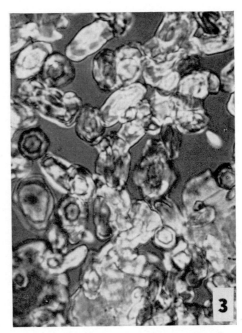

Fig. 2. Later stage of the regenerating membrane, showing boat-shaped and round disks of various sizes; in some of them granules are visible. Light microscope. × 800.

Fig. 3. Still later stage of the crystalline bodies. Partly deformed boat-shaped disks and larger round disks with concentric crystalline layers are beginning to form. At the bottom completely mineralized sheets can be seen. Polarizing microscope. × 700.

The definite calcification of the organic matrix

During the second phase of mineralization the organic crystalline bodies are completely transformed into the inorganic ones. All forms of crystalline bodies finally yield needle clusters or pairs of clusters, the form of which is determined by the initial form of the organic crystal (Fig. 9). The needle clusters (dendrite-like spherulites) are transformed into polyhedral inorganic crystals of calcium carbonate (Aboliņš-Krogis, 1958). The latter join together to form a calcified layer. The carbonate crystals no longer give positive reactions to the presence

D

of organic substances within them. Yet on their surface there can often be observed the remains of organic substances in the form of granules and thick broken fibrils (Figs 4 and 8). Moreover, the minute organic crystalline disks and elongate bodies often join with the polyhedral inorganic crystals, gradually adding to the latter new organic substances (Fig. 5).

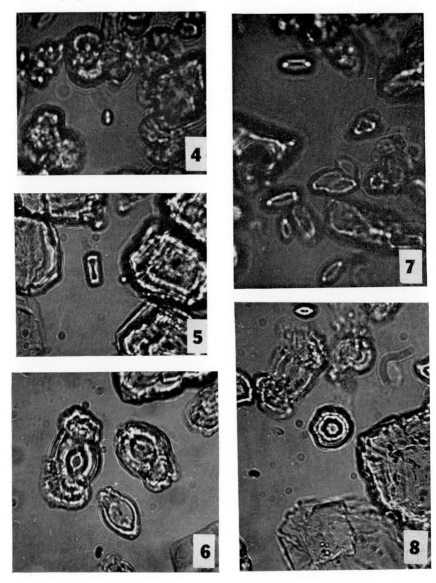

In the shells of molluscs the calcium carbonate is present as calcite and aragonite; the third crystal type of calcium carbonate, vaterite (Mayer, 1931; Levetzow, 1932; Kessel, 1933) is uncommon. Experimental studies were performed by Wilbur and Watabe (1963) concerning the influence of the matrix, temperature and nitrogen supply on the development of the crystal type of calcium carbonate. They found that two aragonitic species laid down only aragonite crystals during regeneration, one calcitic species laid down both calcite and aragonite crystals, and two other species deposited all three crystal types. Species which deposited additional crystal types of calcium carbonate during regeneration or showed divergence from the normal crystal type exhibited an alteration in matrix structure, revealed by the X-ray diffraction method. There are no similar detailed investigations concerning the development of the different crystal types in the regenerating matrix of *H. pomatia*.

I am particularly indebted to Dr. D. Carlström* (personal communication) for his kindness in subjecting some preparations of the shell-regenerating membranes of *Helix* to a preliminary X-ray examination. The results revealed some important facts. Absorption microradiograms showed that even the small crystalline bodies (*ca* $5 \cdot 0$ μ in diameter) were opaque to X-rays—they had a deposition of calcium salts. The diffraction pattern, which was relatively weak, made it possible to suppose the presence of vaterite or calcite in such particles. The powder of larger solitary crystals evidently showed the microdiffraction pattern of calcite. The diffraction pattern of the irregular polycrystalline spherulites was that of aragonite. Obviously, during the mineralization of the organic crystalline structures in the shell-regenerating matrix of *H. pomatia*, all three types of calcium carbonate crystals are present simultaneously.

* The Institute of Medical Physics, Karolinska Institutet, Stockholm, Sweden.

FIG. 4. In the middle is an early boat-shaped disk (out of focus) with two granules in it. Some complex crystalline bodies can be seen.

FIG. 5. Boat-like disk with granules in end positions and rows of smaller granules (out of focus) between them are seen. On the right are larger greatly mineralized crystalline bodies with rows of granular material on their surfaces.

FIG. 6. Large boat-shaped bodies, one of them is double-fan-shaped. There are concentric rows of tightly packed granules corresponding to shape of the crystalline body.

FIG. 7. Several boat-shaped disks showing granules in end positions and fibres between them. Fragments of larger crystalline structures can be seen.

FIG. 8. In the middle is a larger round disk showing concentric rows of granules around the central granulum. At the top are fragments of large boat-shaped bodies (out of focus) and a minute disk. On the right and at the bottom are two polyhedral, completely mineralized crystal sheets with remains of granular organic substance on their surfaces.

FIGS 4–8. Negative phase contrast. × 700–800.

Wilbur and Jodrey (1955) established that carbonic anhydrãse accelerates the formation of bicarbonate and thus also the production of calcium carbonate and development of the shell. Carbonic anhydrase inhibitors markedly reduce the rate of deposition of calcium in the shell of the oyster. The results obtained in this study with 2-benzo-thiazole-sulphonamide on the development of shell regenerates in *Helix* agree with the observations reported by Wilbur and Jodrey (1955).

The elementary calcifying particles, the b-granules

The author is particularly interested in investigating a specific structural feature common to all the organic crystalline bodies in the shell-regenerating matrix (Abolinš-Krogis, 1958, 1960, 1963a–d). All of them contain a set of strongly refractive particles, *ca* $0·2$–$0·4$ μ in diameter, arranged in a pattern characteristic of the peculiar type of body. Free particles are randomly dispersed in the matrix. Since the particles have a similar appearance but different sizes, it is probable that they grow by the addition of new substances. When the light is focused to the depth of the crystalline bodies, the particles can be seen even in the usual light microscope but still better in the phase-contrast microscope. The larger of these particles, which are birefringent and are called the b-granules, probably contain calcium (Abolinš-Krogis, 1960, 1963c), the smaller ones possibly none. The latter are probably identical with the proteinaceous a-granules found in the cells of the hepatopancreas (Abolinš-Krogis, 1960). In the light microscope the minute round disks show the presence of one grain in the centre, whereas the minute boat-like bodies have minimally two grains in their inner part. The growth of the organic crystalline structures is accompanied, or perhaps induced, by the addition of new granules in regular rows around the central grain or in boat-shaped bodies between the grains in the end positions (Figs 4–8). In those cases in which the granules are tightly packed in rows or the rows are not focused, their bulk gives the impression of uninterrupted bands. Yet it is not out of the question that in some cases the granules may have been transformed into layers of fine fibres. The arrangement of the granules in complex crystalline structures often reveals their composition from primary single bodies. A more detailed description of the development of the granular patterns in the organic crystals will be published later (Abolinš-Krogis, in preparation). A working hypothesis has been presented by the author (Abolinš-Krogis, 1963d), according to which the b-granules are the elementary calcifying particles in the mineraliza-tion of the shell. They act as the "calcification centres". It is suggested that the organic crystalline bodies correspond functionally in some way to the segments of the nucleating matrix in the bone and the b-granules to the separate nucleation activating centres of this matrix. The accumulations of the granules in the shell-regenerating matrix apparently occur in accordance with the peculiar spatial patterns, and proceed simultaneously with the formation of the organic crystalline structures and constitute their "visible lattice".

The b-granules are possibly identical with the loose "initial granules" observed by Bévelander and Benzer (1948) in shell-regenerates of the marine molluscs. Wada (1964c) has described two kinds of granules in glass coverslip preparations of the organic matrices: (a) the refractive and faintly metachromatic granules, and (b) the birefringent eosinophil granules. As yet it is not possible to compare these granules with the b-granules, since the latter are refractive or birefringent but did not stain with toluidine blue or eosin.

The intracellular origin of the b-granules

The question as to where in the body of molluscs the substances used for shell-formation are produced has been discussed by many authors. The coverslip insertion method demonstrated that the bulk of the substances is released by the mantle. On the other hand, Wagge (1951) showed that during shell repair calcium carbonate and protein are released from other parts of the shell and from the hepatopancreas; the substances were transported to the damaged area by two kinds of amoebocytes. Using radioactive isotopes Fretter (1952) proved that calcium phosphate is also transported from the hepatopancreas to the area of shell repair by migratory cells. In the hepatopancreas calcium is stored in the crystalline calcium spherites.

Further studies show that in *H. pomatia* the substances required for shell regeneration are distributed from the mantle and the hepatopancreas (Aboliņš-Krogis, 1960, 1963b). The b-granules are released both from excretory and calcium cells of the hepatopancreas, and from the cells of the mantle. The main producer of these elementary calcifying particles is the hepatopancreas (Aboliņš-Krogis, 1960). Experimentally induced shell-repair causes a characteristic activation of the ordinary cell functions of the hepatopancreas in the form of a suddenly increased disintegration and output of various accumulated inclusions and substances from all three types of glandular cells (digestive, excretory and calcium). Within the excretory cells large concrements of heterogeneous structure disintegrate. The released structural elements, the proteinaceous a-granules and the refractive b-granules (some of them birefringent), are extruded basally into the connective tissue. Other b-granules migrate into the neighbouring calcium cells, where they are involved in the formation of new calcium spherites. During the disintegration of the spherites, following the activation of calcium cells, their structure becomes looser. Now it is possible to observe that the spherites contain concentric rows of granules. In the course of disintegration a greater number of loose b-granules appear at the sites of

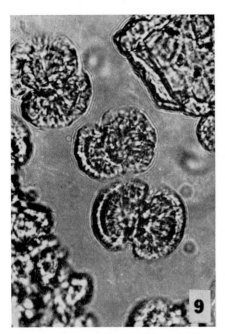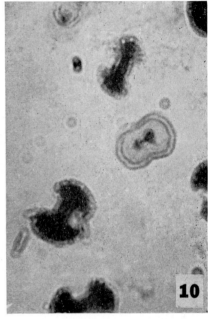

FIG. 9. Fully developed, double-fan-shaped needle clusters. Rows of granules are visible particularly on the ends of the fan radii. Left, a larger aggregate of crystalline bodies; top (right), a polyhedral crystal. Light microscope. × 900.

FIG. 10. Organic remnants of the partly destroyed crystalline bodies, mostly of double-fan type. The sheaves of an organic substance, obviously of fibrous structure, are stained with toluidine blue. At the fan ends of the bodies rows of granules are visible. Other granules are scattered throughout the preparation. In the middle is the remnant of an almost completely destroyed boat-shaped body. The membrane-like matrix is surrounded by rows of a darker-coloured substance. In the centre of the body are the remains of a similar substance (a boat-shaped disk). Light microscope. × 900.

the calcium spherites. Most of them leave the calcium cells at their base and reach the connective tissue. They are transported to the mantle and to the area of shell repair by the migratory cells (amoebocytes and Leydig cells) (Fig. 11).

The histochemical investigations of the hepatopancreas showed that the calcium spherites contained, besides the b-granules, the same groups of substances as the organic crystalline bodies of shell regenerates. The character of the birefringence of the spherulites of the shell-regenerating membrane, including their negative Brewster cross, was similar to that of the spherites in the hepatopancreas (Aboliņš-Krogis, 1958, 1960). Obviously in the shell and the hepatopancreas similar substances are available for the formation of the crystalline bodies in which the presence of a peculiar aggregation of b-granules is necessary for the development of the crystal shape and the progress of mineralization.

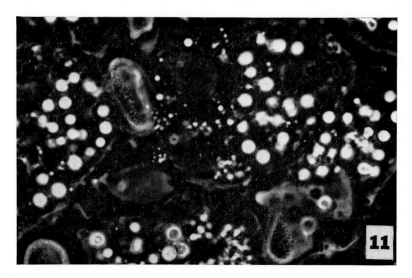

Fig. 11. A transverse section through the hepatopancreas, showing the initial stage of the disintegration of calcium spherites in the activated hepatopancreas of a shell-regenerating snail (1 h after shell damage). There are many opaque calcium spherites. In the middle are b-granules in the places of the disintegrated calcium spherites. The granules are extruded from the calcium cells and enter the still half-empty Leydig cells. Negative phase contrast. × 800.

It was possible to isolate the granules both from the hepatopancreas and from the shell-regenerating membranes (Aboliņš-Krogis, 1963c). The chromatographic analysis showed the presence of calcium, protein-bound amino acids, particularly tyrosine and tryptophan, phenol derivatives and cholesterol within the granules. Polyphenol oxidase of tyrosine type is present in the amoebocytes found in the shell-regenerating membrane and possibly also in the b-granules. All the above-mentioned substances constitute a chemical system capable of hardening the structural proteins and binding the calcium ions (Abolinš-Krogis, 1963c). The method used did not prove whether the acid mucopoly-saccharides were present in b-granules. Of particular interest is the fact that cholesterol was found within them: its presence possibly facilitates the precipitation of calcium ions and the formation of protein fibrils (Abolinš-Krogis, 1963c,d).

Some results obtained by electron microscope studies concerning the intracellular origin of b-granules will now be described. A more detailed description will be published elsewhere (Abolinš-Krogis, in preparation). As was previously mentioned, an important part of the b-granules arise from specific cytoplasmic inclusions, the concrements, in the excretory cells of the hepatopancreas. The material needed for the formation of concrements, which are aggregates of pigment (yellow) bodies, is produced in cells of another type, the digestive cells, and extruded from the latter into the excretory cells (or perhaps the digestive cells transform themselves into the excretory cells). The presence of proteinaceous material, melanin, acid mucopolysaccharides, lipids, cholesterol, pterines and purine derivatives has been demon-strated both in concrements and yellow bodies (Abolinš-Krogis, 1960, 1963a,c).

Electron microscope studies made of tissues fixed in potassium permanganate, formalin and osmium tetroxide from the hepato-pancreas of intact snails showed that the pigment bodies originate in the region of the Golgi complex in the cytoplasm of digestive cells. The Golgi complex of these cells is always well developed, showing parallel arrays of paired smooth membranes, i.e. the Golgi cisternae. Vesicles of small size (microvesicles) are usually scattered around the cisternae. Some of them contain dense osmiophilic material. The vesicles show a tendency to fuse together and form membrane-bounded cyto-plasmic granules of medium electron opacity. Similar granules also arise when paired membranes of the Golgi complex encircle and isolate

FIG. 12. Medium-power electron micrograph of a cross-section of the digestive cells of the hepatopancreas. The Golgi complex is apparent in four cells. The paired Golgi membranes are straight or elliptical. Numerous small vesicles can be seen in their vicinity. Some of them contain osmiophilic material, which is also present in Golgi cister-nae and in the granules within the Golgi region. In some cases the Golgi membranes en-circle small areas of the cytoplasm. Larger, membrane-bounded granules, with less dense contents, are distributed throughout the cytoplasm. In some of them minute dense granules or fibrous material can be seen. The mitochondria appear to be slightly deformed: some of them are closely associated with membrane-bounded granules. A nucleus is seen in the cell at the bottom (right). Fragments of the smooth-surfaced endoplasmic reti-culum appear as tubules or vesicles randomly scattered in the cells. The granular endoplasmic reticulum occurs very rarely. Fixed in potassium permanganate, stained with uranyl acetate. × 18 000.

small areas of the cytoplasm or the mitochondrion. The granules grow by the incorporation of new small vesicles frequently observed on their surface. The membrane-bounded granules of various origins show almost no differences in their fine structure. [Within the larger ones differentiation of the content in one or more concentric layers was often perceptible (Fig. 12).]

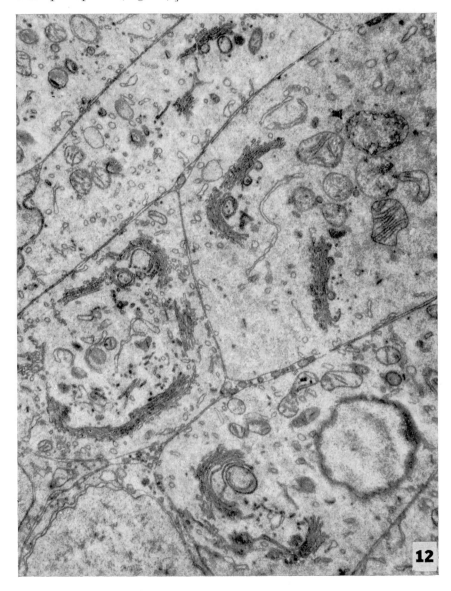

Apart from the membrane-limited granules within the digestive cells varied numbers of extremely dense pigment inclusions are always present in the cytoplasm. These correspond to the previously described yellow bodies visible in the light microscope. They are randomly dispersed within the cell or appear as large clusters accumulated in vacuoles. The electron microscope study of a great number of ultra-thin sections permits us to assume that the pigment inclusions originate from the osmiophilic material concentrated within the Golgi vesicles. This material subsequently accumulates in larger pigment granules and in a number of membrane-bounded granules which are the sites of

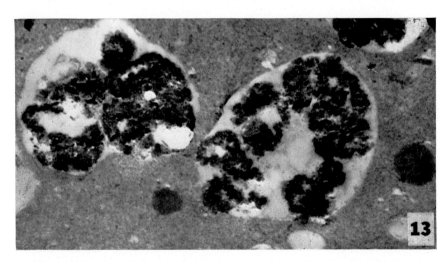

Fig. 13. Electron micrograph of a cross-section of part of a digestive cell. Irregular pigment (yellow) bodies have accumulated within the vacuoles. Two round pigment bodies are in the cytoplasm. Fixed in osmium tetroxide, unstained section. × 18 000.

formation of precursors of pigment (yellow) bodies. The observations showed that these precursor granules were transformed into pigment bodies by incorporation of convenient materials from the cytoplasm. The pigment clusters in vacuoles arise from the aggregation of single pigment bodies. Then, probably by the addition of further available substances, the dense pigmented material attains a looser structure (Fig. 13). The vacuole around it enlarges and the condensation of substances within it proceeds anew. As a result of the condensation, extremely dense spherical bodies, ca $0·4–1·5\,\mu$ in diameter, arise. Finally, by aggregation of these spherical bodies the large concrements are formed. The size of the fully developed concrements varies considerably. All concrements, independently of their size, burst when they

attain the final state of differentiation. Shortly before the rupture the
whole concrement looks heterogeneous. Within a mesh of an electron-
opaque substance, more or less rounded, electron-light areas are visible.
The substance of these areas seems to be fibrous (Fig. 14). In many
cases the remains of the concrements, i.e. the framework of the walls,
are visible in the vacuoles of excretory cells. After the burst the contents

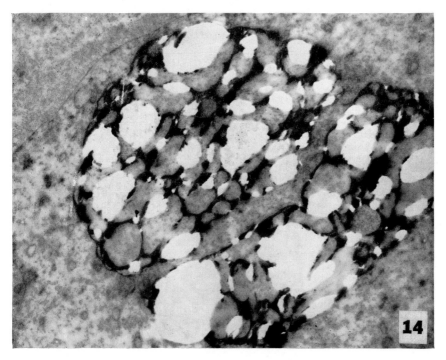

FIG. 14. Electron micrograph of a completely developed concrement (shattered by
section). The heterogeneous structure of the body is clearly visible. In a mesh of electron-
opaque substance more or less rounded, electron translucent bodies are dispersed. The
structure of the rounded bodies is apparently fibrous. Fixed in potassium permanganate,
stained with uranyl acetate. × 10 000.

of light areas reach the cytoplasm. It is supposed that the fibrous
substance of the light areas is related to the fibrous substance of calcium
spherites (Aboliņš-Krogis, 1965) and to the substance of b-granules or
their precursors. Possibly the remaining framework consists of a
complex of lipids and pigments which is extruded from the cell.

  It is one of the aims of further investigation to prove which substances are accumulated in the osmiophilic material of the Golgi

vesicles, in the membrane-bounded granules and within the light areas of concrements. Further, it is important to examine whether some other substances join with the content of the released light areas before it is finally converted to the b-granules or to the organic matrix of spherites. In a paper to be published the structural relations between the matrix of calcium spherites and the b-granules will be discussed. The investigation of the chemical mechanism of incorporation of calcium into the matrix of the b-granules and the organic crystalline structures is being continued.

SUMMARY

1. The process of mineralization of shell-regenerating membranes was studied in *Helix pomatia*.

2. In the first phase of mineralization different types of *organic* crystalline bodies are formed within the organic matrix. In the second phase the organic crystalline bodies are transformed into the inorganic calcium carbonate crystals. Crystal types of calcite and aragonite, and probably also of vaterite, are present in the matrix simultaneously.

3. The organic crystalline bodies contain refractive particles, b-granules, which are arranged in a pattern characteristic of the peculiar type of the body. They probably represent the elementary calcifying particles and act as kind of "calcification centres".

4. The substances required for the shell regeneration are delivered both from the hepatopancreas and from the mantle. They are transported to the area of the damaged shell by the migratory cells.

5. The refractive b-granules, produced mainly in the hepatopancreas, contain substances capable of hardening the structural proteins and binding the calcium ions.

6. Investigation of the ultrastructure of cells of the hepatopancreas shows that both the Golgi complex and mitochondria are involved in the formation of the characteristic yellow pigment bodies which contribute to the formation of the b-granules.

REFERENCES

Abolinš-Krogis, A. (1958). The morphological and chemical characteristics of organic crystals in the regenerating shell of *Helix pomatia*, L. *Acta zool., Stockh.* **39**: 19–38.

Abolinš-Krogis, A. (1960). The histochemistry of the hepatopancreas of *Helix pomatia* (L.) in relation to the regeneration of the shell. *Ark. Zool.* **13**: 159–201.

Abolinš-Krogis, A. (1963a). Some features of the chemical composition of isolated cytoplasmic inclusions from the cells of the hepatopancreas of *Helix pomatia* (L.). *Ark. Zool.* **15**: 393–429.

Abolinš-Krogis, A. (1963b). The histochemistry of the mantle of *Helix pomatia* (L.) in relation to the repair of the damaged shell. *Ark. Zool.* **15**: 461–474.

Abolinš-Krogis, A. (1963c). On the protein stabilizing substances in the isolated b-granules and in the regenerating membranes of the shell of *Helix pomatia* (L.). *Ark. Zool.* **15**: 475–484.

Abolinš-Krogis, A. (1963d). The morphological and chemical basis of the initiation of calcification in the regenerating shell of *Helix pomatia* (L.). *Acta Univ. upsal.* **20**: 1–22.

Abolinš-Krogis, A. (1965). Electron microscope observations on calcium cells in the hepatopancreas of the snail, *Helix pomatia*, L. *Ark. Zool.* **18**: 85–92.

Bernauer, F. (1929). Forschungen zur Kristallkunde. In *"Gedrillte" Kristalle*: 1–102. Johnson, A. (ed.). Berlin: Borntraeger Verlag.

Bévelander, G. and Benzer, P. (1948). Calcification in marine molluscs. *Biol. Bull. mar. biol. Lab.*, *Woods Hole* **94**: 176–183.

Fretter, V. (1952). Experiments with p^{32} and I^{131} on species *Helix*, *Arion* and *Agriolimax*. *Q. Jl microsc. Sci.* **93**: 135–146.

Grégoire, C. (1957). Topography of the organic components in mother-of-pearl. *J. biophys. biochem. Cytol.* **3**: 797–808.

Kessel, E. (1933). Ueber die Schale von *Viviparus viviparus* L. und *Viviparus fasciatus* Müll. Ein Beitrag zum Strukturproblem der Gastropodenschale. *Z. Morph. Ökol. Tiere* **27**: 129–198.

Levetzow, K. G. (1932). Die Struktur einiger Schneckenschalen und ihre Entstehung durch typisches und atypisches Wachstum. *Jena. Z. Naturw.* **66**: 41–105.

McLean, F. C. and Budy, A. M. (1959). Connective and supporting tissues: Bone. *A. Rev. Physiol.* **21**: 69–89.

Mayer, F. K. (1931). Röntgenographische Untersuchungen an Gastropodenschalen. *Jena. Z. Naturw.* **65**: 487–512.

Rånby, B. G. (1957). Nativa och syntetiska polymerers kristallisation. *Svensk kem. Tidskr.* **69**: 61–83.

Sobel, A. E., Laurence, P. A. and Burger, M. (1960). Nuclei formation and crystal growth in mineralizing tissues. *Trans. N.Y. Acad. Sci.* **22**: 233–243.

Wada, K. (1961). Crystal growth of molluscan shells. *Bull. natn. Pearl Res. Lab.* **7**: 703–828.

Wada, K. (1964a). Studies on the mineralization of the calcified tissue in molluscs. II. Experiments by the administration of tetracycline on the mineralization. *Bull. Jap. Soc. scient. Fish.* **30**: 326–330.

Wada, K. (1964b). Studies on the mineralization of the calcified tissue in molluscs. IV. Selective fixation of ^{45}Ca into or onto the metachromatic matter in the processes of shell mineralization. *Bull. Jap. Soc. scient. Fish.* **30**: 393–399.

Wada, K. (1964c). Studies on the mineralization of the calcified tissue in molluscs. VIII. Behaviour of eosinophil granules and of organic crystals in the process of mineralization of secreted organic matrices in glass coverslip preparations. *Bull. natn. Pearl Res. Lab.* **9**: 1087–1098.

Wagge, L. E. (1951). The activity of amoebocytes and of alkaline phosphatases during the regeneration of the shell in the snail, *Helix aspersa*. *Q. Jl microsc. Sci.* **92**: 307–321.

Watabe, N., Sharp, D. G. and Wilbur, K. M. (1958). Studies on shell formation. VIII. Electron microscopy of crystal growth of the nacreous layer of the oyster *Crassostrea virginica*. *J. biophys. biochem. Cytol.* **4**: 281–286.

Wilbur, K. M. and Jodrey, L. H. (1955). Studies on shell formation. V. The inhibition of shell formation by carbonic anhydrase inhibitors. *Biol. Bull. mar. biol. Lab., Woods Hole* **108**: 359–365.

Wilbur, K. M. and Watabe, N. (1963). Experimental studies on calcification in molluscs and the alga *Coccolithus huxleyi*. *Ann. N.Y. Acad. Sci.* **109**: 82–112.

Symp. zool. Soc. Lond. (1968) No. 22, 93–107.

THE MECHANISM OF CALCIFICATION IN THE MOLLUSCAN SHELL

PETER S. B. DIGBY*

*Department of Biology, St. Thomas's Hospital Medical School,
London, England*

SYNOPSIS

Observations on the mussel, *Mytilus edulis*, and other species strongly suggest that calcification of the molluscan shell, like that of crustacean cuticle and mammalian bone, is of electrochemical origin. The calcified part of the shell is always formed beneath the periostracum which consists of quinone-tanned protein and is permeable to salt water. The outer side of the periostracum at the edge of the shell is acid and the inner alkaline. Precipitation of lime is attributed to the action of this alkalinity on the calcium-rich organic colloids. The outer side of the periostracum is electro-negative, the inner positive. Alkalinity and potential are related to mantle activity and probably result from streaming of salt water through periostracum, resulting from suction by the mantle. The acidity and alkalinity are probably due to electrode action within the periostracum, this membrane being strongly semiconducting. Crystal orientation in the prismatic layer is attributed to orientation of the organic matrix and to the sucking of water through the periostracum and between the prisms at the shell edge. Formation of nacre is attributed to the bathing of the inner surfaces of the shell, not subject to suction of sea water through them, with extrapallial fluid, the alkalinity of which is produced at the mantle rim. The flattened form of the crystallites of the nacreous layer is attributed to their deposition in the layer of mucus with a laminar arrangement of the organic matrix caused by sliding movements of the animal within its shell.

INTRODUCTION

Calcification of the shell of molluscs has attracted attention for many years, but the mechanism is still not understood. Recent work is summarized in papers and reviews by Wilbur (1960, 1964). The present paper presents a new theory, that the lime of the calcified shell is precipitated by the alkaline reaction resulting from suction of water through the periostracum by the mantle edge; the alkalinity itself probably resulting from electrode action following internal short-circuiting of streaming potentials through the semiconducting organic complexes in the periostracum.

Theories of the mechanism of calcification

The mechanism of calcification in biological material in general is obscure. The various theories put forward to account for this process are summarized by Digby (1967b). Mostly concerned with bone, the

---

* Present address: Department of Zoology, McGill University, Montreal, Canada.

theories briefly are as follows. The enzyme theory, based on the common association of calcification with phosphatases, phosphorylase and carbonic anhydrase, suggests that the presence of these enzymes determines the site of mineral deposition. The template theory, pointing to the association of first-formed hydroxyapatite with collagen fibres having a particular molecular spacing, suggests that a particular type of organic matrix may cause deposition of inorganic salts of a particular crystalline form by action like that of a seed crystal. It is widely thought that different types of organic matrix may thus lead to correspondingly different inorganic deposits. The crystal poison theory suggests mineral deposition follows enzymatic decomposition of complexes which when present inhibit crystallization: this could possibly account for the association of calcium carbonate deposition with phosphatase, which is otherwise difficult to explain. Serious objections may be raised against all these theories.

Implication of electrode action in the calcification of crab cuticle was suggested by Digby (1964, 1967b). In the crab the outer side of the cuticle is alkaline, the inner side acid; a potential gradient arises from diffusion processes, the outer side being positive. This appears to be due largely to outward diffusion of salt. The cuticle is electronically conducting (semiconducting) because quinone-tanned protein has this property, (Digby, 1961, 1965, 1967b) and the initial lime deposits form around demonstrable centres of electrochemical action. According to the electrochemical theory proposed outward flow of electrons through the semiconducting cuticle causes electrode action on the two sides, the outer thus becoming alkaline and the inner acid. The production of alkalinity in a saturated environment leads to deposition of lime, for around neutrality the solubility of calcium salts is very sensitive to pH. This electrochemical process can account for the changes in calcification through the moult cycle and it also provides the most reasonable explanation for sensitivity to hydrostatic pressure in such forms, this being attributed to hydrogen formed electrolytically in minute quantities on the outer surface (Digby, 1967a). A good test of a theory lies in whether reversing the presumed cause reverses the effect. In the crab this is so; reversing the salinity gradient causes lime to be dissolved from the outer side of the cuticle and deposited on the inner (Digby, 1968).

A piezo-electric theory has been proposed for bone moulding (Bassett, Pawluk and Becker, 1964; Bassett, 1965); calcified bone was shown to form round the cathode of a small mercury cell implanted into the leg of a dog, and small d.c. potentials can arise by rectification of piezo-electric potentials produced by movement.

Calcification of bone can, however, be related to the production of potentials across the growing regions at tne ends of the shaft and below the periosteum by blood flow (Digby, 1966), flow through the fine vessels in these regions producing streaming potentials with the calcifying side positive. The presence of semiconducting complexes in these regions suggests that here also the precipitation mechanism may depend upon the alkalinity resulting from electrode action at the positive side of the semiconducting region. The precise nature of the conduction mechanism and identification of the charge carriers here await further study.

Calcification of the lamellibranch shell

The shell of lamellibranchs such as *Mytilus* consists of three layers, an outer uncalcified organic membrane, the periostracum, and calcified layers beneath; the outer being the prismatic layer with calcite crystals arranged approximately normal or obliquely to the periostracum, and the inner the nacreous layer with crystalline lamellae arranged parallel to the surface (Fig. 1). The calcified layers have a soft organic matrix,

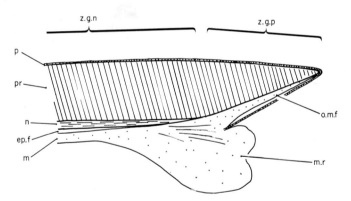

Fig. 1. Diagram of the structure of the edge of a shell of a lamellibranch such as *Mytilus*. *ep.f*, Extrapallial fluid; *m*, mantle; *m.r*, mantle rim; *n*, nacreous layer; *o.m.f*, outer mantle fold; *p*, periostracum; *pr*, prismatic layer; *z.g.n*, zone where shell growth is by deposition of nacreous layer; *z.g.p*, zone where shell growth is by deposition of prismatic layer.

the fibrils of which run parallel to the crystallites. The soft tissues next to the shell constitute the mantle, the edge of which forms a strong muscular band encircling the inner side of the shell and in life more or less adherent to it. A flap of this mantle edge, the outer mantle fold, covers the inner rim of the shell where growth of the prismatic layer occurs. The periostracum is secreted by a fold in the mantle edge from which it passes round the rim of the shell to the outer surface. In the more central parts of the shell a small amount of fluid, the extrapallial fluid, is found between the soft parts and the shell. This fluid is slightly alkaline and supersaturated with lime (Wilbur, 1964). The cells of the mantle, particularly at the edge, secrete mucoprotein and organically bound calcium and iron; the calcified shell is formed by crystallization of lime outside the living cells.

The periostracum consists of quinone-tanned protein (Brown, 1952). This suggests that electrode processes might again be involved in calcification. It is necessary therefore to examine the relation of calcification to possible pH changes across the periostracum, possible electrical potentials across the periostracum and the way in which they are brought about, and the extent to which calcification may result from such potentials, whether by membrane hydrolysis or by electrode action.

MATERIAL

Most of the observations were made on the common mussel, *Mytilus edulis* L. obtained from the Thames Estuary or south coast of England. Additional observations have been made on this species and on the horse mussel, *Modiolus modiolus* L. and a variety of other littoral molluscs from Porsanger Fjord in northern Norway and at the marine station at Kristineberg, Sweden.

CRYSTALLINITY AND PERMEABILITY OF THE SHELL

In the living mussels studied the most recently formed part of the shell at the edge of the current year's growth is soft and breaks easily with a series of parallel fractures, giving a prismatic or fibrous appearance to the broken edge. At the extreme tip new crystals are granular and not yet fused to the rest of the prismatic layer. They show through the periostracum at the rim as a silvery line around the edge. The current year's periostracum may easily be peeled away from such soft newly calcified shell.

When examined from the inside with directional light the rim of the shell below the periostracum is jet black while the nacreous layer over the older parts is pearly. This exposed inner surface of the prismatic layer is permeable. If a shell from which the mollusc has just been removed is partly dried and a small drop of water is placed on the edge of the prismatic layer, the drop is rapidly adsorbed. It can be seen to spread into the prismatic layer, altering its reflectivity. This permeability, presumably arising from the continuous pathways of organic matrix between the crystallites, is largely lost after a few hours or a day or two after removal of the mollusc, probably by recrystallization of the carbonate to form an impervious matrix. In older parts of the shell further recrystallization occurs and these parts then fracture like broken china-ware. These areas on the more central parts of the shell are normally covered with the nacreous layer. In winter the exposed edge of the prismatic layer may, in molluscs from exposed situations, retain their jet black appearance but the mode of fracture of the shell shows that recrystallization of the edge does still in fact occur. Under quiet conditions in estuaries and in an over-crowded aquarium tank a thin coat of nacre may be laid down almost to the edge of the shell, giving it a pearly grey lustre and rendering it very inpervious.

SHELL FORMATION IN RELATION TO ACIDITY AND ALKALINITY

If calcification were brought about by the production of alkalinity which then precipitates lime from an environment, mainly the internal tissues, saturated with the salts concerned at about neutral pH, such localized alkalinity should be detectable. If, however, calcification depended on enzymatic or template processes there would be no obvious need for such pH changes to accompany deposition. Changes of pH are therefore particularly relevant.

Methods and results

The dye Nile blue sulphate used as 1% aqueous solution may be used as a simple pH indicator (Digby, 1965). If an old mussel shell is soaked in sea water or fresh water, stained for 1 min or less with the dye and then replaced in the original medium, the dye adheres slightly to the whole outer surface and stains it blue, indicating a pH in equilibrium with the surrounding water.

If now this process is applied to a living mussel which is in the act of forming shell as shown by the silvery line of new crystals at the rim, and has just been removed from a beaker of sea water in which it has been open and actively filtering, the dye adheres poorly to the periostracum and there assumes a greenish tint at the edge of the shell. This indicates a reaction roughly 0·5 pH unit more acid than the surrounding sea water. If such a specimen freshly removed from a beaker of sea water is held with the rim in contact with a drip of 1% Nile blue, and the rim beneath the drop of dye is nicked with a needle or scalpel blade, a little dye is drawn in beneath the periostracum.

If now the mussel is opened with a knife and the halves examined from the inside, the dye drawn into the space between the infolded edge of the periostracum and the inner surface of the shell is found to have assumed the salmon-red colour of the alkaline form (Fig. 2). This indicates a reaction of the order of pH 9·4, roughly 1·5 pH units alkaline to the sea water. The colour is not due to selective adsorption of the lipophilic red component of the dye by the periostracum because if only a small amount of dye is injected it all assumes the red form, the red colour fading to blue some minutes after opening, indicating dispersal of the alkalinity. Furthermore if fragments of periostracum are equilibrated to sea water before Nile blue staining, they stain only blue.

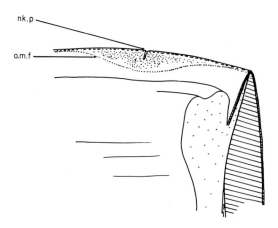

Fig. 2. Diagram of the alkaline reaction shown by the red colour of Nile blue (stippled) sucked into the edge of the shell when the rim is damaged after 1–2 h activity in sea water. Stipple shows how the dye passes round the rim of the shell under the periostracum. nk.p, Nick in periostracum; o.m.f, edge of outer mantle fold retracted from the shell edge at the site of damage, seen through the periostracum.

The mantle edge which normally occupies the space between the infolded periostracum and the calcareous shell retracts at the site of damage, and stains blue. The mantle edge itself does not therefore hold an alkaline fluid either in process of secretion or adsorption. If, after nicking the edge to admit the dye, the mussel is returned to sea water for some minutes before opening, the dye is found to have spread round the edge of the shell beneath the infolded periostracum. Later the dye appears in the extrapallial fluid in the more central parts. Nile blue staining, either in this way or by opening a previously active mussel and applying the dye directly, shows that the alkalinity here is slight.

This is in accord with the observations on the pH of pallial fluid tabulated by Wilbur (1964). In this region also, as at the mantle rim, the soft tissues turn blue with Nile blue and do not obviously secrete or adsorb an alkaline fluid.

The alkalinity on the inner side of the shell edge below the periostracum is very slight or absent in winter, and in summer in mature shells which have practically ceased growth. Neither is it present in mussels which are quiescent after removal from water for some hours.

The distribution of alkalinity and movements of the dye suggests that an alkaline fluid is produced at the shell edge at the inner side of the periostracum. The alkalinity arises not from the mantle edge but from some action involving the periostracum which thus also becomes acid on the outer side. The alkaline fluid then becomes distributed over the inner surface of the shell, its alkalinity dissipating on the way. The deposition of lime in the shell, greater at the edges and less in the more central regions, would be in keeping with simple precipitation brought about by contact between this alkaline fluid and tissues more or less saturated with calcium salts.

THE PRODUCTION OF ACIDITY AND ALKALINITY BY THE PERIOSTRACUM

Since acidity and alkalinity are produced on the two sides of the periostracum, this membrane was examined further. In general appearance it closely resembles the quinone-tanned epicuticle of Crustacea; this latter is semiconducting and exhibits electrode effects (Digby, 1961, 1965). The techniques applied were similar to those used in the study of crustacean material.

Methods and results

Whole living specimens of small mussels or fragments of shell of larger specimens were stained with the Nile blue brine technique (Digby, 1965, 1967b). This consists of soaking the specimens in brine for some minutes, rinsing in fresh water, staining in 1% aqueous Nile blue for 30 sec to 1 min, rinsing in fresh water and observing in fresh water. Certain membranes, notably crustacean cuticle and the periosteum and growing regions of mammalian bone thus develop a purple to salmon pink colour, indicating an alkaline reaction. The tissues become electro-negative owing to more rapid outward diffusion of sodium ions. There is strong evidence to suggest that the alkalinity is due to cathodic action resulting from the short-circuiting of this potential through semiconducting complexes in the organic material, alkali being thus produced electrolytically on the outer surface (Digby, 1965, 1967a,b).

When applied to young mussels the periostracum adsorbs dye heavily, the outer surface of the membrane becoming bronze then purple. Inner parts of the periostracum exposed by damage become purple more readily. A purple reaction is also given by the outer surface of the prismatic layer from which the periostracum has been stripped. The prismatic layer itself adsorbs very little dye. The colours cannot readily be seen in black shells, but in the pale variety, and in other species such as *Cardium* and *Tellina*, no purple is developed, the colour remaining blue. The periostracum thus shares with other calcifying membranes, crab cuticle and bone periosteum, the property of producing an alkaline reaction on outward diffusion of salt. The mucus between the mollusc and its shell also gives this reaction.

Observations on living mussels suggested that suction of sea water through the periostracum, rather than salt diffusion, might be involved. Small perspex cells were therefore made in such a way that fragments of shell or periostracum 0·5 cm square could be held or sealed in position while sea water or saline was forced through, the surface being observed through a microscope or binocular (Fig. 3).

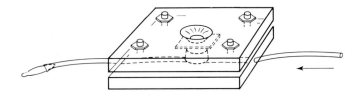

Fig. 3. Perspex cell approx. 2·5 × 3·5 cm used for applying pressure to one side of piece of shell or periostracum while the upper surface is observed by a binocular or microscope. A small piece of shell is shown mounted in the centre.

Pressures of 1–3 atm were applied to sea water or saline beneath such mounted fragments and drops of 1% Nile blue were pipetted on to the upper surface at intervals, each time being left for 15 sec then replaced by sea water or saline. With this treatment periostracum mounted with the normal outer side uppermost became strongly bronze-purple then purple-red. When this treatment was applied to fragments of shell edge, mounted inner side up, a markedly alkaline reaction appeared. After maintaining 1–3 atm for 10–20 min with several applications of Nile blue, a network of minute acicular crystals of the red alkaline form of the dye formed flat on the surface of the prismatic layer. In no cases could definite flow of fluid through pores be seen. The amount of flow is small and the pores are possibly sub-microscopic.

An alkaline reaction can thus arise by forcing fluid through the permeable or slightly porous organic material. In view of the properties of similar membranes in Crustacea it appears likely that the alkalinity here arises as an electrode effect following the short-circuiting of a streaming potential through organic complexes in the material.

POTENTIALS ACROSS THE PERIOSTRACUM OF *Mytilus* AND THEIR PRODUCTION

Electrical potential and the separation of acidity and alkalinity on the two sides of a membrane are clearly related. Potentials were there-fore investigated in order to trace the origin of the alkalinity on the inner side of the shell. The values presented are preliminary in that they have so far been made in November and March, at the end and at the very beginning of the period of shell growth and with electrodes in positions where the magnitudes of the potentials measured are likely to be only a fraction of those generated. The measurements made so far are, however, relevant.

Methods and results

Potentials were measured as in crustacea (Digby, 1965) by a Vibron vibrating-capacitor electrometer (Electronic Instruments Ltd., Rich-mond, Surrey) driving a pen recorder. Silver–silver chloride electrodes with saturated KCl-filled capillaries were used, the tips dipping into droplets of saline or sea water placed on the surface to be measured. Mussels were supported in position with plasticine in a dish of sea water, the level being arranged so as just to cover the opening of the shell (Fig. 4A). Potentials were measured between electrodes on the outer surface, one being placed as reference on the umbo where the perio-stracum was worn away, or on an area close to the rim from which the periostracum had been removed, and the other on the surface of the periostracum in various positions. The potentials measured are thus substantially those across the periostracum although smaller than the true values if they are generated at some distance from the elec-trodes.

Measurements on old shells, soaked for some hours in sea water, showed little or no potential difference between electrodes with tips on the outer surface of the periostracum and the surface of the prismatic

layer respectively. Measurements on shells from which an active living
mussel had been removed a few minutes previously showed marked
potentials of 1–3 mV with the outer side of the periostracum negative.
This could be increased by placing a drop of acid on the surface and
decreased or reversed by a drop of alkali. It is therefore a diffusion
potential arising from the junction of acid periostracum and alkaline
prismatic layer.

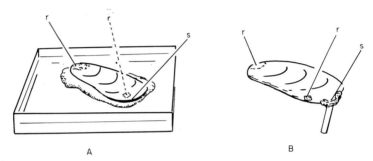

A   B

FIG. 4. A. Arrangement for measuring potentials on the shell of a living mussel.
The rim of the mussel is at water level. Capillary electrodes on flexible mounts follow
the shell movement. Reference electrode, *r*, is shown on the umbo; an alternative position
near the shell edge where the periostracum has been removed is shown with a broken line.
"Search" electrode, *s*, is on the rim. B. Arrangement for measuring potentials generated
by suction applied to the inner surface of the wet shell, through glass tubing fixed with
wax.

Measurements on shells containing living mussels showed similar
potentials. These, however, increased during activity of the mantle
edge. A few minutes after placing a mussel in the dish it commonly
started to open. Opening was preceded and accompanied by fluctuations
of 0·5–1·0 mV above the resting potential. On tapping the dish the
mussel shut and potentials declined (Fig. 5A). With prolonged periods
of activity the level of the resting potential rose (Fig. 5B). The maximum
negative potential occurred on the extreme rim of the shell, comparable
potentials were recorded over the last few millimetres of the shell edge,
corresponding to the extent of the exposure of the prismatic layer
inside.

Consideration of possible ways in which such potentials might be
brought about suggested suction by the mantle edge might be involved.
Further measurements of potential were therefore made on shells
subjected to suction from the inner side. Shells from which the animals
were freshly removed were partially dried and pieces of glass tubing
were attached with sealing wax. They were soaked again and if treated
carefully the tube remained adherent (Fig. 4B). When subjected to

suction of 25 in. Hg potentials of 4–5 mV were produced, the outside of the periostracum becoming negative to the prismatic layer as in living specimens (Fig. 5C). In the specimens shown there is an appreciable lag, possibly because the tube was attached a little in from the edge of the shell. Some specimens have given considerably greater potentials. The shells used were partly recrystallized; experiments with winter shells with a thin covering of nacre produced only a

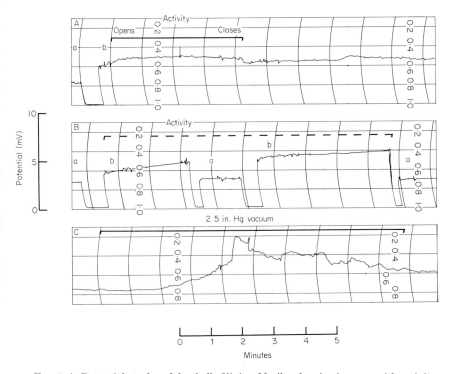

Fig. 5. A. Potential at edge of the shell of living *Mytilus* showing increase with activity as the shell opens, decrease again when the shell closes. At *a*, electrode zero is obtained, both electrodes being together on the umbo. At *b*, one electrode is on the rim. B. Potential at edge of the shell of living *Modiolus* arising from internal activity while the valves are closed. This causes potential of the shell edge to rise progressively. *a*, Electrode zero; *b*, potential of rim. *c*. Potential at edge of the empty shell of *Modiolus* when subject to suction of 25 in. Hg from the inner side of the shell as in Fig. 4B.

fraction of 1 mV, the shells being impervious. Under natural conditions the permeability of the shell edge is probably considerably greater than in the shells used.

Observations on living mussels showed that suction is indeed exerted by the mantle edge. The movement of dye into the shell edge when the periostracum is nicked has been described; if the shell edge is damaged when above water air is immediately drawn in. In further experiments small holes were drilled in the edge of the shell with fine twist drills. When the bit penetrated the inner side of the prismatic layer, a drop of saline or dye around it was immediately drawn in. Suction can therefore be applied to the shell by the mantle edge; the magnitude of the suction pressure and its variation with activity has yet to be determined.

DISCUSSION

The mechanism of formation of acid and alkaline reactions

The theory of streaming potentials is considered in standard texts of physical and surface chemistry (Adam, 1958; Adamson, 1960; Davies and Rideal, 1961). In the present context they arise by suction of salt water through a protein membrane which allows positive ions to pass through more readily than negative ions. The region to which the saline is flowing thus becomes relatively positive. The effects are closely comparable to those arising when the salt diffuses down a density gradient across such a membrane, as in the crab. Following production of such a potential, acidity and alkalinity arise at the two surfaces. As discussed elsewhere (Digby, 1965, 1967b) this could perhaps be a Donnan effect, that is, a movement of hydrogen and hydroxyl ions, or an electrode effect. The latter would involve charge transport by electronic means through the semiconducting material with electrode action at the two surfaces. Critical investigation of the nature of the charge carriers is necessary to settle the question but, as in crustacean cuticle and mammalian bone, the coincidence of calcification with tissues which give the red and purple alkaline colours of Nile blue on outward diffusion of salt and which are semiconducting gives strong support to the view that electrode action is involved. As in the other cases discussed, although the specific conductivity by electronic conduction may be low, the potential gradient is produced across a membrane which is thin and polarization effects may be expected to be slight, so the electronic component of the current which flows may be expected to be large.

As in bone, the system is in some ways comparable to that of a storage battery. A cathodic deposit is accumulated while on charge, while suction is proceeding and when charging ceases by the mollusc becoming inactive a potential of the same sign is produced by diffusion of hydrogen ions from acid periostracum to alkaline prismatic layer, the potential declining slowly until equilibrium is reached.

The mechanism of shell formation

It has been shown that the outside of the growing edge of the shell is acid, the inner alkaline. The outer is negative, the inner positive. It has also been shown that these pH and potential differences are brought about by the activity of the mollusc and that they can also be produced by suction of sea water through the shell and periostracum, the latter being semiconducting. A degree of suction exerted by the mantle edge on the shell is obvious, although precise values have not yet been measured. It so far appears likely that the calcium carbonate of the shell is precipitated by the alkalinity produced initially at the growing edge, and that the alkalinity is brought about by suction exerted by the mantle edge on the inner side of the shell. It is tempting to view these conclusions in the light of the reduced pressures found in the shells of cuttlefish and their allies (Denton and Gilpin-Brown, 1961, 1966; Denton, Gilpin-Brown and Howarth, 1961, 1967). Whether or not the lamellibranch mechanism is entirely muscular or whether there is some osmotic mechanism as described for cuttlefish remains to be seen; the latter so far appears unlikely for if it were so the abstracted fluid would be expected to be alkaline and the mantle edge does not give such an alkaline reaction with Nile blue; the alkaline fluid finds its way to the extrapallial space. The muscular mechanism indicated at once suggests that this may in fact be the main reason for the mantle edge being so strongly muscular.

The interpretation which thus appears most likely is shown in Fig. 6. The structure of the shell is compatible with controlled crystallite

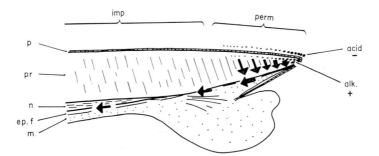

Fig. 6. Summarized interpretation of mechanism of growth of lamellibranch shell as suggested by experiments reported. Broad arrows indicate movement of alkaline fluid, developed during activity, from shell edge towards extrapallial cavity, precipitating prismatic layer in permeable region and nacre where prismatic layer has become impermeable by recrystallization. *Acid*, acid reaction; *alk*, alkaline reaction; *imp*, prismatic layer impermeable; *perm*, prismatic layer permeable. Other abbreviations as Fig. 1.

orientation by fibril direction in the organic matrix and direction of flow of fluid. In the prismatic layer the flow of fluid is through the shell edge along the length of the crystallites, and deposition of lime occurs on the inner ends where alkalinity and organically bound calcium meet. So long as flow through the matrix between the prisms continues the diffusion channels are kept open, shortly after flow ceases the channels become blocked. Nacre owes its characteristic form to its deposition in a matrix containing many planes at which gliding movement occurs with the activity of the mollusc. The way in which the shell becomes impermeable following cessation of activity of the mollusc for a while clarifies the production of annual and disturbance rings; in each case cessation of suction through the shell results in the edge becoming impermeable and with resumption of activity a new start in prismatic layer formation has to be made at the extreme edge of the shell, giving rise to a prominent growth ring.

In the molluscs therefore, as in mammalian bone and crustacean cuticle (Digby, 1966, 1967b), it is suggested that although enzymes may have a part to play in facilitating calcification they are indeed no more than biological catalysts, the process being driven by potentials produced in the ways described. Since most of the salts found in biological calcified structures are precipitated from saturated solutions by alkalinity, it follows that the precise mineralogical structure of the deposits is not strictly relevant to the main issue of calcification and although the template theory is not disproved, the need for it disappears. Further it follows that the activities of the various cells implicated in shell formation are incidental, providing the matrix, the enzymes and the salts for precipitation, but not themselves being responsible for the crystallization of the lime to form the shell.

ACKNOWLEDGEMENTS

My thanks are due to the Director, Dr. B. Swedmark, and the staff of the marine station at Kristineberg, Sweden, for facilities and to St. Thomas's Hospital Medical School for expenses in connexion with this work. I am grateful to the Trustees of the Central Research Fund of the University of London for a grant for some of the equipment used.

REFERENCES

Adam, N. K. (1958). *Physical chemistry.* Oxford: Clarendon Press.
Adamson, A. W. (1960). *Physical chemistry of surfaces.* New York: Interscience.

Bassett, C. A. L. (1965). Electro-mechanical factors regulating bone architecture. In *Calcified tissues 1965*: 78–89. Fleish, H., Blackwood, H. J. J. and Owen, M. (eds). Berlin: Springer-Verlag.

Bassett, C. A. L., Pawluk, R. J. and Becker, R. O. (1964). Effects of electric currents on bone *in vivo*. *Nature, Lond.* **204**: 652–654.

Brown, C. H. (1952). Some structural proteins of *Mytilus edulis*. *Q. Jl microsc. Sci.* **93**: 487–502.

Davies, J. T. and Rideal, E. K. (1961). *Interfacial phenomena*. London: Academic Press.

Denton, E. J. and Gilpin-Brown, J. B. (1961). The buoyancy of the cuttlefish, *Sepia officinalis* (L.) *J. mar. biol. Ass. U.K.* **41**: 319–342.

Denton, E. J. and Gilpin-Brown, J. B. (1966). On the buoyancy of the pearly nautilus. *J. mar. biol. Ass. U.K.* **46**: 723–759.

Denton, E. J., Gilpin-Brown, J. B. and Howarth, J. V. (1961). The osmotic mechanism of cuttlebone. *J. mar. biol. Ass. U.K.* **41**: 351–364.

Denton, E. J., Gilpin-Brown, J. B. and Howarth, J. V. (1967). On the buoyancy of *Spirula spirula*. *J. mar. biol. Ass. U.K.* **47**: 181–191.

Digby, P. S. B. (1961). Mechanism of sensitivity to hydrostatic pressure in the prawn, *Palaemonetes varians* Leach. *Nature, Lond.* **191**: 366–368.

Digby, P. S. B. (1964). The mechanism of calcification in crustacean cuticle. *J. Physiol., Lond.* **173**: 29P–30P.

Digby, P. S. B. (1965). Semi-conduction and electrode processes in biological material. I. Crustacea and certain soft-bodied forms. *Proc. R. Soc.* (B) **161**: 504–525.

Digby, P. S. B (1966). Mechanism of calcification in mammalian bone. *Nature, Lond.* **212**: 1250–1252.

Digby, P. S. B. (1967a). Pressure sensitivity and its mechanism in the shallow marine environment. *Symp. zool. Soc. Lond.* No. 19: 159–188.

Digby, P. S. B. (1967b). Calcification and its mechanism in the shore-crab. *Carcinus maenas* (L.) *Proc. Linn. Soc. Lond.* **178**: 129–146.

Digby, P. S. B. (1968). Mobility and crystalline form of the lime in the cuticle of the shore crab, *Carcinus maenas*. *J. zool., Lond.* **154**: 273–286.

Wilbur, K. M. (1960). Shell structure and mineralization in molluscs. In *Calcification in biological systems*: 15–40. Sognnaes, R. F. (ed.). Washington, D.C.: Am. Ass. Adv. Sci.

Wilbur, K. M. (1964). Shell formation and regeneration. In *Physiology of Mollusca*: 243–282. Wilbur, K. M. and Yonge, C. M. (eds). New York and London: Academic Press.

Symp. zool. Soc. Lond. (1968) No. 22, 109–134.

A REVIEW OF THE BIVALVED GASTROPODS AND A DISCUSSION OF EVOLUTION WITHIN THE SACOGLOSSA

E. ALISON KAY

*General Science Department, University of Hawaii,
Honolulu, Hawaii, U.S.A.*

SYNOPSIS

The members of the sacoglossan opisthobranchiate family Juliidae bear a bivalved shell. They are associated with the alga *Caulerpa*, and the family is distributed throughout the Indo-west-Pacific to Victoria, Australia, the west coast of the Americas, and in the Caribbean. Fossil records date from the Eocene of the Paris Basin.

The opisthobranch features of the bivalved gastropods include their slug-like bodies, head with grooved rhinophores, mantle cavity with a lamellate gill, vas deferens sunk into the haemocoele, and ascus-like radula. The bivalved shell develops from a veliger larva: the threefold mantle secretes the valves which develop as projections from each side of the larval shell. An adductor muscle forms from a portion of the larval retractor, joining right and left shell valves.

Despite the anomalous feature of the bivalved shell, the Juliidae are comparable with other shelled sacoglossans, *Cylindrobulla*, *Volvatella*, *Lobiger* and *Oxynoe*. Mantle cavity, genitalia and nervous system are similar in all, the slight differences which are exhibited being associated with the degree of shell formation and habit. In *Cylindrobulla* and *Volvatella* a horizontally oriented muscle runs between the ventral edges of the shell.

Within the Sacoglossa the shelled tectibranchs comprise largely monotypic families and genera, while the nudibranchs—*Elysia*, *Stiliger*, etc.—have speciated more actively. The possibility that the Juliidae, which represent a creeping radiation in the adaptively radiating sacoglossans, have given rise to the creeping nudibranch forms is suggested. The restricted association of the tectibranchs with the algal genus *Caulerpa* is to be contrasted with the habits of the nudibranchs on a wide variety of algae. The occurrence of zooxanthellae and zoochlorellae in the tissues of a number of nudibranchs suggests the possibility that these organisms provide an essential nutriment which has released the group from its original dependence on *Caulerpa*.

INTRODUCTION

Since the discovery in Japan by Kawaguti and Baba in 1959 of "a curious sea slug having a two-valved shell and a small helicoid shell at the apex of the left valve", molluscan literature has been enriched by some thirty reports wherein the original discovery has been elaborated, species previously referred to the class Bivalvia have been transferred to the Gastropoda, and new species described.

The peculiarities of the anomalous mollusc, *Tamanovalva limax*, have stimulated a reassessment of the family Juliidae, long referred to the Bivalvia, and led to recognition of the widespread distribution and long geological history of this group of sacoglossan opisthobranchs. While their bivalved shell invites comparison with that of the Bivalvia, the slug and its valves also motivate a comparison of the habits and structures of other tectibranch sacoglossans. The diversity of form and habit of the tectibranch sacoglossans brings into focus questions of evolution within the Sacoglossa: the origin of the nudibranch sacoglossans from their shelled ancestors, and radiation in food habits within the order.

SYSTEMATIC REVIEW

Kawaguti and Baba's description of *Tamanovalva limax* stimulated Cox and Rees' (1960) comment on the strong resemblance of *Tamanovalva* to the figures of *Edentellina* Gatliff and Gabriel, 1911, from Australia, and Burn (1960a) reported within a few weeks living specimens of *Edentellina* as bivalved gastropods. The resemblance of *Edentellina* to some shells from the Eocene of the Paris Basin had been noticed by Hedley (1920), whereupon Keen (1960), reviewing the nomenclature of the Paris Basin shells, suggests that *Edentellina* and *Tamanovalva* are synonyms of the genus described from France, *Berthelinia* Crosse, 1875. Reference to *Berthelinia* led Keen (1960) to review the curious pelecypod family Prasinidae (now known as Juliidae) to which Crosse and Fischer (1887) refer not only *Berthelinia* but also *Julia* Gould, 1862. The result was the recognition of the Juliidae as a family of sacoglossan opisthobranchs bearing a bivalved shell with a more or less well-developed heterostrophic protoconch on the left valve posterior to the midline and with a subcentral adductor muscle scar on the interior of the shell. Observations of the habits of living animals of *Julia*, *Tamanovalva* and *Edentellina* indicate that all of the recent genera are associated with the algal genus *Caulerpa*.

Subfamily Juliinae

Included in this subfamily are small (2–10 mm) cordate, solid, porcelaneous green shells with a heavy hinge bearing a prominent tooth-like knob on the right valve and a deep socket in the left valve (Fig. 6 I). A minute, heterostrophic protoconch of about one and one-quarter whorls is present in young stages and living animals but is absent in beachworn specimens. The adductor muscle scar may have a central constriction or be divided into two parts, one above the other.

A single genus, *Julia*, is currently recognized. The type species, *J. exquisita* Gould, 1862, was described from the Hawaiian Islands and is widely distributed in the Indo-west-Pacific; *J. borbonica* (Deshayes,

1863) from Réunion appears to be a synonym (Smith, 1885; but see also Beets, 1944, 1949; Boettger, 1963, for another opinion). Shells examined from Mauritius, the Seychelles, Réunion, Ceylon, New Caledonia, the New Hebrides, the Kermadecs, Lord Howe Island, and the Marshall Islands fall well within the range of variation exhibited by the Hawaiian shells. *J. japonica* Kuroda and Habe, 1951, from southern Japan and *J. equatorialis* Pilsbry and Olsson, 1944, from the west coast of the Americas (Lower California to Peru) may represent geographic subspecies of *J. exquisita*, the Japanese shells being somewhat more solid, larger (7–10 mm compared with 5–7 mm in the Indo-west-Pacific shells), and lighter in color, while the American shells are smaller, ranging in length from 2–3·8 mm. Kawaguti and Yamasu (1966) report *J. japonica* feeding on *Caulerpa ambigua* and depositng its eggs on *C. okamurai*.

Another recent species is *J. cornuta* (De Folin, 1867), known only from the type specimens described from Mauritius. It differs from *J. exquisita* in having two prominent horns projecting from the umbones.

There are, in addition, several fossil records of *Julia*. Three species have been described from the Miocene of Europe, *J. douvillei* (Cossmann and Peyrot, 1914) and *J. girondica* (Cossmann and Peyrot, 1914) from the Aquitaine of France, and *J. lecointreae* (Dollfuss and Dautzenberg, 1901) from the Helvetian near Frankfurt. In the Americas *J. floridiana* Dall, 1898, is found in the Miocene of Florida, and *J. gardnerae* Woodring, 1925, in the Miocene of Bowden, Jamaica. Boettger (1963) has described *J. borneensis* from the Upper Miocene of Java.

Subfamily Bertheliniinae

The shells of the Bertheliniinae are distinguished by their lenticular shape, thin, elastic texture, weak hinge with sometimes obsolete teeth, a central, undivided, circular muscle scar, and retention of the protoconch in the adult shell (Figs 1 and 6H).

There is some question as to the appropriate generic and subgeneric groupings within the subfamily. Keen and Smith (1961) recognize *Berthelinia* Crosse, 1875, and *Midorigai* Burn, 1960b, as genera, and subdivide *Berthelinia* into *Berthelinia* s.s. with three fossil species and the recent *B. schlumbergeri* Dautzenberg, 1895, from Madagascar; the fossil subgenera *Analomya* Cossmann, 1888, and *Ludovicia* Cossmann, 1888; and the recent genera *Tamanovalva* and *Edentellina*. This arrangement was followed by Baba (1961b) and Boettger (1963), although Baba (1961a) earlier treated all the Keen and Smith (1961) subgenera as genera. Burn (1965, 1966) recognizes four genera, *Midorigai*,

E

Berthelinia which he restricts to fossil species with two-whorled proto-conchs, *Tamanovalva* with recent species having one and one-half whorled protoconchs, and *Edentellina*.

Since there are no firm foundations on which to evaluate the sub-familial divisions, I will follow the precedent of Keen and Smith (1961), recognizing *Berthelinia* as an encompassing genus, and of Boettger (1963) restricting *Berthelinia* s.s. to fossil forms, but I consider *Midorigai* a subgenus of *Berthelinia*.

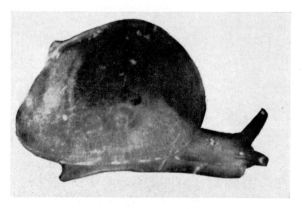

Fig. 1. *Berthelinia pseudochloris* Kay, 1964. (Photo: J. Poulter.)

The shells of *Tamanovalva* are ovate-trigonal, the posterior narrow and rounded; the protoconch is elevated and erect; the radular teeth are laterally microscopically denticulate and the tip is simple. Three species may be recognized.

One species group centers around Kawaguti and Baba's *B. limax* and includes *B. chloris* Dall, 1918, from California, *B. pseudochloris* Kay, 1964, from Hawaii, *B. fijiensis* Burn, 1966, from Fiji, and *B. corallensis* Hedley, 1920, from Queensland. *B. limax* is associated with *Caulerpa okamurai* (Kawaguti and Baba, 1959), *B. chloris* and *B. pseudochloris* occur on *C. racemosa* (Kay, 1964), and *B. pseudochloris* has also been found on *C. sertularioides* (Keen and Smith, 1961). Because of the variability in shell shape displayed by the bivalved gastropods, all except *B. corallensis* may eventually be considered geographic sub-species of *B. chloris*; *B. corallensis* has thus far been recorded only from beachworn specimens and for the present cannot be fruitfully discussed.

B. caribbea Edmunds, 1963, from the Caribbean, is distinguished anatomically by its heart-shaped buccal mass, more concentrated nervous system, lack of oral lobes, and its habit on *Caulerpa verticillata*.

B. babai (Burn, 1965) (= *B. typica* Burn, 1960b), from Victoria, Australia, has radular teeth with a bifid tip; it is recorded from *C. scapelliformis*. *B. schlumbergeri* is known only from a single valve described from Madagascar. Although the protoconch resembles that of the fossil subgenus, it may be a juvenile specimen, and adults would resemble other recent species.

Midorigai Burn, 1960b, is represented by the type species, *M. australis*, from Victoria, Australia (Burn, 1960b). Its obese shape, lobate and large oral lobes, and the occurrence of two adductor muscles distinguish the subgenus. *M. australis* has been recorded from three species of *Caulerpa*.

Edentellina Gatliff and Gabriel, 1911, is apparently also monotypic, represented by *E. typica* Gatliff and Gabriel, from Victoria, Australia. The shell is elongate-oval, the protoconch small and inclined horizontally, and the radula is smooth with a bifid tip. On the basis of this combination of characters, *Edentellina* is subgenerically distinct from *Midorigai* and *Tamanovalva* but Burn (1966) suggests that other as yet undescribed anatomical characters require it being placed in a third subfamily of the Juliidae.

Two fossil species of *Berthelinia* are described from the Eocene of the Paris Basin by Crosse (1875), and *B. burni* Ludbrook and Steel, 1961, comes from the Upper Pliocene of South Australia. *Analomya* Cossmann and *Ludovicia* Cossmann are monotypic, from the Eocene of the Paris Basin.

Distribution and fossil history

The recent and fossil histories of the Juliidae are summarized in Figs 2 and 3. The recent members of the family reflect the distribution of the algal genus *Caulerpa* which occurs not only throughout the Indo-west-Pacific, but as far south as Victoria, Australia, in the Mediterranean, and in the Caribbean. It is noteworthy that neither *Julia* nor *Berthelinia* has been reported from the Mediterranean. While recent *Julia* are probably absent from there, as they are from the Caribbean, it is possible that *Berthelinia* has thus far been overlooked in the Mediterranean. The description of *B. schlumbergeri* from Madagascar indicates that *Berthelinia* also occurs in the Indian Ocean although living animals have not yet been found.

The fossil history of the Juliidae shows that the *Bertheliniinae* are older than the Juliinae. The Juliinae may have arisen from the Bertheliniinae in the Paris Basin, and, in a way similar to that postulated by Fleming (1957) for some pectens, the Miocene ancestors of both subfamilies entered the Indian Ocean by a mid-Tertiary sea route,

E. ALISON KAY

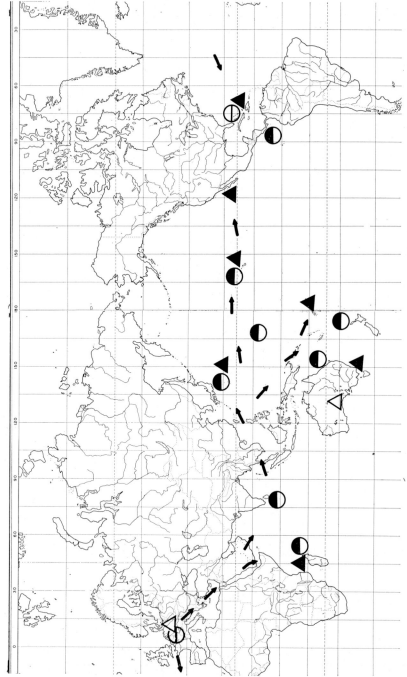

Fig. 2. Distribution of recent and fossil Juliidae. ⊖ Fossil *Julia*; △ fossil *Berthelinia*; ◑ recent *Julia*; ▲ recent *Berthelinia*.

and then moved eastward through the Pacific, *Berthelinia* migrating farther south to South Australia, and both genera reaching across the Pacific to the west coast of the Americas. Since several elements of the Caribbean and Florida Tertiary faunas are related to those of the Miocene of Europe (Woodring, 1928), a similar migration may have occurred across the Atlantic, with *Julia* dying out in the Upper Miocene of Florida and Jamaica but with *Berthelinia* persisting in the Caribbean.

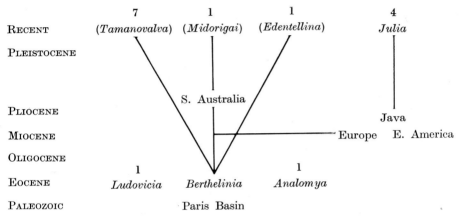

Fig. 3. Diagram summarizing recent and fossil genera and subgenera of the Juliidae. Numbers indicate species.

An alternative hypothesis to account for the occurrence of both *Berthelinia* and *Julia* on the west coast of the Americas is to postulate that the west-American forms were derived from those in the eastern Atlantic, but because of the similiarities of the west American *Julia* and *Berthelinia* and those of the Pacific, the former hypothesis is more attractive.

ANATOMICAL AND CONCHOLOGICAL FEATURES OF THE JULIIDAE

External features and habit

The bivalved gastropods are small, usually less than 10 mm in length, and green like their host plant *Caulerpa*. They differ from species to species in details of color pattern: *B. limax*, *B. chloris*, *B. pseudochloris*, and *B. babai* are minutely speckled with cream, in *B. babai* the mantle contains patches of red (Burn, 1960b), and in *Edentellina typica* there are horizontal, parallel black lines on the mantle (Burn, 1965).

The opisthobranch features of these animals are apparent as they crawl about on *Caulerpa*, projecting anteriorly and posteriorly from the confines of the bivalved shell. The head is bilaterally symmetrical with paired, auriculate, grooved rhinophores and small oral lobes. The eyes are on short projections on the midline of the neck. The foot is narrow, anteriorly hamate, and extends posteriorly as a slender tail; the sole is medially grooved, tending to curl laterally about the axes of *Caulerpa*.

Internal anatomy

When the valves are separated the unorthodox features of *Berthelinia* and *Julia* are easily seen. A smooth, horizontal adductor muscle (Figs 4 and 7A) runs across the anterior third of the visceral mass,

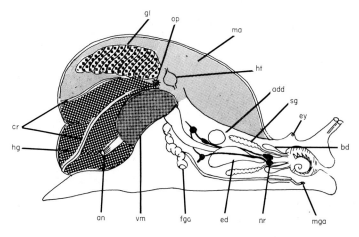

Fig. 4. *Berthelinia limax*: diagrammatic representation of the whole animal with the right valve removed and the right mantle turned back. *add*, adductor; *an*, anus; *bd*, buccal diverticulum; *cr*, ciliated ridges; *ed*, esophageal diverticula; *ey*, eyes; *fga*, female genital aperture; *gl*, gill; *hg*, hypobranchial gland; *ht*, heart; *ma*, right mantle; *mga*, male genital aperture; *nr*, nerve ring; *op*, osphradium; *sg*, salivary gland; *vm*, visceral mass.

uniting the right and left shell valves, and posteriorly a foot retractor rises from the root of the tail. The mantle is divided ventrally into two symmetrical lobes, and the visceral mass is uncoiled, lying in the hollow of the left valve.

If the right valve is removed as in Fig. 4, the mantle cavity is seen between the visceral mass and the right half of the mantle, opening ventrally on the surface of the body. The gill (*gl*) is lamellate, forming a wide belt extending somewhat diagonally across the right mantle from near the midline to the mantle skirt overhanging the foot. There is a small osphradium (*op*) anterior to the gill in the ventral portion of the mantle cavity, and posteriorly there is a pair of well-developed ciliated ridges (*cr*), one on the right mantle, the other on the visceral mass. The heart and kidney are embedded dorsomedially in the right mantle above the gill. The hypobranchial gland (*hg*) is pigmented and diffuse, lying over the posterior surface of the visceral mass. The greater portion of the visceral mass is composed of the digestive gland and the hermaphrodite gland. The pallial portions of the gonoduct are embedded in the visceral mass, and the female genital aperture opens medially on the right side of the foot (*fga*). The vas deferens is sunken into the haemocoele of the foot, with the penis opening on the right side of the head (*mga*). Anteriorly the head-foot bears an oviducal groove.

In details of the alimentary system the bivalved gastropods are clearly sacoglossan. The buccal mass is pyriform with the ascus on the ventral surface containing a single row of articulating teeth, each tooth bladelike with a simple or bifid apex, smooth, or with a lateral series of microscopic denticles. There are long, paired diverticula (*bd*) projecting from the dorsal surface of the buccal mass. The esophagus is thin-walled, with paired salivary glands (*sg*) lying on the ventral surface and a dorsal diverticulum (*ed*). The esophagus passes into the stomach which is a rather ill-defined chamber communicating with the digestive gland by small ducts. The rectum is short and the anus (*an*) opens into the mantle cavity ventrally, posterior to the female genital aperture but anterior to the hypobranchial gland.

The nervous system is streptoneurous but the cerebral and pleural ganglia (*nr*) are concentrated around the esophagus and the pedal ganglia lie beneath the cerebrals and pleurals.

Origin of the bivalved shell

The origin of the bivalved shell and its musculature have been described by Kawaguti (1959) and Kawaguti and Yamasu (1960c). The larva at hatching is a typical veliger with velar lobes and operculum and is about 250 μ in diameter and 150 μ in height (Fig. 5A). Within three or four days after hatching the marginal portion of the larval shell begins to extend, forming projections on each side (Fig. 5B), and the adductor muscle becomes visible near the center of the shell, running obliquely toward the margin of the right mantle. As development proceeds the hinge line becomes visible, gradually becoming more prominent as the valves develop, and the adductor reaches the right valve (Fig. 5C). During development the muscle attachment gradually shifts its position, the original left attachment slipping from the larval shell onto the left valve, and there is also a slight ventral movement of the right attachment. Shortly after metamorphosis, after the velar cilia have been cast off, the velum retractor begins to function as a head retractor which runs somewhat obliquely to the interior of the shell and the animal can now retract into its bivalved shell.

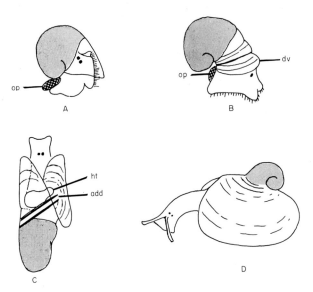

Fig. 5. *Berthelinia limax*: diagrammatic representation of the development of the shell (after Kawaguti and Baba, 1959; Kawaguti and Yamasu, 1960c). A, Veliger larva; B, development of the right valve; C, appearance of adductor; D, metamorphosed snail. *add*, Adductor; *ht*, heart; *op*, operculum.

COMPARISON WITH THE BIVALVIA

The bivalved gastropods are comparable with the Bivalvia only in terms of the shell; the musculature and mantle cavity which might conceivably also be reminiscent of the Bivalvia are structurally and functionally gastropod. Comparison of these structures points up the similarities and differences.

Yonge (1953) has suggested that the primitive shell of the Bivalvia evolved by a bending of the ancestral cap-shaped shell of the early gastropods and lateral compression accompanied by a division of the mantle into two lobes, each with its own center of calcification. The calcified valves consist of three layers, an outermost periostracum, and inner and outer calcareous layers which are secreted by the threefold mantle. The periostracum originates from a groove between the outer and middle folds, the outer calcareous layers are produced by the outer fold, and the inner calcareous layers by the general pallial surface (Owen, Trueman and Yonge, 1953). The calcareous valves are joined by a passively elastic uncalcified ligament, composed of the same layers as are the valves, but produced by a connecting neck of tissue between the two mantle lobes, the mantle isthmus which represents the original dorsal surface of the mantle (Owen et al., 1953).

The shell of the bivalved gastropods is clearly similar to that in the Bivalvia. As described above, the valves are secreted by two lobes of the mantle of the larval gastropod. The mantle is threefold, the outer-most forming the outer calcareous layer, and there is a periostracal groove between the outer and middle layers (Kawaguti and Yamasu, 1961). But while in the Bivalvia the middle fold is primarily sensory and well-developed, and the inner fold muscular, in *Berthelinia* the middle fold is small, and the inner fold consists of a layer of thick cells, some of which are ciliated (Kawaguti and Yamasu, 1961). A portion of the inner fold projects into the mantle cavity as the lamellar-like gill. These modifications may be associated with the gastropod nature of *Berthelinia*, the reduced middle mantle fold perhaps due to the well-developed sensory structures on the head which are absent in the Bivalvia, and the lack of the muscularized inner fold associated with a differently functioning mantle cavity.

As in the Bivalvia, the valves of the bivalved gastropod shell are joined by a connecting ligament which is essentially the primary ligament of Owen et al., (1953). Kawaguti and Yamasu (1961) have described the three uncalcified layers, and the fused mantle beneath. In the bivalved gastropods, the greater portion of the ligament beneath the periostracum consists of the outer layer, and the inner layer is much reduced.

In the Bivalvia and the bivalved gastropods the passive elasticity of the ligament is overcome by adductor muscles which open and close the shell valves. In the Bivalvia there are typically two adductors, and, in addition, the shell and the mantle are attached by pallial muscles near the shell margin; Yonge (1953) suggests that the adductors have arisen by a fusion of some of the pallial muscles.

There is but a single adductor in bivalved gastropods and pallial muscles are absent. The adductor in *Berthelinia* develops from a portion of the larval retractor (= head retractor of Kawaguti and Yamasu, 1960c). The larval retractor in other gastropods appears to be the left member of primitively paired retractors; in *Haliotis* and other proso-branchs the left retractor functions in guiding torsion during larval life but disappears, and the right member of the pair becomes the

functional columellar muscle (Crofts, 1955). Kawaguti and Yamasu
(1960a) point out that the fine structure of the adductor of *Berthelinia*
studied with the electron microscope is gastropod, comparable with the
retractor of the buccal mass of prosobranchs rather than with muscles
in the Bivalvia.

Mention should also be made of the mantle cavity. In the bivalved
gastropods it is limited in extent to the right side of the body and the
gill is single and has no skeleton. The inhalant current is produced by
ciliary action of the gills and mantle walls; the exhalant current is
directed by ciliary currents on the posterior ciliated ridges. The mantle
cavity of the Bivalvia extends the width of the body, contains paired,
skeletonized ctenidia, and the pallial currents are variously modified
and developed in association not only with the respiratory current but
with ciliary feeding.

COMPARISON WITH OTHER SACOGLOSSA

The structural features common to the members of the Sacoglossa
are the suctorial feeding apparatus with the sacoglossan radula and
concentrated nervous system. The opisthobranchs which exhibit these
features include both tectibranch and nudibranch forms. The tectibranchs can be considered as comprising five families, the Cylindrobullidae, Volvatellidae, Juliidae, Oxynoidae and Lobigeridae. There is but
a single genus each in the Cylindrobullidae, Volvatellidae and Lobigeridae, and apparently no more than three or four species in each
genus. In the Juliidae and Oxynoidae there are two or three genera,
some of which may have up to eight species. The nudibranch sacoglossans also comprise at least five families and there appears to be a
similar small number of genera in each family; the number of species in
the nudibranch genera is far greater than in the tectibranchs however,
more than thirty species having now been described for *Elysia* and
perhaps fifteen species each in *Stiliger* and *Hermaea*.

Four tectibranch sacoglossans (Fig. 6) will be compared with the
bivalved gastropods: *Cylindrobulla* (C and D), *Volvatella* (= *Arthessa*
Evans, 1950) (E and F) and *Oxynoe* (A) with bulloid shells in which the
margins meet midventrally, and *Lobiger* (B) with a cap-shaped shell,
the margins of which are free on all sides. In *Cylindrobulla* the apex of
the shell is inrolled, in *Volvatella* the coiled portion of the shell is sunken
in the left posterior margin and the posterior margins are produced as
a spout, and in *Oxynoe* the apex is barely coiled and the margins of the
shell meet only posteriorly. In all the genera the shell is thin and
elastic.

Body form and habit

In body form the tectibranch sacoglossans display a diversity of head form and structures and variously developed foot and parapodia which may be associated with the habits of the animals. The bivalved gastropods with their bilaterally symmetrical form, long, narrow foot extending the length of the animal, creeping habit, well-developed

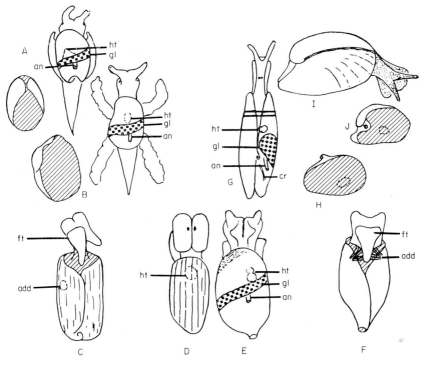

Fig. 6. Tectibranch sacoglossans (diagrammatic). A, *Oxynoe* and shell; B, *Lobiger* and shell; C, *Cylindrobulla*, ventral view (after Marcus and Marcus, 1956); D, *Cylindrobulla*, dorsal view (after Marcus and Marcus, 1956); E, *Volvatella*, dorsal view (after Kay, 1961); F, *Volvatella*, ventral view (after Kay, 1961); G, *Berthelinia*; H, *Berthelinia* shell; I, *Julia exquisita* Gould; J, *Julia exquisita* shell. *add*, adductor muscle; *an*, anus; *cr*, ciliated ridge; *ft*, foot; *gl*, gill; *ht*, heart.

head with paired, auriculate rhinophores and oral lobes, and median, well-developed eyes are most reminiscent of the elysiomorphs. Their small size enables them to remain on the axes of *Caulerpa*. *Lobiger* and *Oxynoe* resemble the bivalved gastropods in color, simulating that of *Caulerpa*, but are larger, and their development of paired parapodia (single, inrolled lobes in *Oxynoe*, paired lateral extensions in *Lobiger*) enables them to move more freely within the mass of *Caulerpa* in which

Lobiger tends to "wallow" and from which *Oxynoe* can swim (Eliot, 1906). The parapodia in *Lobiger* are apparently utilized in defensive action (Gonor, 1961a), but they also enable the animal to flap about in the water although the animals do not do so unless disturbed. *Lobiger* has rhinophores, oral lobes, and the long tapered foot of the bivalved gastropods, but the eyes are lateral. The foot in *Oxynoe* is also elysiomorphic but is apparently more muscular and utilized in swimming (Eliot, 1906); as in *Lobiger* the eyes are widely separated, but there are no oral lobes.

Cylindrobulla and *Volvatella* diverge even more noticeably from the elysiomorphic habit, the foot in both being subtriangular and shorter than the head and the head is flattened, with median, sunken eyes (Fig. 6C–F). In *Volvatella* paired rhinophores and oral lobes are distinguishable in living animals but are much flattened compared with those of the former three tectibranchs, and in preserved material resemble a cephalic shield (Evans, 1950; Burn, 1966). Marcus and Marcus (1956) have described *Cylindrobulla* as possessing a cephalic shield divided into anterior and posterior lobes by a deep median furrow. In color, too, *Cylindrobulla* and *Volvatella* differ from the other tectibranchs: *Cylindrobulla* is described with an ivory cephalic shield, brown visceral mass and pale yellow shell (Marcus and Marcus, 1956) and some species of *Volvatella* are orange, others white (Evans, 1950; Burn, 1966). The flattened head and sunken eyes of *Cylindrobulla* and *Volvatella* also suggest a different habit, that of burrowing. Marcus and Marcus (1956) report *Cylindrobulla* from "muddy algae", while Gonor (*fide* Baba, 1966) describes *Volvatella* in Fiji from *Caulerpa*. My own observations suggest that *Volvatella* normally occurs deep within the algal mat formed by *Caulerpa*.

Shell

The external features of the shells are more difficult to assess in terms of habit adaptations than is general body form. All that can be said is that there is an apparent reduction in the shell among these forms. *Cylindrobulla* with its distinctively inrolled shell (Fig. 6C) represents the most coiled form of the shell in the Sacoglossa, while *Volvatella*, *Oxynoe*, and *Lobiger* (Fig. 6A, B and E) have progressively less coiled and more reduced shells. Although the bivalved shells of *Berthelinia* and *Julia* do not at first glance appear to fit the sequence, their single shells (Fig. 6H and J) approximate those of *Lobiger*. It is interesting that Gould (1862), Deshayes (1863), and Hedley (1920) all comment on the resemblance of the single shells to those of opisthobranchs.

Despite its externally aberrant form, the bivalved gastropod shell is clearly the homologue of the bulloid sacoglossan shells when musculature is considered. In *Cylindrobulla* (Fig. 7B) and *Volvatella* (Fig. 6F) a horizontally oriented muscle runs diagonally between the two ventral edges of the shell. Marcus and Marcus (1956) describe the contraction of the muscle in *Cylindrobulla* as causing an overlapping of the shell edges, and assume the alternative contraction and relaxation of the muscle would result in an effective respiratory current.

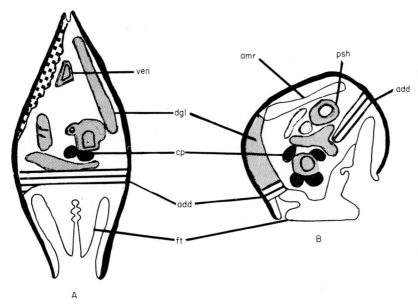

Fig. 7. A, *Berthelinia limax*: diagrammatic cross-section at the level of the adductor (after Baba, 1961b); B, *Cylindrobulla*: diagrammatic cross-section at the level of the adductor (after Marcus and Marcus, 1956). *add*, adductor; *amr*, anterior recess of mantle cavity; *cp*, cerebropleural ganglia; *dgl*, digestive gland; *ft*, foot; *psh*, penial sheath; *ven*, ventricle.

Mantle cavity

The mantle cavity is remarkably similar among the tectibranch sacoglossans, and the few differences which are noticeable may be associated with the varying degrees of shell development and coiling of the visceral mass.

In *Berthelinia* and *Cylindrobulla* the pallial cavity is deeper than in the other three genera, depth in the former genus being associated with lateral compression of the visceral mass and in the latter with the coiled shell. In *Volvatella*, *Oxynoe* and *Lobiger*, which have broad, almost uncoiled visceral masses, the mantle cavity is spacious and shallow, horizontally reaching the full width of the visceral mass. The anterior reaches of the cavity are delimited by the anterior mantle commissure which in *Cylindrobulla* projects from the left side and produces an anterior recess (Fig. 7B, *amr*) (Marcus and Marcus, 1956) and in *Lobiger* separates an anterior chamber from the pallial cavity proper (Gonor, 1961a). The anterior mantle commissure is present in *Berthelinia* in histological cross-sections figured by Baba (1961b), although it has not been described, and is presumably also present in *Volvatella* and *Oxynoe*. In the posterior regions of the mantle cavity there are ciliated ridges which in *Cylindrobulla* parallel the gill around the coils of the visceral whorls (Marcus and Marcus, 1956), and in *Berthelinia*, *Volvatella* and *Oxynoe* lie on the right mantle wall and the visceral mass. There are no ciliary ridges in *Lobiger*, but the posterior borders of the mantle cavity are upturned in a siphon-like arrangement posteriorly. Fretter and Graham (1962) note the presence of ciliated ridges in the mantle cavities of a number of opisthobranchs with deep mantle cavities where they give rise to the excurrent stream of water; Fretter (1960) suggests these ciliated bands compensate for the poor ciliation and small size of the gill.

In all the tectibranch sacoglossans the gill is similar to that described in *Berthelinia*, consisting of projecting, ciliated folds of the mantle wall. The osphradium is small, anterior to the gill, and the hypobranchial gland is posterior to the gill as a diffuse structure covering the visceral mass.

Despite the differences in the posterior regions of the mantle cavity, the ciliary currents and patterns of the inhalant and excurrent water flows are similar in *Berthelinia* and *Lobiger*, the latter described by Gonor (1961a). Water is drawn in anteriorly in a current directed along the anterior half of the aperture. The stream passes over the osphradium and is then moved posteriorly across the gill lamellae by ciliary currents originating on the surface of the gill. Posteriorly the current passes over the hypobranchial gland, and mucus produced by both the gill and the gland entangle fine particles, preventing them from accumulating and clogging the respiratory surface (Gonor, 1961a). The excurrent stream is directed in *Berthelinia* by the ciliated ridges and in *Lobiger* by the excurrent siphon (Gonor, 1961a). The feces from the mid-dorsal anus of *Lobiger* and the more ventrally and laterally placed anus of *Berthelinia* are thus both directly in the path of the excurrent flow. As in the case of other tectibranchs, the result of the ciliary currents produced by the ridges and the sipohn is that of replacing the inhalant by an exhalant force as the main driving power in the mantle cavity.

The heart and kidney are embedded in the mantle in all the tecti-branch sacoglossans. In *Cylindrobulla* the ventricle is posterior to the auricle (Marcus and Marcus, 1956), which is the arrangement charac-teristic of prosobranchs; in the bivalved gastropods the heart lies on the midline with the auricle to the left and the ventricle to the right, and in the other genera the usual opisthobranch pattern of an anterior ventricle and posterior auricle is present.

Alimentary system

The hallmark of the sacoglossans is their suctorial method of feeding and its associated buccal apparatus consisting of a uniseriate radula of articulating teeth which pass below the odontophore into a spiral ascus, anterior esophagus with a dorsal diverticulum, well-developed salivary glands, small stomach opening freely into the digestive gland, and short intestine with the anus opening mid-dorsally. Fretter (1939) has discussed the functioning of the sacoglossan gut in *Elysia*: the radular teeth puncture the plant cells, the fluid contents of which are sucked into the buccal cavity and esophagus by muscular contractions and expansions of the buccal mass. The esophageal pouch apparently serves as a pump, moving the fluid through the stomach to the digestive gland and the food is then directed freely through the stomach to the wide openings of the digestive gland. The short intestine may be associated with the nature of the food, little fecal matter being formed from plant juices.

Essentially the same arrangement is exhibited in the tectibranch sacoglossans, although there is more emphasis on the musculature of the buccal bulb, the anterior buccal glands may be absent, and diverti-cula or "crops" are present on the buccal mass. In the bivalved gastro-pods the buccal diverticula are muscular and paired, and internally there are rugose folds covered with cilia (Baba, 1961b). There are no buccal diverticula in *Cylindrobulla* (Marcus and Marcus, 1956), but there are unpaired diverticula in *Volvatella*, *Oxynoe* and *Lobiger*, which in section show paired lumina (Evans, 1950; Baba, 1961b). The homologies of these diverticula cannot be discussed until detailed studies are made of the buccal apparatus of the sacoglossans.

The esophageal diverticula are variously developed in the tecti-branch sacoglossans, as they are in the nudibranch forms: in the bivalved gastropods the diverticulum forms an extremely long, oval pouch (Fig. 4, *ed*); in *Cylindrobulla*, *Lobiger*, and *Oxynoe* it is a shorter, more circular structure, and there is no esophageal diverticulum in *Volvatella*. Fretter (1939) suggests that the dorsal esophageal diverti-culum, which is also present in bulloid tectibranchs, is an innovation in the opisthobranchs and not the homologue of the esophageal pouches of the prosobranchs.

The lateral position of the anus in the bivalved gastropods differs from the mid-dorsal position in the other genera, a consequence of the lateral compression of the visceral mass.

Genital system

Ghiselin (1965) suggests that the Sacoglossa are primitively oodiaulic, illustrating with *Cylindrobulla* which does not have a vas deferens completely separated from the pallial gonoduct (Marcus and Marcus, 1956). In all the other genera the system is triaulic. Since there are no detailed comparative treatments of the genital ducts of the tectibranch sacoglossans, no attempt will be made here to assess the various features which have been described.

The position of the genital apertures and the presence of an oviducal groove in the bivalved gastropods and *Volvatella* do require comment, however. In the bivalved gastropods and *Volvatella* the male and female apertures are widely separated, and an open oviducal groove leads from the female aperture to the edge of the right rhinophore. Kawaguti and Yamasu (1960b, 1966) have described ovulation in *Berthelinia* and *Julia*, where the eggs, coated with mucus in the membrane gland, pass down the large oviduct toward the female orifice, emerge at the surface of the body, and then move forward in the external oviducal groove to its exit near the mouth where they are deposited in the egg mass. Presumably the same sort of thing happens in *Volvatella*.

The homologies of the oviducal groove are not immediately apparent. External seminal grooves are characteristic of the Cephalaspidea and the Anaspidea, and are interpreted as retentions of the primitive open seminal grooves of the prosobranchs (Ghiselin, 1965); the oviducal groove in the Juliidae and *Volvatella* may thus be but a modification of the sperm groove.

In *Lobiger* and *Oxynoe* there is neither an oviducal nor a seminal groove, and the male and female apertures are close together on the right surface of the foot.

Nervous system

Although the nervous system of opisthobranchs is typically euthyneurous, various degrees of streptoneury are exhibited within the group, and this is perhaps nowhere better demonstrated than in the Sacoglossa where the elysiomorphs display an extreme degree of euthyneurous concentration and varying degrees of streptoneurous euthyneury are evident among the tectibranch sacoglossans.

In *Cylindrobulla, Volvatella,* and the bivalved gastropods the cerebropleural ganglia are dorsal to the esophagus, united by a short commissure and linked with almost fused pedal ganglia which lie below the esophagus. The paired visceral cords are long and not completely untwisted; they bear the supra-intestinal ganglion dorsal to the alimentary canal, the infra-intestinal ganglion below, a visceral ganglion and a genital ganglion. In *Oxynoe* and *Lobiger* the central ganglia are more concentrated and the visceral loop is so short that all of the principal ganglia are in the anterior nerve ring above the esophagus as they are in *Elysia* and *Hermaeina* (Gonor, 1961b).

While the functional aspects of the varying degrees of streptoneury and euthyneury are unknown, it is obvious that there occurs, along with a reduction in the shell and the uncoiling of the visceral mass, in the *Volvatella–Oxynoe–Lobiger* sequence a progressively more concentrated euthyneurous condition.

EVOLUTION IN THE SACOGLOSSA

The discovery of the bivalved gastropods was heuristic for it stimulates further discussion on the course of evolution in the Sacoglossa. There are two aspects to this question: one, on the origin of the nudibranch forms such as the Elysiidae and Stiligeridae, has long been discussed; but there has thus far been no speculation on the second, the evolution of feeding habits within the order.

Speculation on the evolution of the nudibranch elysiomorphs and stiligerids from shelled ancestors has largely centered on consideration of the nervous system and the progressive reduction of the shell displayed among the tectibranchs. Boettger (1963), primarily on the basis of the nervous system, envisions a central stem with the lowest side branch, the Volvatellidae, terminal and the Juliidae as a slightly higher but also terminal branch, followed by a dichotomy of the central stem. One dichotomy originates with the Oxynoidae and leads to the Elysiidae and Placobranchidae; the other starts as Lobigeridae, develops into the Phyllobranchidae and culminates with the Stilligeridae, Oleidae, and Limapontidae. Morton (1963) and Gonor (1961a) utilize the progressive reduction of the shell in their schemes: a *Volvatella–Lobiger–Limapontiidae–Stiliger* sequence (Morton, 1963) and a *Volvatella–Cylindrobulla–Oxynoe–Lobiger* arrangement leading toward the shell-less bilateral condition (Gonor, 1961a).

Consideration of the form and habits of the tectibranch sacoglossans suggests another approach. The adoption of the suctorial habit by the primitive sacoglossans opened up an unexploited niche, an event which Wright (1963) suggests leads to evolution by adaptive radiation. In the Sacoglossa such an adaptive radiation is reflected by the different habits of the tectibranchs: burrowing by *Cylindrobulla* and *Volvatella*, "wallowing" and swimming by *Lobiger* and *Oxynoe*, and creeping by the Juliidae. The burrowers are apparently the oldest of the stock, retaining most of the primitive features of early opisthobranchs such as a coiled shell, twisted nervous system, and even, in the case of *Cylindrobulla*, the prosobranch auricle-ventricle arrangement of the heart. The "wallowers" and creepers radiated somewhat later, as is reflected by the more concentrated nervous systems and closed genital systems of *Lobiger* and *Oxynoe* and the bilateral symmetry of the Juliidae.

The creeping habit and bilateral symmetry of the nudibranch elysiomorphs suggest that the elysiomorphs are perhaps derivatives of the creeping radiation. The Juliidae are already bilaterally symmetical, the visceral mass is uncoiled, and only loss of the thin and elastic shell of *Berthelinia*, fusion of the mantle to the foot, sinking of the visceral hump into the cephalopedal mass, and concentration of the nervous system are needed to produce a reasonable facsimile of an elysiomorph. The plausibility of such a course of evolution hinges around the question of the derivation of the parapodia in the elysiomorphs. Is it possible that they are mantle derivatives rather than epipodial derivatives, as Thompson (1962) has demonstrated is the case for the dorsal integuments of the dorids and tritonids and suggested for the notaspideans?

The discovery of living bivalved gastropods on the algal genus *Caulerpa* has confirmed earlier suggestions that the tectibranch sacoglossans are all restricted nutrionally to that alga. That *Lobiger* and *Oxynoe* feed exclusively on *Caulerpa* has now been demonstrated, and there is a strong suggestion that *Volvatella* is also restricted to that genus. The nudibranch sacoglossans, however, exhibit a variety of food habits: *Elysia* has been reported from various species of the Codiaceae (MacNae, 1954), *Hermaeina* on *Rhizoclonium* and *Urospora* (Gonor, 1961b), the Limapontiidae on *Cladophora* and *Vaucheria* (Gascoigne, 1956), *Stiliger* on *Chaetomorpha* (Rao, 1937, *fide* Gonor, 1961b), and *Alderia* on *Vaucheria* (Evans, 1953).

The association of the more primitive tectibranch forms with a variety of species of *Caulerpa* suggests that it is not the structure of the alga which has facilitated the feeding habit but that *Caulerpa* provides a nutrient not found in other algae. The question must thus be raised as to how the nudibranch forms have been able to utilize other algal genera as their food.

It would seem to be of more than passing interest that many of the nudibranch sacoglossans, notably *Elysia* (Ostergaard, 1955), *Placo-branchus* (Kawaguti, 1941), *Tridachia* (Yonge and Nicholas, 1940), and members of the Phyllobranchidae (personal observation) contain in their tissues quantities of zooxanthellae or zoochlorellae. Yonge and Nicholas (1940) discussed the occurrence of zooxanthellae in the tissues of *Tridachia* at Dry Tortugas, and suggested that since there was no evidence that the animal normally consumes them they might be of value by removing waste products of metabolism produced within the parapodia. Kawaguti (1941) found them in the inner surface of the parapodia and in the digestive gland of *Placobranchus* and suggests they may function in gaseous exchange or serve as food. The role of zooxanthellae in providing "growth substances" in corals has been suggested by Prosser and Brown (1961), similar roles have been postu-lated for symbiotic bacteria in the guts of the suctorial hemiptera (Goodchild, 1966), and of course there is the general thesis that, from termites to man, internal floras convert certain dietary components into essential nutrients. The possibility is therefore suggested that the release of the tectibranch sacoglossans from a restricted diet of *Caulerpa* was made possible by the association with zooxanthellae or zoochlorellae which provided the nutrient or growth substance lacking in algae other than *Caulerpa*.

Whatever the answers may eventually be shown to be, the Saco-glossa provide us with an almost classic case of an evolutionary program suggested by Mayr (1963): the earlier basal stock of tectibranch saco-glossans radiated adaptively, forming monotypic genera and families, while the modern nudibranch forms appear to be speciating more actively, without much adaptive radiation.

SUMMARY

1. The Juliidae comprise a family of sacoglossan opisthobranchs bearing a bivalved shell with a more or less well-developed hetero-strophic protoconch on the left valve and with a subcentral adductor muscle scar on the interior of the shell. All recent genera are associated with the algal genus *Caulerpa*.

2. Two subfamilies are recognized: the Juliinae with the single genus *Julia* and the Bertheliniinae with the recent genera *Berthelinia* (which includes the fossil subgenera *Berthelinia* s.s., *Analomya*, and *Ludovicia* and the recent *Tamanovalva* and *Midorigai*) and *Edentellina*.

3. The recent members of the Juliidae reflect the distribution of their host plant, *Caulerpa*, and are found throughout the Indo-west-Pacific, in Victoria, Australia, the Caribbean, and the Mediterranean. Fossil

Berthelinia, Ludovicia and *Analomya* have been found in the Eocene of the Paris Basin, fossil *Berthelinia* in the Upper Pliocene of South Australia, and fossil *Julia* occur in the Miocene of Europe, Florida, Jamaica and Java.

4. Opisthobranch features of the bivalved gastropods include the slug-like body, bilaterally symmetrical head with grooved rhinophores, mantle cavity on the right side of the body, lamellate gill, vas deferens sunken into the haemocoele, and sacoglossan alimentary system with an ascus containing a single row of articulating teeth. The nervous system is streptoneurous but the cerebral and pleural ganglia are concentrated around the esophagus.

5. The anomalous features of the bivalved gastropods are their bivalved shell, adductor muscle, and mantle divided into two lobes.

6. The larva at hatching is a typical veliger, and the shell valves develop as projections from each side of the larval shell. The adductor muscle develops near the center of the larval shell, gradually extends to the right valve, and slips ventrally from the larval shell to the left valve.

7. The bivalved gastropods are comparable with the Bivalvia only in terms of the shell: the mantle is threefold, the outermost forming the outer calcareous layer and there is a periostracal groove between the outer and middle layers. The ligament is similar to the primary ligament of the Bivalvia. The adductor muscle is single and in fine structure resembles gastropod muscle rather than adductor muscles of the Bivalvia.

8. The Juliidae are comparable in terms of mantle cavity, alimentary system, nervous system, and genitalia with other tectibranch sacoglossans. The adductor muscle in the bivalved gastropods is the homologue of a muscle joining the ventral edges of the shell in *Cylindrobulla* and *Volvatella*.

9. The tectibranch sacoglossans may be considered as a group which have radiated adaptively, *Cylindrobulla* and *Volvatella* forming the more primitive burrowing forms, *Lobiger* and *Oxynoe* forming a "wallowing" and swimming radiation, and the bivalved gastropods forming the creeping radiation. The possibility is suggested that the nudibranch sacoglossans have evolved from the creeping radiation.

10. The association of the more primitive tectibranch sacoglossans with a variety of species of *Caulerpa* suggests that it is not the structure of the alga which has facilitated the feeding habit but that *Caulerpa* provides a nutrient not found in other algae. The occurrence of zoochlorellae and zooxanthellae in the tissues of such nudibranchs as *Placobranchus, Elysia,* and *Tridachia* which feed on a variety of algae suggests

the possibility that the release of the tectibranch sacoglossans from a restricted diet of *Caulerpa* was made possible by the algal association which provided the essential nutrient.

ACKNOWLEDGEMENTS

I am especially grateful for the help and advice of Dr. A. M. Keen who also provided several references which have been utilized in this study. I would like to thank Dr. Siro Kawaguti for his hospitality and for providing the means of my seeing *Berthelinia limax* in its type locality, and Dr. A. J. Bernatowicz for his helpful criticism of the manuscript.

REFERENCES

Baba, K. (1961a). On the identification and the affinity of *Tamanovalva limax*, a bivalved sacoglossan mollusc in Japan. *Publs Seto mar. biol. Lab.* **9**: 37–61.

Baba, K. (1961b). The shells and radula of *Berthelinia*, a bivalved sacoglossan genus. *Venus* **21**: 389–401.

Baba, K. (1966). Gross anatomy of the specimens of the shelled sacoglossan *Volvatella* = (*Arthessa*) collected from Okino-Erabu Island, Southern Kyushu, Japan. *Publs Seto mar. biol. Lab.* **14**: 197–205.

Beets, C. (1944). Die Lamellibranchiaten-Gattung *Julia* Gould. *Geol. en Mijn.* **6** (3–4): 28–31.

Beets, C. (1949). Additional observations on the genus *Julia* Gould. *Geol. en Mijn.* **11** (1): 22–24.

Boettger, C. R. (1963). Gastropoden mit zwei Schalenklappen. *Verh. dt. zool. Ges.* **1962**: 403–439. (*Zool. Anz.* Suppl. **26**).

Burn, R. (1960a). A bivalve gastropod. *Nature, Lond.* **186**: 179.

Burn, R. (1960b). Australian bivalve gastropods. *Nature, Lond.* **187**: 44–46.

Burn, R. (1965). Rediscovery and taxonomy of *Edentellina typica* Gatliff and Gabriel. *Nature, Lond.* **206**: 735–736.

Burn, R. (1966). The opisthobranchs of a caulerpan microfauna from Fiji. *Proc. malac. Soc. Lond.* **37**: 45–65.

Cossmann, M. (1888). Catalogue illustré des coquilles fossiles de l'Eocene des environs de Paris. Pelecypodes (Suite). *Annls Soc. r. malacol. Belg.* **22**: 3–214.

Cossmann, M. and Peyrot, A. (1914). Conchylologie neogenique de l'Aquitaine. Suite (1). *Act. Soc. linn. Bordeaux* **68**: 5–210.

Cox, L. R. and Rees, W. J. (1960). A bivalve gastropod. *Nature, Lond.* **185**: 749–751.

Crofts, D. R. (1955). Muscle morphogenesis in primitive gastropods and its relation to torsion. *Proc. zool. Soc. Lond.* **125**: 711–750.

Crosse, H. (1875). Description du nouveau genre, *Berthelinia*. *J. Conch., Paris* **23**: 79–81.

Crosse, H. and Fischer, P. (1887). Observations sur le genre *Berthelinia*. *J. Conch., Paris* **35**: 305–311.

Dall, W. H. (1898). Contributions to the Tertiary fauna of Florida, with especial reference to the Miocene Silex beds of Tampa and the Pliocene beds of Calossahatchie River. Part IV. *Trans. Wagner Free Inst. Sci.* **3**: 1–200.

Dall, W. H. (1918). Description of new species of shells, chiefly from Magdalena Bay, Lower California. *Proc. biol. Soc. Wash.* **31**: 5–8.

Dautzenberg, P. (1895). De l'existence du genre *Berthelinia* Crosse à l'epoque actuelle. *Bull. Soc. zool. Fr.* **20**: 37–38.

De Folin, L. (1867). *Les fonds de la mer.* **1** Paris.

Deshayes, G. P. (1863). Catalogue des Mollusques de l'Ile de la Réunion (Bourbon). In *Notes sur l'Ile de la Réunion (Bourbon)*. Maillard, L. (ed.). Paris: Dentu.

Dollfuss, G. and Dautzenburg, P. (1901). Nouvelle liste des Pelecypodes et des Brachiopodes fossiles du Miocene moyen du Nord-ouest de la France. *J. Conch., Paris* **49**: 229–280.

Edmunds, M. (1963). *Berthelinia caribbea* n. sp., a bivalved gastropod from the West Atlantic. *J. Linn. Soc. Lond. (Zool.)* **44**: 731–739.

Eliot, C. N. (1906). Nudibranchs and tectibranchs from the Indo-Pacific. II. *J. Conch., Lond.* **11**: 298–315.

Evans, T. J. (1950). A review of Pease's genus *Volvatella*, together with a preliminary report on a new sacoglossan genus. *Proc. malac. Soc. Lond.* **28**: 102–106.

Evans, T. J. (1953). The alimentary and vascular systems of *Alderia modesta* (Loven) in relation to its ecology. *Proc. malac. Soc. Lond.* **29**: 249–258.

Fleming, C. A. (1957). The genus *Pecten* in New Zealand. *Paleont. Bull., Canberra* **26**: 1–69.

Fretter, V. (1939). On the structure of the gut of the sacoglossan nudibranchs. *Proc. zool. Soc. Lond.* (B) **110**: 185–198.

Fretter, V. (1960). Observations on the tectibranch *Ringicula buccinea* (Brocchi). *Proc. zool. Soc. Lond.* **135**: 537–549.

Fretter, V. and Graham, A. (1962). *The functional morphology and ecology of British prosobranch molluscs.* London: Ray Society.

Gascoigne, T. (1956). Feeding and reproduction in the Limapontiidae. *Trans. R. Soc. Edinb.* **63**: 129–151.

Gatliff, J. H. and Gabriel, C. J. (1911). On some new species of Victorian marine Mollusca. *Proc. R. Soc. Vict.* (N.S.) **24**: 187–192.

Ghiselin, M. T. (1965). Reproductive function and the phylogeny of opisthobranch gastropods. *Malacologia* **3**: 327–378.

Gonor, J. J. (1961a). Observations on the biology of *Lobiger serradifalci*, a shelled sacoglossan opisthobranch from the Mediterranean. *Vie Milieu* **12**: 381–403.

Gonor, J. J. (1961b). Observations on the biology of *Hermaeina smithi*, a sacoglossan opisthobranch from the West Coast of North America. *Veliger* **4**: 85–98.

Goodchild, A. J. P. (1966). Evolution of the alimentary canal in the Hemiptera. *Biol. Rev.* **41**: 97–140.

Gould, A. A. (1862). Descriptions of new genera and species of shells. *Proc. Boston Soc. nat. Hist.* **8**: 280–284.

Hedley, C. (1920). Concerning *Edentellina*. *Proc. malac. Soc. Lond.* **14**: 74–76.

Kawaguti, S. (1941). Study on invertebrates associating unicellular algae I. *Placobranchus ocellatus* von Hasselt, a nudibranch. *Japan Soc. Prom. Sci. Res. Tokyo* **1941**: 307–308.

Kawaguti, S. (1959). Formation of the bivalve shell in a gastropod *Tamanovalva limax. Proc. Japan Acad.* **35**: 607–611.

Kawaguti, S. and Baba, K. (1959). A preliminary note on a two-valved sacoglossan gastropod *Tamanovalva limax* n. gen., n. sp. from Tamano, Japan. *Biol. J. Okayama Univ.* **5**: 177–184.

Kawaguti, S. and Yamasu, T. (1960a). Electron microscopic study on the adductor of a bivalved gastropod, *Tamanovalva limax. Biol. J. Okayama Univ.* **6**: 61–69.

Kawaguti, S. and Yamasu, T. (1960b). Spawning habits of a bivalved gastropod. *Tamanovalva limax. Biol. J. Okayama Univ.* **6**: 133–149.

Kawaguti, S. and Yamasu, T. (1960c). Formation of the adductor muscle in a bivalved gastropod, *Tamanovalva limax. Biol. J. Okayama Univ.* **6**: 150–159.

Kawaguti, S. and Yamasu, T. (1961). The shell structures of the bivalved gastropod with a note on the mantle. *Biol. J. Okayama Univ.* **7**: 1–16.

Kawaguti, S. and Yamasu, T. (1966). Feeding and spawning habits of a bivalved gastropod, *Julia japonica. Biol. J. Okayama Univ.* **12**: 1–9.

Kay, E. A. (1961). A new opisthobranch mollusc from Hawaii. *Pacif. Sci.* **15**: 112–113.

Kay, E. A. (1964). A new species of *Berthelinia* and its associated sacoglossans in the Hawaiian Islands. *Proc. malac. Soc. Lond.* **36**: 191–197.

Keen, A. M. (1960). The riddle of the bivalved gastropod. *Veliger* **3**: 28–30.

Keen, A. M. and Smith, A. (1961). West American species of the bivalved gastropod genus *Berthelinia. Proc. Calif. Acad. Sci.* **30** (2): 47–66.

Kuroda, T. and Habe, T. (1951). Nomenclatorial notes. In *Illustrated catalogue of Japanese shells.* **1** (13) Kuroda, T. (ed.)

Ludbrook, N. H. and Steel, T. M. (1961). A late Tertiary bivalve gastropod from South Australia. *Proc. malac. Soc. Lond.* **34**: 228–230.

MacNae, W. (1954). On four sacoglossan molluscs new to South Africa. *Ann. Natal Mus.* **13**: 51–64.

Marcus, E. and Marcus, E. (1956). On the tectibranch gastropod *Cylindrobulla. Anais Acad. bras. Cienc.* **28**: 117–128.

Mayr, E. (1963). Speciation and systematics. In *Genetics, paleontology and evolution*: 281–298. Jepsen, G. L., Mayr, E. and Simpson, G. G. (eds). New York: Atheneum.

Morton, J. (1963). The molluscan pattern: evolutionary trends in a modern classification. *Proc. Linn. Soc. Lond.* **174**: 53–72.

Ostergaard, J. M. (1955). Some opisthobranchiate Mollusca from Hawaii. *Pacif. Sci.* **9**: 110–136.

Owen, G., Trueman, E. R. and Yonge, C. M. (1953). The ligament in the Lamellibranchia. *Nature, Lond.* **171**: 73–75.

Pilsbry, H. A. and Olsson, A. A. (1944). A west American *Julia. Nautilus* **57**: (3): 86–87.

Prosser, C. L. and Brown, F. A. Jr (1961). *Comparative animal physiology.* Philadelphia: W. B. Saunders.

Smith, E. A. (1885). Report on the Lamellibranchiata collected by H.M.S. *Challenger*, during the years 1873–1876. *Rep. Scient. Results Voy. Challenger 1873–1876.* Zoology **13** (1): 1–341.

Thompson, T. E. (1962). Studies on the ontogeny of *Tritonia hombergi* Cuvier (Gastropoda Opisthobranchia). *Phil. Trans. R. Soc.* (B)**245**: 171–218.

Woodring, W. P. (1925). Miocene mollusks from Bowden, Jamaica. Pelecypods and Scaphopods. *Publs Carnegie Instn* No. 366.

Woodring, W. P. (1928). *Contribution to the geology and paleontology of the West Indies*. Miocene mollusks from Bowden, Jamaica. Part II. Washington: Carnegie Inst.

Wright, S. (1963). Adaptation and selection. In *Genetics, paleontology and evolution*: 365–389. Jepsen, G. L., Mayr, E. and Simpson, G. G. (eds). New York: Atheneum.

Yonge, C. M. (1953). The monomyarian condition in the Lamellibranchia. *Trans. R. Soc. Edinb.* **67**: 443–478.

Yonge, C. M. and Nicholas, H. M. (1940). Structure and function of the gut and symbiosis with zooxanthellae in *Tridachia crispata* (Oerst.) Bgh. *Pap. Tortugas Lab.* **32** (15): 289–301.

Symp. zool. Soc. Lond. (1968) No. 22, 135–149.

THE BIOLOGY OF INTERSTITIAL MOLLUSCA*

BERTIL SWEDMARK§

Station Biologique de Roscoff, Finistère, France

SYNOPSIS

The specific ecological factors in the interstitial water space of marine sand make definite biological demands on the organisms which are to colonize this biotope successfully.

The space factor restricts the fauna to species of very small size. In fact, the fauna comprises a selection of the smallest species of the various phyla of invertebrates, differing in size from about 0·3 mm in the smallest to 3 mm in the largest species. Protozoa and Metazoa are of the same size in this environment.

In both littoral and sublittoral sand or shell-sand, restratification is going on incessantly as a consequence of turbulence and the action of waves. This makes the interstitial environment a dynamic one, making adaptation in the organisms inhabiting it, in the form of cuticular reinforcement or other arrangements for mechanical protection, necessary.

The Mollusca are represented in the interstitial fauna by a series of Solenogastres, which, as yet, have been only incompletely studied. There are also a few prosobranch gastropods. Most numerous are the opisthobranchs, with a dozen or so species belonging to the order Acochlidiacea, and some species belonging to the genera *Philinoglossa* and *Pseudovermis*.

Owing to the smallness of the organisms, and the consequent smaller number of cells available for morphological differentiation, the production of gametes is low in the interstitial molluscs. Cutaneous fertilization occurs, which was known previously only in the lower Metazoa and some Archiannelida. Several features of the biology of their reproduction may be interpreted as adaptation to provide guarantees for fertilization or protection of the larvae during the developmental period.

INTRODUCTION

Since Remane's work during the 1920s (references in Remane, 1952) studies of interstitial fauna of marine sand have made many interesting and valuable contributions to systematic zoology, besides opening up a profitable field for ecological research.

The specific ecological factors in interstitial space make strict demands on the organisms colonizing this biotope. This is manifested in often very striking morphological and biological adaptations to the environment found in these species. It is this interesting adaptation and other features of the microfauna in marine sand that have, during recent decades, stimulated marine biologists to study these animals.

---

* The paper was illustrated by extracts from a 16 mm film by J. Dragesco and B. Swedmark: "Adaptations biologiques de la Microfaune des sables marins", Roscoff, 1957.

§ Present address: Kristineberg Zoological Station, Fiskebäckskil, Sweden.

Since the interstices in marine sand are small, the fauna living there is a microfauna or meiofauna. The average upper limit of interstitial organisms is usually considered to be 3 mm, and it is, therefore, not astonishing that Protozoa and Metazoa in this biotope are of the same size.

The microfauna of marine sand consists of species belonging to practically all classes of invertebrates. Groups of animals with normally very small bodies, ciliates, turbellarians, nematodes, ostracods, copepods, etc., are represented by a great number of species. In addition, a large number of aberrant forms from the other groups of invertebrates, Cnidaria, polychaetes, echinoderms, ascidians, etc., are also present. These aberrant sand-microforms usually deviate considerably from the systematic group to which they belong, and they may even represent quite new types of organization. Owing to their small size, features characteristic of regressive evolution are often found in their organization, and examples of neoteny occur.

There are also groups of animals which are, as yet, known only in the interstitial fauna. To these belong the order Macrodasyoidea among the Gastrotricha, and also Mystacocarida, a group of crustaceans discovered by Pennak and Zinn in 1943.

If the space factor limits the fauna to comprise only microforms, the structure of the interstitial space also leads to certain shapes of body being especially adaptable to life in the interstices of marine sand. It is rather obvious, for instance, that elongated forms fit into this biotope very well, and that also flat forms have good prospects of living there.

Various groups of Mollusca are also represented in this microfauna. As a matter of fact, many of the interstitial species of molluscs had been described scientifically long before the interstitial fauna had been defined ecologically and before their specific biotope had aroused any special interest in biologists.

SOLENOGASTRES

In 1886, Marion and Kowalevsky described a Solenogastres, 2 mm long, a single specimen of which they had found on a stem of *Balanophyllia*, and which was given the name *Lepidomenia hystrix*. This species became of some importance for, among other things, the phylogenetical discussion of this group of animals. In 1954, rather great numbers of this species were found at Marseilles, in shell-sand, so-called *Amphioxus* sand. Since then, I have found several species of Solenogastres, 2–3 mm long, in the same biotope, shell-sand, in other regions

(Roscoff, Friday Harbor). Everything seems to suggest that the normal biotope of *Lepidomenia hystrix* and other as yet undescribed micro-forms of Solenogastres is the interstitial space of sand.

The method of locomotion of interstitial Solenogastres, cilia—gliding, and their armament of scales and epidermal spicules, which provide mechanical protection, give these species striking similarity to macrodasydoid gastrotrichs (fam. Lepidodasyidae), a group of specifi-cally interstitial animals.

The features of the reproduction of interstitial Solenogastres are unknown. Some species are bisexual. The number of gametes formed by each individual for one reproduction situation is small. This is associated with the small size of body and the low number of cells in general. Judging from observations of animals with large egg cells during the greater part of the year, it may be assumed that breeding continues throughout the year. The interstitial Solenogastres studied usually had a maximum of four or five egg cells, and such a low production implies continuous breeding.

PROSOBRANCHIA

The gastropods are represented in the microfauna of marine sand by both prosobranchs and opisthobranchs.

The species of prosobranchs most studied is *Caecum glabrum* (Montagu) (Fig. 1A) which, thanks to its small size, 1–2 mm, and its tubular, slightly curved shell, fits well into the interstices between the grains of sand. *Caecum glabrum* is found on the Atlantic coast of Europe and in sublittoral shell-sand on the coasts of the Mediterranean, often associated with Amphioxus. It feeds on diatoms.

Like all other very small species, also among other groups of animals, *Caecum* has a relatively important extension of cilia, and in its anatomy shows many features of reduction (it is without gills, salivary gland, hypobranchial gland, visceral ganglion).

A modern biography of *Caecum glabrum* has been written by Götze (1938), who also studied the reproduction of the species. The species is monosexual and produces a small number of eggs at a time, which, by spawning, are equipped with a cocoon each and develop into pelagic larvae.

OPISTHOBRANCHIA

Order Acochlidiacea

Immediately before and around the turn of the century, Kowalevsky (1901a) studied an interesting material of small marine shell-less

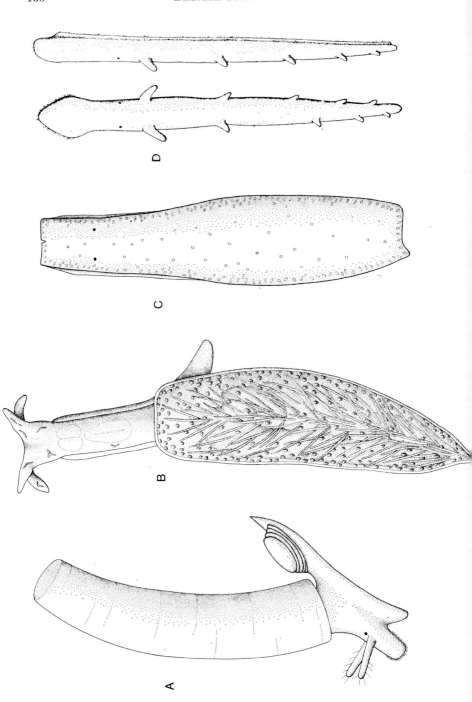

molluscs from the eastern Mediterranean, which he assigned to the genus *Hedyle* Bergh. He described four species from different regions: *H. milaschewitchii*, *H. glandulifera* and *Parhedyle tyrtowii*.

These hedylids discovered earliest of all were found in sublittoral shell-sand. It is remarkable that Kowalevsky (1901a) in his work on hedylids was the first to describe a method of separating small organisms, which cannot be seen with the naked eye, from samples of sand.

Later, the number of species and genera was increased in works published by Hertling (1930), Odhner (1937, 1952), Marcus (1953) and Swedmark (1968).

The systematics was studied by Thiele (1931), Odhner and others. Odhner (1937) placed these forms in a special order, Acochlidiacea. Today twelve species are known.

Interstitial Acochlidiacea have a length of body varying from 0·8 mm in the smallest (*Hedylopsis loricata* Swedmark) to 3–4 mm (*Hedylopsis spiculifera* Kowalevsky). No great variations occur in the shape of the body. The head-foot complex may be cylindrical in some species and dorso-ventrally flattened in others according to the development of the foot. There are cylindrical or slightly flattened tentacles in the anterior part and in most species also a pair of rhinophores. In most species the foot is shorter than the visceral sac. Most species are elongated, and the shape of the body is well adapted to the interstitial environment.

All species are colourless, as is usual among interstitial organisms which live in an environment where the light is weak.

One typical feature of marine sand and shell-sand is the great amount of movement. Under the influence of turbulence and waves, restratification is going on incessantly, and as a consequence the interstitial organisms live in a very dynamic environment which is changing continuously. In molluscs, as in other animal forms, this situation makes special demands, which can be traced in morphological and biological adaptations. As far as the adaptation of the Acochlidiacea to the mobility of the environment is concerned, the great contractile capacity of the species may be interpreted as such adaptation. Another characteristic, which is of importance in this connexion, is the cutaneous spicules which occur in all species.

The cutaneous spicules are very small in several of the species and in such species they cannot be of any mechanical protective significance. In the genus *Hedylopsis*, the species *H. brambelli* (Fig. 1B) *H. spiculifera*

Fig. 1. Interstitial molluscs. A, *Caecum glabrum*; B, *Hedylopsis brambelli*; C, *Philinoglossa helgolandica*; D, *Pseudovermis papillifer*.

and *H. loricata* form an interesting morphological series in respect of
the arrangement of the spicules. In all species the spicules are of the
same type: they are rod-shaped, straight or slightly curved. Their
density is greatest in the visceral sac, and in the main they are oriented
along the longitudinal axis of the animal. The spicules are thinly distributed in the largest species, *H. spiculifera* (3–4 mm); in *H. brambelli*
(2 mm) they are more dense and cover more of the organism, while in
H. loricata (0·8 mm) they are so dense that the visceral sac has a
constant form. The cutaneous spicules of *H. loricata* serve the same
purpose as a shell.

The importance of the calcareous spicules as mechanical protection
for interstitial organisms has been mentioned in other connexions
(Swedmark, 1964). It is also a striking phenomenon that in the interstitial fauna spicules are found on representatives of animal forms in
which spicules do not otherwise occur. One example of this is the turbel-
larian *Acanthomacrostomum spiculiferum* Papi and Swedmark, known
only from sublittoral shell-sand banks.

As in many interstitial organisms belonging to other groups of
animals, numerous examples of reduction in anatomy, related to small
size and small number of cells, can be found in Acochlidiacea. Only the
biological aspects of some of these reductions will be considered here.

Organs of *light reception* are lacking in nearly half of the species
(*Hedylopsis brambelli, H. loricata, Microhedyle glandulifera, Unela
remanei*). *Hedylopsis spiculifera*, which is the largest species, has the
largest eyes. In this species, the diameter of the pigment cup is approximately 50 μ, which is four or five times greater than in the other
species with eyes.

In many interstitial animals the eyes are degenerated or absent.
It must be assumed that the dark environment does not require
very differentiated organs of light reception. However, many species,
in the first place those in tidal areas, make rhythmical vertical migrations. It has been assumed (Swedmark, 1955) that these migrations
may be of importance when the animal is seeking food, particularly
for the large and important category of interstitial species living on
diatoms to which some of the molluscs belong. Boaden (1963) maintains
that the light factor plays an important role in controlling the distribution of interstitial animals. Gray (1966), in his studies of Archiannelida,
demonstrated in light/shade choice experiments the presence of
epidermal light receptors in eyeless species with the help of an apparatus
in which the animals could choose between different degrees of light,
and he showed that vertical migrations were conditioned at least
partly by the animal's reaction to light.

Regression in the structure and function of the *sexual apparatus* is manifested most in the small number of gametes produced, particularly by the smallest species. The sperms of *Hedylopsis suecica*, which have a spiral structure, were studied by Franzén (1955). According to my own studies, sperms of the same type also occur in *H. brambelli* and *Microhedyle lactea*. The eggs are rich in yolk and fewer than fifty are produced on each reproduction occasion.

Most of the Acochlidiacea are monosexual but the genus *Hedylopsis* is hermaphroditic.

Fertilization also takes place in the hermaphroditic species, probably by cross-fertilization. Evidence for this is the occurrence of spermatophores in at least one of these species, *H. brambelli*.

Spermatophores were first observed in Acochlidiacea by Hertling (1930) in *Microhedyle lactea*, which is monosexual, but were interpreted by the German author as relics of a copulatory organ. Marcus (1953) established the presence of spermatophores in *Ganitus*, a genus he discovered on the coast of Brazil. I have observed that spermatophores are present in *M. lactea*, *M. milaschewitchii* and *Hedylopsis brambelli*. In *M. lactea* they are thin sacs, almost as long as the visceral sac. The spermatophores observed in *H. brambelli* are about half as long, while *M. milaschewitchii* have very small spermatophores. The formation of spermatophores has not been observed, but it may be assumed that they are formed from an accessory gland in the vicinity of the germinal duct. During the reproduction period, the spermatophores are attached to the animal to be fertilized. One or more spermatophores are attached to each animal, usually on the visceral sac, but not infrequently on the head-foot complex, then most commonly in the middle of the complex. The spermatophores are completely full of spermatozoa. After a few days the distal point is empty, and after a time the whole spermatophore is emptied, beginning with the distal end. In the meantime, the sperms have penetrated through the skin of the receiving animal. On histological sections, changes can be observed in the contact part between the epidermis and the base of the spermatophore. The nuclei of the epidermal cells have either disappeared or altered, which implies that an autolysis of cells may have occurred under the influence of the sperms to enable the latter to penetrate the skin. They can then be seen in histological sections on their way to the fertilization centre.

The Acochlidiacea probably represent the first examples of cutaneous fertilization without copulation in molluscs. Otherwise this method of fertilization is found in some small Turbellaria and Archiannelida, for example.

It may be assumed that, in animals with a limited production of gametes, spermatophores are a certain guarantee for fertilization, and thereby are a factor in the perpetuation of the population. In the interstitial sand fauna, the formation of spermatophores has been observed in some gastrotrichs and archiannelides.

The formation of spermatophores and cutaneous fertilization may occur in most Acochlidiacea, but copulation has also been observed. Odhner (1937) has described a copulatory organ in *H. suecica*.

The development of larvae of *M. lactea* and *H. spiculifera* has been followed to an advanced stage at Roscoff. Spawning occurs some time after fertilization. The eggs are laid in transparent, gelatinous cocoons which are attached to grains of sand or fragments of shells. The cocoons contain a maximum of fifty eggs, usually fewer. In the cocoons, the eggs develop into veliger larvae (Fig. 2) which are hatched out when the cocoon is dissolved or perforated. After that the larvae pass through a free-swimming phase, during which they live on interstitial plankton, primarily small flagellates. Attempts were made in the laboratory to make larvae of both species react to light. However, light did not seem to affect the orientation of the larvae. This absence of reaction to light makes it likely that the great majority of free-swimming larvae remains in the interstitial environment and never moves upwards into the water. Such a situation is probably of importance in the perpetuation of the population within the often rather restricted areas of the sublittoral shell-sand banks.

The duration of the breeding period is naturally of importance for these low-producing species. It has also been demonstrated that in the species studied breeding goes on throughout the greater part of the year.

Thus, in Acochlidiacea, as in many other interstitial animals, interesting adaptations in the sphere of reproductive biology, which must be seen partly on the background of the small size of the animals, but which may also be related to specific conditions, take place.

Order Philinoglossacea

Hertling (1932) described *Philinoglossa helgolandica* (Fig. 1G) from Amphioxus sand at Heligoland. This species and two others discovered later form the systematic group Philinoglossacea, which Odhner (1952)

FIG. 2. Development stages of *Hedylopsis spiculifera* (Kowalevsky). A–D, Early cleavages; E, gastrulation; F, early differentiation of the veliger larva; G, veliger larva after hatching; H, larval shell. *a*, Anus; *dg*, digestive gland; *f*, foot; *k*, kidney; *m*, mouth; *op*, operculum; *rm*, retractor muscles; *sh*, larval shell; *st*, stomach; *stc*, statocyst; *v*, velum.

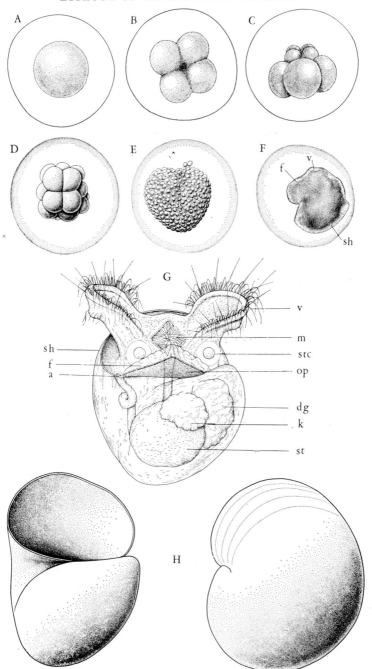

F

FIG. 2.

considered should be an order. Like other interstitial opisthobranchs, the adult animals have no shells.

The bodies of these animals are elongated and have rectangular cross-sections; the head is without appendices. Generally speaking, they are reminiscent of the *Philine*, and certain similarities to this genus can also be discerned in the structure of the radula.

Spicules are absent, but the epidermis has numerous glands which produce mucus of importance for the great adhesiveness of the animal.

The *Philinoglossa* are hermaphroditic. Sexually matured animals are found at Roscoff during a great part of the year. How fertilization is effected is unknown, but the characteristic features found in the reproductive biology of the Acochlidiacea are present here, too.

The eggs, fifty or so in number, are spawned in spherical cocoons (Fig. 3G) which adhere to grains of sand. They pass through the typical cleavages (Fig. 3A–D), and develop into veliger larvae (Fig. 3F) with shells. The veliger larva is similar to that of the Acochlidiacea, but can be distinguished by the black larval kidney.

Aeolid Opisthobranchia

The most interesting adaptation to the interstitial environment in respect of shape of body is found in *Pseudovermis* (Fig. 1D), an aeolid opisthobranch which, at the beginning of the century, was described by Kowalevsky (1901b). They are elongated vermiform, with an acorn-like head. This shape of body is found in certain representatives of other groups of animals, e.g. in the interstitial polychaete *Psammodrilus balanoglossoides*, and it allows the animals to penetrate through the uneven interstices.

Liver papillae are well developed in *Pseudovermis papillifer*, described by Kowalevsky, but in species described later (Marcus and Marcus, 1955; Fize, 1963) they are more or less reduced.

The liver papillae of *Pseudovermis* species sometimes contain nematocysts, and it must be assumed that they feed on, among other things, interstitial Cnidaria.

The animals are hermaphroditic, and their reproductive biology is similar to that of the Acochlidiacea. Part of the larval development has been studied in *Pseudovermis axi* at Roscoff. At spawning, a couple of dozen eggs are laid in a transparent, stalked, gelatinous cocoon (Fig. 4), which is attached by the flattened base of the stalk to a fragment of shell or grain of sand. The eggs, which are about 70 μ in diameter, pass

FIG. 3. Development stages of *Philinoglossa helgolandica* Hertling. A–D, Early cleavages; E, gastrulation; F, veliger larva before hatching; G, cocoon. See Fig. 2 for key.

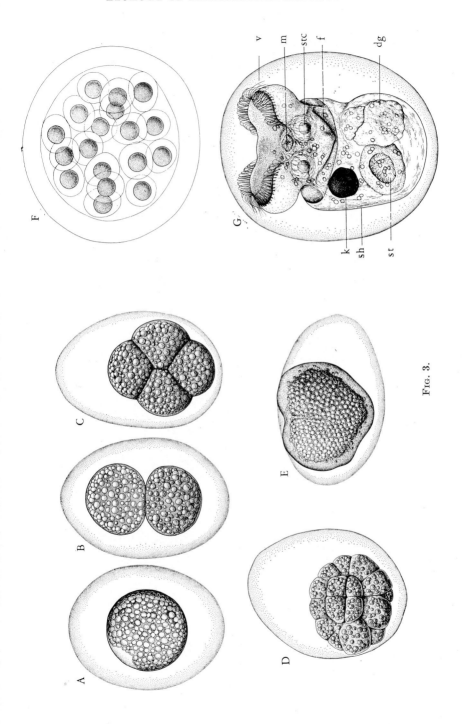

Fig. 3.

through a process of cleavage typical of Opisthobranchia. After about two days, at 13–15°C, gastrulation occurs and the differentiation of the veliger larvae begins. The thin, transparent shell of the larvae is the same

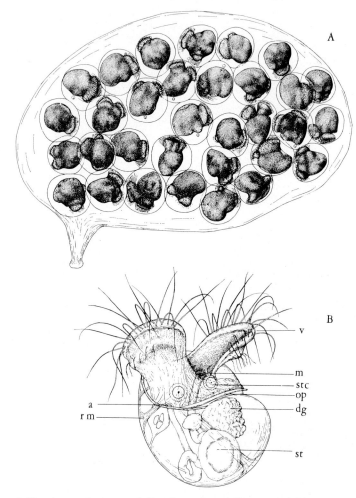

Fig. 4. Development stages of *Pseudovermis axi* Marcus and Marcus. A, Stalked cocoon with young veliger larvae; B, veliger larva after hatching. See Fig. 2 for key.

shape as that of the Acochlidiacea. At this temperature, hatching occurs after more than a week. As with the pelagic larval stage of Acochlidiacea, the veliger larva of *Pseudovermis axi* does not seem to react to light. Swimming movement of the larvae is in an elongated spiral, but, in laboratory situations at least, the larvae remain passive for long periods, attached to the substratum. The larvae clearly take their food from the interstitial microplankton. In the laboratory, it was just as impossible to cultivate larvae up to metamorphosis as it was with Acochlidiacea. The difficulty seems mainly to have been in the low egg production: it is impossible to get enough larvae to elaborate the cultivation experiment. This is a general difficulty inherent in research on the life history of marine interstitial fauna.

Animals with filled gonads have been observed at Roscoff during the greater part of the year, which suggests that the breeding period is of long duration.

If, to sum up, the species of interstitial forms taken from various systematic groups of molluscs, sketched here, are compared, many common features in morphology and biology can be found. Several of these features are common to molluscs and other interstitial Metazoa. The most typical of these tendencies are as follows.

(1) Small size.

(2) Regressive characteristics in the anatomy, which may affect many systems of organs. All are without shells, but larval shells occur.

(3) Specialized features, in shape of body and anatomy, such as vermiform tendencies, strong development of spicules and mucus-producing epidermis glands.

(4) Extension of body cilia more important than in non-interstitial molluscs.

(5) Striking adhesive ability.

(6) Striking contractability, especially in species without cutaneous differentiation for mechanical protection.

(7) The limited production of gametes is compensated by a prolonged reproduction period. Adaptation in the biology of reproduction is of importance to ensure fertilization (hermaphroditism, copulation, spermatophores, cutaneous fertilization). Tendencies towards stationary larval development (weak photic reaction, relatively weak development of larval organs of locomotion, brief free-swimming period), which counteract dispersion of the larvae and keep most of them within the territory of the population.

The interstitial molluscs are difficult to classify, for in their organization they often deviate greatly from other molluscs. Systematists have met with great difficulties with these forms, and in most cases the discussion of their relation to the classical orders of molluscs is speculative. Intermediate forms from other biotopes, which would undoubtedly facilitate classification, are unknown too. This is true not only of molluscs in the interstitial fauna, but of all aberrant forms of Metazoa in this environment.

This circumstance suggests that the interstitial fauna is of ancient origin. Here, in this environment, selection has favoured small organisms, and the regressive evolution has led to the number of cells available for the formation of the anatomic structures characteristic of the animal in question being small. Frequently the organs have become simplified, rudimentary and, in extreme cases, eliminated. Neoteny may be one of the mechanisms active in the regressive evolution which have given rise to these forms of the size of Protozoa.

It is these changes, regressions and specializations caused by the slight variations in the environment which have gradually eliminated the original features of the organisms developed there, and which have given rise to the difficulties of interpreting their relationships.

Thus, the interstitial fauna offers interesting but difficult objects of study for the systematist and the phylogeneticist.

For the student of ecology, the interstitial environment and its unique conditions of life, as well as the organisms it has evolved, is a sphere of work offering great possibilities, and in England, not least, important contributions have been made in this field.

In physiology and experimental biology of these often simplified organisms, whose life functions are based on few cells, should provide much of interest, but experimental methods must be developed to penetrate the small dimensions characteristic of these organisms and their environment.

REFERENCES

Boaden, P. J. S. (1963). Behaviour and distribution of the archiannelid *Trilobodrilus heideri*. *J. mar. biol. Ass. U.K.* **43**: 239–250.
Fize, A. (1963). Contribution à l'étude de la microfaune des sables littoraux du Golfe d'Aigues-Mortes. *Vie Milieu* **14**: 669–774.
Franzén, A. (1955). Comparative morphological investigations into the spermatogenesis among Mollusca. *Zool. Bidr. Upps.* **30**: 399–455.
Götze, E. (1938). Bau und Leben von *Caecum glabrum* (Montagu). *Zool. Jb.* (Syst.) **71**: 55–122.
Gray, J. S. (1966). The response of *Protodrilus symbioticus* (Giard) (Archiannelida) to light. *J. Anim. Ecol.* **35**: 55–64.
Hertling, H. (1930). Über eine Hedylide von Helgoland. *Wiss. Meeresunters.* N.F. **18** (5): 1–10.
Hertling, H. (1932). *Philinoglossa helgolandica*, ein neuer Opisthobranchier. *Wiss. Meeresunters.* N.F. **19** (2): 1–9.
Kowalevsky, A. (1901a). Etude anatomique sur le genre *Pseudovermis*. *Mém. Acad. Sci. St. Pétersb.* (Sci. math. phys. nat.) **12** (4): 1–28.

Kowalevsky, A. (1901b). Les Hedylides, étude anatomique. *Mém. Acad. Sci. St. Pétersb.* (Sci. math. phys. nat.) **12** (6): 1–32.

Kowalevsky, A. O. and Marion, A. F. (1887). Contributions à l'histoire des Solenogastres ou Aplacophores. *Annls Mus. Hist. nat. Marseille* **3**: 1–77.

Marcus, E. (1953). Three Brazilian Sand-Opisthobranchia. *Bolm Fac. Filos. Ciênc. Univ. S Paulo* (Zool. Ser.) No. 18: 165–203.

Marcus, E. and Marcus, E. (1955). Über Sand-Opisthobranchia. *Kieler Meeresforsch.* **11**: 230–243.

Marion, A. F. and Kowalevsky, A. O. (1886). Organisation du *Lepidomenia hystrix*, nouveau type de Solénogastre. *C. r. hebd. séanc. Acad. Sci., Paris* **103**: 757–759.

Odhner, N.Hj. (1937). *Hedylopsis suecica* n. sp. und die Nacktschneckengruppe Acochlidiacea (Hedylacea). *Zool. Anz.* **120**: 51–64.

Odhner, N.Hj. (1952). Petits opisthobranches peu connus de la côte méditerranéenne de France. *Vie Milieu* **3**: 136–147.

Pennak, R. W. and Zinn, D. J. (1943). Mystacocarida, a new order of Crustacea from intertidal beaches in Massachusetts and Connecticut. *Smiths. misc. Colln* **103**: (9) 1–11.

Remane, A. (1952). Die Besiedelung des Sandbodens im Meere und die Bedeutung der Lebensformtypen für die Okologie. *Verh. dt. zool. Ges.* **1951**: 327–339. (*Zool. Anz.* Suppl. **16**).

Swedmark, B. (1955). Recherches sur la morphologie, le développement et la biologie de *Psammodrilus balanoglossoides*, Polychète sédentaire de la microfaune des sables. *Archs. zool. exp. gén.* **92**: 141–220.

Swedmark, B. (1964). The interstitial fauna of marine sand. *Biol. Rev.* **39**: 1–42.

Swedmark, B. (1968). Sur deux espéces nouvelles d'Acochlidiaceés (Mollusques Opisthobranches) de la fauna interstitielle marine. *Cah. Biol. mar.* **9** (2). (in press).

Thiele, J. (1931). *Handbuch der systematischen Weichtierkunde* I–III. Jena: Gustav Fischer.

Symp. zool. Soc. Lond. (1968) No. 22, 151–166.

THE FEEDING MECHANISM AND BEHAVIOUR OF THE OPISTHOBRANCH *MELIBE LEONINA*

ANNE HURST*

Friday Harbor Laboratories, Friday Harbor, Washington, U.S.A.

SYNOPSIS

Adult *Melibe* live in sheltered eel grass beds, swimming by lateral flexions of the body only when disturbed. Younger stages probably bring about distribution and are more active. *Melibe* has a distinctive method of feeding. It has no radula and captures free-swimming prey with a unique apparatus, the oral hood. This is an umbrella-shaped anterior flange of the body wall and oral veil. The hood is spread widely and closes around any small object which touches it. By reduction of the enclosed cavity, movements of the lips and opening of the mouth, prey is forced into the mouth. Borne on the oral hood are rhinophores and two series of sensory tentacles. Within the hood are sinuses, haemal sacs and blood vessels to the margins and the lips. The state of turgidity of these affects the shape of the hood, and blood may be temporarily confined in the area by means of a physiological valve. Series of muscles running obliquely, circularly and dorso-ventrally also affect the shape and movements of the hood. They allow its angle to be changed, and opening, closing and swallowing movements. Innervation of the feeding apparatus is cerebral, including nerves to the rhinophores, lips, oral tube, hood and tentacles. All these (excepting the separate rhinophore nerves) are interconnected, with a particularly well-marked network around the edge of the hood. The multiplicity of routes for nerve impulses within the hood achieves quick reactions and efficient co-ordination. Intracellular recordings and stimulation of nerves showed that the giant cells of the cerebral ganglia are probably important in co-ordinating feeding movements. There appeared to be direct connexions between the cerebral giant cells and the largest nerves of the hood. Direct electrical stimulation of nerves and muscles supplying the hood caused reproduction of some integral feeding movements.

INTRODUCTION

Melibe leonina is a nudibranch mollusc of the family Tethyidae and is found on the Pacific coast of North America. It is well known for its unusual appearance (Figs 1 and 2) and behaviour, described by several authors including Gould (1852), Bergh (1892, 1904), Heath (1917), Agersborg (1919, 1921, 1923), O'Donoghue (1921, 1922) and MacFarland (1966). Its feeding habits and the apparatus used for capturing prey are very unlike those of its near relatives. Whereas most opisthobranchs obtain food by means of a radula, operated by a complex buccal mass, *Melibe* has no radula or buccal mass and feeds with the aid of a much enlarged and modified oral veil. This forms a specialized anterior hood-like structure which is equipped with sensory,

---

* Present address: Zoology Department, University of Reading, England.

muscular and vascular elements well suited for efficient capture of
free-swimming prey. The anatomical and physiological adaptations
necessary for the use of such a feeding apparatus are considerable.
The rather spectacular way in which feeding occurs has not been described in detail previously and the mechanism of the feeding apparatus
has not been investigated.

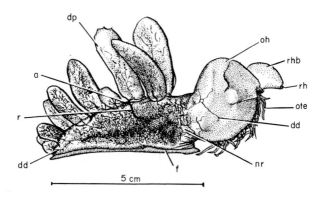

Fig. 1. Right side view of *Melibe* with the oral hood closed. One dorsal papilla
(second right) is regenerating after loss. *a*, Anus; *dd*, digestive duct; *dp*, dorsal papilla;
f, foot; *nr*, neck region; *oh*, oral hood; *ote*, outer tentacle; *r*, rectum; *rh* rhinophore;
rhb, rhinophore bearing process.

MATERIAL, ECOLOGY AND BEHAVIOUR

Present specimens of *Melibe* were collected off the San Juan Islands
and Vancouver Island where their distribution is localized. The adults
live in sheltered, sublittoral eel grass beds but are occasionally found on
kelp, floats or pilings. Groups may be found at any time of year amongst
their egg masses which are easily distinguished from those of other eel
grass opisthobranchs (Hurst, 1967). According to the depth of the eel
grass beds the animals were taken by hand-netting or by plankton-net
hauls. In the latter method the disturbance caused *Melibe* to swim and
its presence was indicated by a characteristic smell produced by a
glandular secretion from the body wall (Agersborg, 1921). The adult is
not primarily pelagic, contrary to the suggestion of O'Donoghue (1921),
and probably only swims when accidentally dislodged from the eel
grass. Young specimens, however, may sometimes be collected using a
nightlight, to which they swim actively. Very young adults are much
more active than the older ones and swim much more readily. It is
probable that distribution is achieved during the young stages.

The initiation of swimming depends on whether or not the anterior part of the foot is firmly attached to the substratum. When *Melibe* is about to swim, it first removes this part of the foot from its attachment, then folds the whole foot longitudinally so that the right and left halves of the sole meet. The body becomes compressed laterally while the dorsal papillae (Figs 1, 2, *dp*, 3) elongate and flatten considerably, each pair being held closely together in the midline. Thus the animal becomes

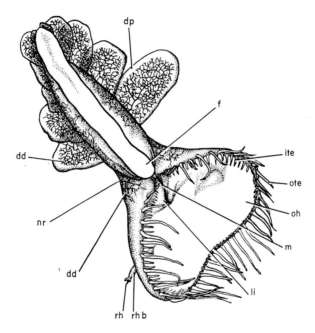

Fig. 2. Ventral view of *Melibe* with the oral hood opened. One dorsal papilla of the right side has been dropped. *dd*, Digestive duct; *dp*, dorsal papilla; *f*, foot; *ite*, inner tentacle; *li* lip; *m*, mouth; *nr*, neck region; *oh*, oral hood; *ote*, outer tentacle; *rh*, rhinophore; *rhb*, rhinophore bearing process.

more streamlined and as deep as possible dorso-ventrally so that it presents the maximal lateral surface area. This allows greater efficiency in propulsion since swimming is caused by lateral flexions of the body in alternate directions. The rate and degree of bending determine the speed of progress. In vigorous swimming the most posterior dorsal papillae touch the tip of the rhinophore (*rh*) or rhinophore bearing process (*rhb*) of the appropriate side during each flexion. The hood is semi-contracted and plays no major part in propulsion. *Melibe* swims upside down, not back upward as Agersborg (1921, 1923) described.

In settling, the anterior part of the foot attaches first followed by partial opening and shutting of the hood before attachment spreads back gradually to the rest of the foot. The body bulges over the sides of the foot as in Fig. 1 and is no longer laterally compressed. The hood opens widely.

Undisturbed *Melibe* remain attached to the eel grass by means of the long muscular foot which may clasp the edges of the blades or creep on their flat surfaces. Here adults may feed, copulate and lay eggs. They are also able to crawl on the surface film. This rarely occurs in the field but is common in the laboratory. *Melibe* can be made to crawl very rapidly by repeated mechanical stimulation of the posterior part of the foot. The anterior tip is attached after great elongation of the complete sole, and the hind end is then quickly brought forward by longitudinal contraction of the pedal muscles. The posterior part is then

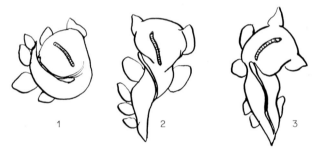

Fig. 3. Three successive stages in swimming, taken from a moving film of *Melibe*.

attached as closely as possible to the anterior end. This "galloping" progress has been described by Agersborg (1923). More vigorous stimulation usually causes swimming.

Melibe feeds on small free-swimming crustaceans such as gammarids, caprellids and copepods which are abundant in eel grass beds. Its colour varies from greyish to brownish according to the staple diet in a given area and can change within a few days. Prey is captured by means of the oral hood as Agersborg (1919, 1921) has mentioned. When feeding begins the hood is spread widely, usually towards any prevailing current, although in young animals its angle is constantly changed. Swift raising of the hood causes a water current to flow under and into it, possibly sweeping small animals in also. When such a small object touches the underside or tentacles (Figs. 1, 2, *ite, ote*) there is an instant closing reaction. The lateral edges of the hood come together so that it closes in the ventral midline with a longitudinal closure. The inner series of tentacles (*ite*) interlock within the resulting cavity enclosed by the hood, while the outer tentacles (*ote*), interdigitating, trail outside it. The prey is now enclosed by the hood which is raised up and back so that it becomes indented postero-dorsally. (Figure 1 shows the early stages of indentation.) Simultaneously the hood contracts bringing the prey into contact with the bulging circular area of the lips (Fig. 2, *li*) immediately surrounding the mouth. Where the lips are touched there is a local contraction followed by bulging movements. These manoeuvre the prey towards a median dorsal groove between the lips, which leads to the mouth. As the mouth opens food is sucked or pressed in and the flat median surfaces of the lips close together so that only a narrow longitudinal slit remains between them.

ANATOMY AND FUNCTIONING OF THE FEEDING APPARATUS

The oral hood of *Melibe* has been loosely called the head, which is not a precise description since some of the cephalic structures (notably the nerve ring and eyes) are not a part of the hood. It is an expanded flange of the antero-dorsal body wall combined with a much enlarged oral veil to form an anterior food-catching structure. This is approximately circular extending forwards and surrounding the mouth (Figs 2, 4, *m*) which lies on its concave ventral surface close to the posterior margin. The mouth is not in the centre of the hood as in Gould's (1852) figure. The lips surround the mouth (Fig. 2, *li*, *m*) forming a lobe on either side of it. They can be closed together longitudinally. The inner face of each lobe (Fig. 4, *ili*) is flattened, while the outer face (*oli*) bulges, particularly when the lip is filled with blood. When this occurs the inner faces of the lobes meet. Between them is a median dorsal groove (*dgr*) leading to the mouth and oral tube (*ot*).

The oral hood is well equipped with sensory structures. The rhinophores (Figs 1, 2, *rh*) have migrated forward onto its mid-dorsal surface and are situated on a pair of leaf-like processes (*rhb*) which are oriented at right angles to the antero-posterior axis of the body. Each of the two small rhinophores is retractile into a small pocket on the outermost tip of the appropriate process. Two series of sensory tentacles also form an integral part of the hood. These comprise an inner and an outer series. There are over 100 small inner tentacles (Figs 2, 5, *ite*) extending circumorally around the innermost part of the margin of the hood. They are uniform in size except for ten to twelve very small ones situated mid-ventrally, posterior to the mouth. An outer series of longer, larger tentacles (*ote*) extends around the edge of the hood except in the region posterior to the mouth. There are forty-six (occasionally

a few more) of these, somewhat less than Heath (1917) described.
The same number of tentacles occurred in both young and old animals
of adult form.

The musculature of the oral hood fulfils several functions. These
include changing the shape of the hood while preventing distortion,
producing feeding activity comprising opening, closing and changing
the direction and angle at which the hood is held. Such functions may
only be accomplished in conjunction with movements of the blood and
are governed by the cerebral ganglia. These provide several pairs of
cerebral nerves, the distribution of which allows rapid reactions and
successful integration of impulses.

The hood is made up of a sponge-work of sinuses and haemal sacs
(of the type described in *Philine* by Hurst, 1965) amongst which run
many muscular trabeculae from the dorsal to the ventral wall. Ramify-
ing between these are muscles, nerves and blood vessels. Two main sets
of muscles radiate out on each side to supply the tentacles. They cross
each other obliquely, one set running longitudinally (Fig. 4, *lm*) the
other circularly (*dcm*). Supplying the first of these sets some of the
larger longitudinal muscle bands of the right side of the body wall
come together in the neck region, continue forward, then separate out
again dorsal to the lips. Here they cross over to supply the opposite,
left, side of the hood. Originating from the left side of the body wall
are the longitudinal muscles of the right side of the hood. Where the
right and left sets of longitudinal muscles cross, above the lips, they
interweave with each other. This area is the region of indentation
important in swallowing prey. The second set of muscles (*dcm*) running
around and across the hood on each side also originates from muscles
of the body wall but radiates within the hood on the same side of the
body. Around the rim of the hood are well-marked circular muscles
(*cmr*) forming a marginal sphincter.

The hood is supplied by two large blood vessels (Fig. 5, *ora*) running
adjacent to its rim, one on each side. These supply blood to the tentacles and also branch to the inner area of the hood. Small paired labial
arteries (*laa*) supply the lips.

Maintaining the hood in an outspread condition (or achieving this
position) depends largely on the state of turgidity of the blood-filled
spaces and sacs within it. Their expansion is confined and directed by
the dorso-ventral muscular trabeculae. Changes in the direction, angle
or degree of curvature of the hood are achieved by differential contraction of the longitudinal and diffuse circular muscles running within it.
Contraction of the longitudinal muscles alone causes the tentacles to
tip inwards. Closure depends on several factors. The result of contraction

of the circular muscles of the rim is reduction of the perimeter of the hood, while contraction of the longitudinal and diffuse circular muscles tends to reduce its area. When the oral arteries are full they help to keep the hood margins stretched out. Thus when prey touches the underside of the hood, the longitudinal muscles contract, tipping the tentacles

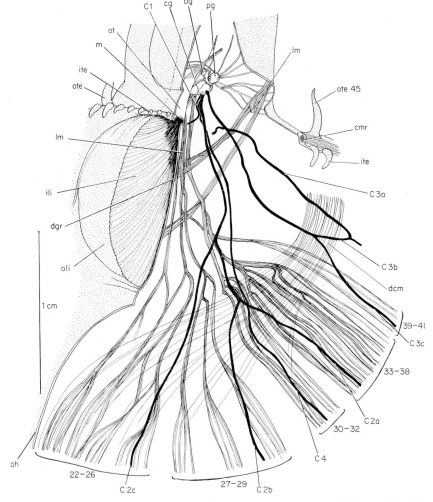

Fig. 4. Ventral dissection of the neck region and proximal part of the oral hood in *Melibe*. The stippled regions represent the unremoved outer part of the body. *bg*, Buccal ganglion; C 1–4, cerebral nerves; *cg*, cerebral ganglion; *cmr*, circular muscles of the rim; *dcm*, diffuse circular muscles; *dgr*, dorsal groove; *ili*, inner face of lip; *ite*, inner tentacle; *lm*, longitudinal muscle; *m*, mouth; *oh*, oral hood; *oli*, outer face of lip; *ot*, oral tube; *ote*, outer tentacle; *pg*, pedal ganglion; 1–45, outer tentacles.

inward, the circular muscles of the rim contract, closing the margins together, but these can only fold maximally in the median line, proximally and distally (because of the turgid oral arteries invading the rest of the margin), so that the lateral edges of the hood close longitudinally. This encourages efficient interlocking of the tentacles, so that the prey cannot escape. In facilitation of swallowing, the enclosed cavity is reduced by further contraction of the longitudinal muscles, followed

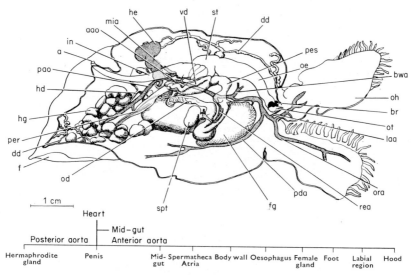

Fig. 5. Diagram of *Melibe* in side view with the right side of the body wall removed and the hood cut in half medially and pinned out. The main route taken by the blood vessels is shown below the diagram. *a*, Anus; *aao*, anterior aorta; *br*, brain; *bwa*, artery to body wall; *dd*, digestive duct; *f*, foot; *fg*, female gland; *hd*, hermaphrodite duct; *he*, heart; *hg*, hermaphrodite gland; *in*, intestine; *laa*, labial artery; *mia*, mid-gut artery; *od*, oviduct; *oe*, oesophagus; *oh*, oral hood; *ora*, oral artery; *ot*, oral tube; *pao*, posterior aorta; *pda*, pedal artery; *per*, penial retractor; *pes*, penial sac; *rea*, reproductive system artery; *spt*, spermatheca; *st*, stomach; *vd*, vas deferens.

by indentation of the posterior part of the hood. This indentation is an inevitable result of antagonistic pulls at the cross-over of the right and left sets of longitudinal muscles. The cross-over region cannot move backwards due to the interweaving of the longitudinal muscle fibres with others running at right angles in the neck region, and due to the presence of a complex system of connective tissue sheets in this area (Fig. 8). The effect of this system is to stabilize the structures in the neck region and to limit the position and movement of blood within the area. Figure 8 illustrates part of the neck region complex.

Entering the oral hood are four pairs of cerebral nerves (Figs 4, 6, 7, 8, C 1–4), the numbering of which follows that of Heath (1917). Each C1 is a small nerve leaving the anterolateral surface of the cerebral ganglion (Fig. 7, C1, *cg*) and innervating the oral tube and the lips.

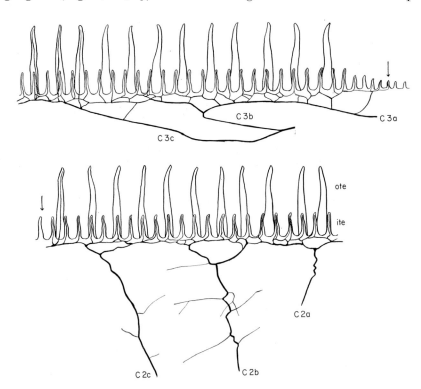

FIG. 6. Diagram of the inner and outer series of tentacles of the left side of the oral hood of *Melibe* seen from the ventral side. The nerve network supplied to the rim by C3 is shown in the upper row (proximal quarter of the hood), that supplied by C2 in the lower row (distal quarter of the hood). Arrows indicate the median line. The diagram shows nerves stained in a single preparation, thus the individual nerves of some of the outer tentacles are not drawn. C2–3, Cerebral nerves; *ite*, inner tentacles; *ote*, outer tentacles.

It forms an elaborate network here with numerous junctional enlargements or small ganglia. One marks the junction with a nerve branching from C3 (Figs 4, 8, C1, C3). There is also a dorsal commissure between left and right C1, lying within the oral tube wall. Each C2 innervates a distal quarter of the hood. On each side C2 runs within a muscular sheath with C4 (Fig. 8, C2, C4) and although these nerves are in close proximity at least half their length, there are no connexions between them. Each C2 branches twice (Fig. 4, C2) to give three main nerves, each supplying a small group of tentacles (Fig. 6, C2a, b, c, *ite*, *ote*). From each of the three branches leave fine nerves supplying other regions of the hood, and some of these join each other. A well-marked nerve network, with frequent ganglia at the junctions, surrounds the perimeter of the hood (Fig. 6). Nerves from this network go to each of the tentacles and provide an elaborate series of connexions between the branches of C2 and C3 of both sides. After giving a branch to join C1, each C3 supplies the proximal quarter of the hood, with three main nerves joining the marginal network (Figs 4, 6, C3, a, b, c, *ite*, *ote*). C4 (Figs 4, 7, 8) is an entirely separate nerve on each side, which arises from a tentacular lobe (Fig. 7, *tel*) of the cerebral ganglion (following Heath's 1917 terminology). It innervates the rhinophore (Figs 1, 2, *rh*) and the process bearing it (*rhb*).

The interconnexions between pairs C1, C2, C3 allow nerve impulses to take a variety of possible routes. This allows short circuits to be made, short cuts taken and it is likely that not all impulses are relayed through the brain. This is reflected in the behaviour of *Melibe* which shows rapid responses to stimuli and quick co-ordination of its reactions. The frequency of interconnexions is also notable between the nerves of the rest of the body. The use of the rhinophores is independent of the movements of the hood which is further indicative of its quite separate innervation.

EXPERIMENTAL INVESTIGATION OF THE FUNCTIONING OF
THE FEEDING APPARATUS AND DISCUSSION OF THE RESULTS

Experiments were carried out on the blood system and the muscles and nerves of the hood, with the aim of clarifying their functions during feeding. The description of the feeding mechanism described above was based on both anatomical and experimental studies. Only healthy specimens were used in experiments, although several contained two or three parasitic nematodes living within the well-vascularized body walls. These were most often within the hood, but some occurred in the foot or in the lateral body walls—there was no apparent adverse effect on the host.

Intravitam injection of methylene blue into the ventricle or the aortae of *Melibe* allowed the major blood vessels to be filled and traced (Fig. 5). It was noted that besides filling the vessels, the stain also filled large numbers of haemal sacs. These thin-walled, narrow-necked vesicles subsequently resembled blue bubbles hanging from the connective tissue sheets (Fig. 8, *cts*) and muscular sheaths (*msh*) surrounding the nerves of the neck region and were also crowded between the muscular meshwork of the body wall. There was sometimes a delay before the haemal sacs filled, but the only way to fill a large quantity of them experimentally was by injection into one of the blood vessels or the heart.

Some interesting results were obtained by injecting into the anterior aorta in the neck region, just before the beginnings of the labial, oral and pedal arteries. Using large numbers of *Melibe* it was found that in many cases nearly all the stain reached and was retained by either the oral hood or the foot and the walls of the main part of the body. The fact that the animals seemed able to confine the stain in one of these regions was circumstantial evidence for the presence of a physiological valve in the neck region, affecting blood flow in the vessels concerned. This might take the form of one or more sphincters, each able to reduce the lumen of a particular blood vessel. Such a mechanism has been postulated in *Monodonta* (Nisbet, 1953) and in *Philine* (Hurst, 1965).

The presence of such a physiological valve in *Melibe* would be functionally advantageous in both feeding (during which the oral hood is turgid at least part of the time) and swimming (when the body must be turgid). Similarly the haemal sacs are probably important in such activities as feeding, where at times (for instance in closure of the hood) large amounts of blood may flow to a new area. The sacs may be able to accommodate this fluid, thus reducing the possibility of backflow through the heart, and also provide a means of restricting blood to an advantageous or useful area. A similar function has been suggested for the haemal sacs in *Philine* (Hurst, 1965). The presence of haemal sacs, especially in large numbers, seems to be associated with specialized functions involving sudden considerable changes in shape.

When methylene blue was injected into the hearts of undamaged whole animals, which were then left undisturbed, the blue colour first appeared (by transparency) in the neck region and base of the hood. As swimming commenced, the stain appeared in the dorsal papillae, indicating that blood is redistributed during the onset of a new activity, such as the change from crawling to swimming.

As an experimental animal for neurophysiological investigation *Melibe* has some undoubted advantages but some concurrent disadvantages. Some pilot experiments were carried out, including intracellular recording from an isolated brain preparation, stimulation of nerves in both isolated preparations and *in situ* within the specimen. The most obvious advantages of *Melibe* for such experiments are these.

(a) The oral hood demonstrates a few clear-cut activities.

(b) The oral hood is innervated by only a few nerves, all cerebral, none of which supplies any other part of the body.

(c) The oral hood is a thin structure, its nerves being easy to reach with minimal damage.

(d) *Melibe* does not produce large quantities of mucus, unlike many nudibranchs.

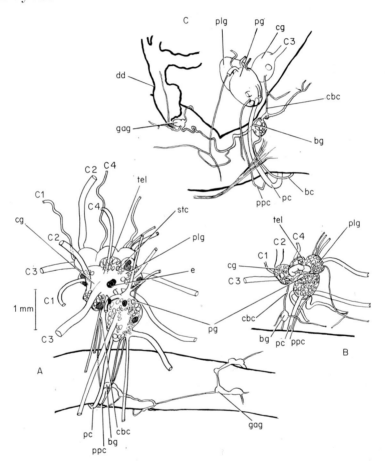

Fig. 7. Three views of the nerve ring of *Melibe*, showing the layout of the giant nerve cells. Those cross-hatched have been used in intracellular recordings. A, Left side view of the complete nerve ring; B, left side view of the right half of the nerve ring after removal of the left half; C, right side view of the nerve ring to show the buccal ganglion in greater detail. *bc*, Buccal commissure; *bg*, buccal ganglion; C 1–4, cerebral nerves; *cbc*, cerebrobuccal connective; *cg*, cerebral ganglion; *dd*, digestive duct; *e*, eye; *gag*, gastric ganglion; *pc*, pedal commissure; *pg*, pedal ganglion; *plg*, pleural ganglion; *ppc*, parapedal commissure; *tel*, tentacular lobe; *stc*, statocyst.

(e) The brain is easy to reach, being instantly revealed by a cut into the neck region.

(f) The brain has some large giant cells (not more than 90 μ) which are situated uniformly from animal to animal.

(g) The blood is largely confined to haemal sacs and thus leakage loss is not great.

Some of the disadvantages are as follows.

(a) The oral hood is very mobile and it is difficult to keep electrodes in place in a whole animal.

(b) The nerves of the hood are so interconnected that the route of an impulse is difficult to predict or follow.

(c) Dissection of the brain is difficult due to its small size and the presence of intricate connective tissue and muscular sheaths around the nerves.

(d) An isolated brain preparation does not live long due to its small size and to the length of time needed to dissect it out adequately cleared and undamaged.

For intracellular recording, the brain was isolated by careful dissection, clearing away all muscles and connective tissue, without pinching or pulling any of the nervous material. The nerves were then cut at a short distance (3–4 mm) from the main ganglia and the whole preparation was removed from the specimen and placed in a dish of slowly circulating sea water. Each nerve was sucked individually into a separate polythene tube (of diameter only slightly larger than that of the nerve) by means of a hypodermic syringe filled with sea water. Inserted into each polythene tube was a silver electrode. Thus each nerve could be separately stimulated and the preparation was also held in position in the dish by suction exerted on the nerves by the spaced out polythene tubes. (The method is that used by Mr. A. O. D. Willows of the University of Oregon.) A conventional glass capillary micro-electrode was introduced to any specific giant nerve cell of the brain for intracellular recording.

Recording from the giant cells indicated in Fig. 7 revealed that most had an endogenous rhythm and might function as pacemaker cells. This is common in nudibranchs as described by Willows (1965). There appeared to be direct or near axonal connexions between some of the largest giant cells of the cerebral ganglia and the large cerebral nerves C2 and C3 supplying the hood. This was indicated by a quick response, shown by changed spike activity of the giant cell being investigated, after stimulation of the cerebral nerve. A response was noted in giant cells of the right side to stimulation of the appropriate nerves of the same side, but concurrent responses in giant cells of the left side indicated the probable presence of connexions between the giant cells of left and right cerebral ganglia. The results suggest that the cerebral giant cells play an important role in the control and co-ordination of feeding movements.

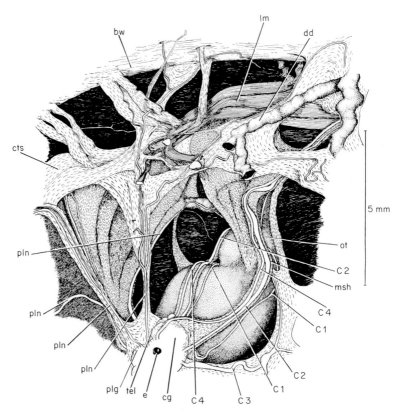

Fig. 8. Dissection of the dorsal part of the neck region of *Melibe*, seen from the right side. All areas with dashed lines represent connective tissue sheets, some of which have been partially removed during dissection. The gut is stippled. *bw*, Body wall; C 1–4, cerebral nerves; *cg*, cerebral ganglion; *cts*, connective tissue sheet; *dd*, digestive duct; *e*, eye; *lm*, longitudinal muscle; *msh*, muscular sheath; *ot*, oral tube; *plg*, pleural ganglion; *pln*, pleural nerve; *tel*, tentacular lobe.

Direct stimulation of the nerves and muscles of the oral hood was achieved by hooking a pair of fine chlorided silver electrodes under the selected structure after making a small incision through the hood's surface. In response to low voltage stimulation (usually 10–15 V

5–20/sec) contractions of the longitudinal and diffuse circular muscles were individually obtained, also contractions of the circulars of the rim. Movements of the tentacles occurred, including tipping inwards (longitudinals contracting), reduction in length (intrinsic muscles of the tentacles contracting), coming closer together (circulars of the rim contracting), besides partial closure of the whole hood. Stimuli applied to one branch of C2 or C3 first affected the immediate area, but subsequently other parts of the hood reacted. Similar results were obtained whatever the direction of the current applied, but the timing of the response varied. This emphasizes the multiplicity of possible routes that an impulse might take. Most of the muscular contractions which together form part of the feeding mechanism were individually invoked at least partially, but due to the essential interaction between muscle and blood movements (which could not be produced) and the crude nature of stimulation, none of the major feeding movements were completed normally during experiments.

Acknowledgements

The author is greatly indebted to Dr. R. L. Fernald for the use of facilities at the Friday Harbor Laboratories and to the National Science Foundation for financial support. Thanks are also due to Mr. A. O. D. Willows who lent the electronic apparatus for intracellular recording and stimulation of nerves and gave much helpful advice in the use of it. The film from which Fig. 3 was taken was also made with the co-operation of Mr. Willows.

References

Agersborg, H. P. K. (1919). Notes on *Melibe leonina* (Gould). *Publs Puget Sound mar. biol. Stn* **2**: 269–277.

Agersborg, H. P. K. (1921). Contribution to the knowledge of the nudibranchiate mollusk, *Melibe leonina* (Gould). *Am. Nat.* **55**: 222–253.

Agersborg, H. P. K. (1923). The morphology of the nudibranchiate mollusc *Melibe* (syn. *Chioraera*) *leonina* (Gould). *Q. Jl microsc. Sci.* **67**: 507–592.

Bergh, L. S. R. (1892). Die Nudibranchiata holohepatica porostomata. *Verh. zool.-bot. Ges. Wien* **42**: 1–16.

Bergh, L. S. R. (1904). Nudibranchiata kladohepatica (*Melibe pellucida*). In Semper, C. *Reisen im Archipel der Philippinen, Wiss. Resultate* **9** (6): 1–55.

Gould, A. A. (1852). Mollusca and shells. In *United States exploring expedition during the years 1838, 1839, 1840, 1841, 1842, under the command of Charles Wilkes, U.S.N.* **12**: 1–150. Philadelphia: Lippincott.

Heath, H. (1917). The anatomy of an eolid, *Chioraera dalli. Proc. Acad. nat. Sci. Philad.* **69**: 137–148.

Hurst, A. (1965). Studies on the structure and function of the feeding apparatus of *Philine aperta* with a comparative consideration of some other opisthobranchs. *Malacologia* **2**: 281–347.

Hurst, A. (1967). The egg masses and veligers of thirty Northeast Pacific opisthobranchs. *Veliger* **9**: 255–288.

MacFarland, F. M. (1966). Studies of opisthobranchiate mollusks of the Pacific coast of North America. *Mem. Calif. Acad. Sci.* **6**: 1–546.

Nisbet, R. H. (1953). *The structure and function of the buccal mass in some gastropod molluscs. 1.* Monodonta lineata (*da Costa*). Ph.D. Thesis, Univ. Lond.

O'Donoghue, C. H. (1921). Nudibranchiate mollusca from the Vancouver Island region. *Trans. R. Can. Inst.* **8**: 147–209.

O'Donoghue, C. H. (1922). Notes on the nudibranchiate mollusca from the Vancouver Island region. III. Records of species and distribution. *Trans. R. Can. Inst.* **14**: 145–167.

Willows, A. O. D. (1965). Giant nerve cells in the ganglia of nudibranch molluscs. *Comp. Biochem. Physiol.* **14**: 707–710.

Symp. zool. Soc. Lond. (1968) No. 22, 167–186.

THE BURROWING ACTIVITIES OF BIVALVES

E. R. TRUEMAN

Zoology Department, The University, Hull, England

SYNOPSIS

Bivalve molluscs dig into sand by means of a series of steps, each termed a "digging cycle", which continue until the animal is beneath the surface. Digging cycles consist of six different phases of activity, similar in all bivalves so far investigated, and involve the integration of pedal protraction and retraction with the opening and closing of the valves, much of the musculature of the body playing a part in each cycle.

The hinged shell acts as the basis of a fluid-muscle system which allows the strength of adduction to be used in digging. The fluid-muscle system consists of two separate fluid-filled chambers, the haemocoele and the mantle cavity, adduction generating high pressures in each equally and simultaneously. In the haemocoele this pressure gives rise to the characteristic dilated form of the foot which ensures a secure pedal anchorage so that at retraction the shell is drawn down. From the mantle cavity the pressure produces powerful jets of water which assist movement of the shell by loosening the adjacent sand. Subsequently the foot is protracted with probing movements by means of the intrinsic pedal musculature at relatively low hydrostatic pressures, while the shell is held still by the elastic ligament pressing the valves open against the substrate (shell anchor).

The hinge teeth function to maintain contact between the valves dorsally during digging, when the valves are gaping ventrally. The possibility that the tissues adjacent to and between the teeth contain tactile receptors is considered and the nervous co-ordination of digging is discussed.

INTRODUCTION

The process of burrowing in the Bivalvia has been studied by means of visual observations and kymograph recordings in respect of the Veneracea (Quayle, 1949; Ansell, 1962), the Lucinacea (Allen, 1958) and the Solenacea (Drew, 1907; Fraenkel, 1927; Pohlo, 1963). These descriptions are conveniently summarized by Morton (1964).

Electronic techniques of recording the activity of aquatic inverte-brates (Hoggarth and Trueman, 1967), based on the measurement of pressure and impedance changes, together with the analaysis of ciné film of the initial stages of burrowing have recently led to a fuller under-standing of the dynamics of burrowing (Trueman, 1966a). Records have been obtained of the pressure changes in the sand into which a bivalve was burrowing (Fig. 1A), together with synchronous recordings of valve movement by the determination of the impedance changes between electrodes (e) attached to the valves and the downward pull of the animal by means of isotonic or isometric myographs (my). Internal fluid pressures were recorded by cannulation through the shell into the pericardial or mantle cavities, or in *Ensis* by the insertion of a hypodermic needle into the pedal haemocoele (Trueman, 1966c, 1967a).

Burrowing consists of a step-like series of movements, involving the integration of the muscular system of the whole body, which continues until a stable position in the substrate is attained. The activity from the commencement of burrowing until the final position is reached is termed the "digging period" (Ansell, 1962). The events occurring in respect of each downward movement into the substrate (vertical movement, Fig. 2A) are referred to as a "digging cycle" (Trueman, Brand and Davis, 1966a) and are repeated many times during a digging period.

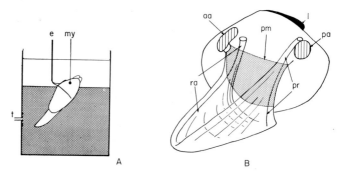

A   B

Fig. 1. A. Diagram of equipment used to record digging activity. The bivalve is suspended from a myograph (*my*) and has electrodes (*e*) attached to the valves to record their movements. The pressures produced externally in the sand (shaded area) are recorded by means of a pressure transducer attached to the tube (*t*) which is protected by a coarse nylon mesh. B. Generalized diagram of a bivalve with the foot extended showing the principal musculature involved in burrowing. *aa*, Anterior adductor; *l*, opisthodetic ligament; *pa*, posterior adductor; *pm*, protractor; *pr*, posterior retractor; *ra*, anterior retractor.

The anatomy of the foot may be conveniently outlined before discussing the results of experimental work. When extended in digging the foot of the Bivalvia is relatively large in size and consists of two parts, dorsally a viscero-pedal region and ventrally a muscular region into which the haemocoel extends. The latter region is typically compressed and blade-like, being adapted in shape for rapid penetration of the substrate as in members of the Tellinacea. The pedal musculature generally consists of three pairs of shell muscles, the anterior and posterior retractors and the protractors (Fig. 1B), and the transverse muscles (Fig. 6). The fibres of the retractor muscles extend through the ventral region of the foot, forming a geodetic network. Muscle fibres arranged in this fashion are not mutually antagonistic (Chapman,

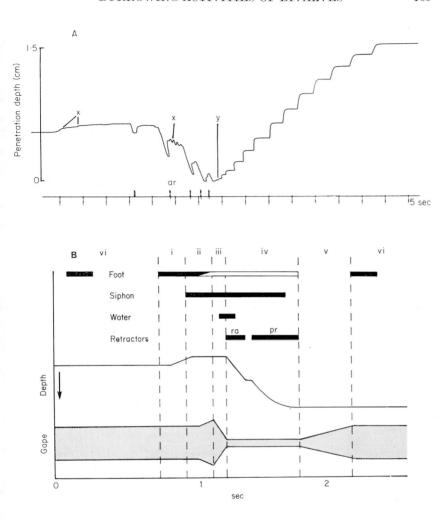

Fig. 2. Diagrams showing (A) complete digging period of *Donax vittatus* and (B) the principal activities involved in a single digging cycle of a generalized bivalve. The recording (A) was obtained by a thread attached from the posterior end of the shell to an isotonic myograph and shows the two phases of the digging cycle, namely the probing of the foot (X) to obtain a secure anchorage which is then followed (at Y) by a succession of digging cycles. Upstrokes represent movement into the sand and the horizontal lines static periods while the foot is probing. The adduction-retractions (ar) occurring during the early part of the digging period are marked above the time trace (5 sec) by visual observation. B. Stages of the digging cycle (i–vi) as described in the text; gape, refers to the angle of opening of the valves; depth, to the movement of the shell into the sand (arrow); retractors, to the contraction of the anterior (ra) and posterior (pr) retractor muscles; water, to the ejection of water from the mantle cavity at adduction; siphon, to the period of closure of the siphons; foot, to pedal probing or extension (■) and dilation (□).

1958) for although they may allow change of shape by the shortening of one or other set of fibres, additional antagonistic muscles are required for lengthening. Apart from the protractor muscles which, in many bivalves, may act as circular muscles around the upper part of the foot, e.g. *Macoma* (Trueman *et al.*, 1966a), there are little or no circular muscles as occur in the Annelida. In their place the retractors are opposed by intrinsic transverse pedal muscles (Fig. 6, *tm*) which run across the haemocoele and are inserted into the connective tissue of the basement membrane of the epithelium. Contraction of these transverse muscles can cause pedal protraction, with elongation of the retractors, in the same manner as contraction of the circular muscles may extend a polychaete worm. In the Bivalvia the blood in the pedal sinus, restricted by Keber's valve from flowing out of the foot, functions as the fluid of the fluid-muscle system, whereas in a worm the coelomic fluid is involved.

Differential tension in the retractor muscles allows postural control of the foot. This is particularly apparent during digging (Fig. 3, iv) when the contraction of first the anterior and secondly the posterior retractor muscles imparts a rocking movement to the shell. This effect is reduced in bivalves with thin, more streamlined shells and is at a minimum in *Ensis*, where the elongation of the shell and specialization of the body form make very rapid movement through sand possible. In this genus the anterior retractors are relatively reduced and act only as pedal protractors (Morton, 1964), the shell being pulled downwards by the contraction of the powerful posterior retractors.

THE DIGGING PROCESS

Digging period

The digging period of a bivalve exposed on the surface of sand exhibits two phases. During the first the foot is extended sideways and downwards probing rhythmically into the sand, and digging cycles generally occur at rather long or irregular intervals (Fig. 2A, *X*). The second phase of the digging period follows when the foot has penetrated the substrate sufficiently to obtain an anchorage firm enough to allow the shell to be lifted from a horizontal to a more vertical position (*Y*). The initial probing of the foot is particularly well observed in recordings of valve movements (Fig. 4A) where probing (*P*) is at first continuous but is subsequently interspersed by a series of adductions of the valves (*ad*) indicative of successive digging cycles. A series of step-like digging cycles gradually take the animal beneath the sand, the interval (horizontal line, Fig. 2A) between them increasing during the digging period. The reasons for this retardation and the termination of digging are discussed below.

Digging cycle

A digging cycle consists of a number of co-ordinated activities, involving much of the musculature of the animal, which are repeated

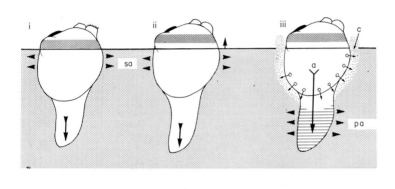

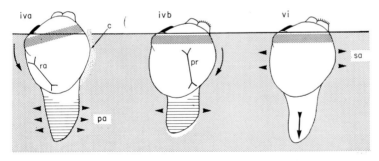

FIG. 3. Series of diagrams of a generalized bivalve burrowing into sand (shaded area) at different stages of the digging cycle (i–vi as Fig. 2). The dotted band across the valve indicates movements of the animal with reference to the surface of the sand and the horizontal shading on the foot the region of the pedal anchor (*pa*). ◄——— Movement of the shell; ◄———○ water ejection from the mantle cavity loosening the sand around the shell (*c*); ◄———◄ probing and extension of the foot; ◄———< hydrostatic pressure produced by adduction (*a*) of the valves causing pedal dilation; >———< contraction of anterior (*ra*) and posterior (*pr*) retractor muscles; arrowheads indicate pedal (*pa*) and shell (*sa*) anchors. Further information in the text.

in the same sequence for each cycle. Both the overall pattern of the digging period and the sequence of the digging cycle are common to many burrowing bivalves. Representative genera that have been recently investigated include *Nucula*, *Glycymeris*, *Anodonta*, *Cardium* *Tellina*, *Donax*, *Mercenaria*, *Mactra*, *Mya* and *Ensis* (Trueman,

1966a,c,d, 1967a; Trueman *et al.*, 1966a; Ansell and Trueman, 1967a, and 1967b).

The digging cycle is best understood by reference to Figs 2B and 3 which are derived from the analysis of ciné film supplemented by recordings such as those shown in Figs 4 and 5. It consists of the

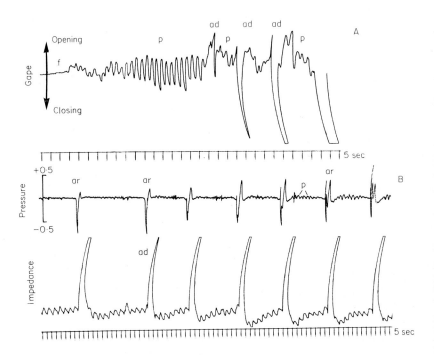

Fig. 4. Recordings of the commencement of burrowing of *Mercenaria mercenaria* (from Ansell and Trueman, 1967a). A. Impedance record (a.c. coupled) of valve movements (gape). At *f* the foot is extended onto the sand making a series of rapid probes which are detected as valve movements (*p*). Subsequently the first four adductions (*ad*) of the digging period show increasing amplitude with penetration and anchorage is obtained. B. Pressure recording from adjacent sand (pressure, in cm of water throughout) and simultaneous impedance recording (impedance) from the pericardium showing both heart beat and adduction. Pressure record exhibits at first negative and subsequently complex positive wave forms at successive adduction-retractions (*ar*) as depth of burial increases due to ejection of water from the mantle cavity into the sand.

following stages, numbered i–vi to correspond to previous descriptions (Trueman *et al.*, 1966a).

(i) The foot makes a major probe downwards tending to raise the shell if penetration is not easily achieved.

(ii) Siphons close, preventing water from passing out at adduction (iii). The probe continues to maximum pedal extension and pedal dilation may commence.

(iii) Rapid adduction of the valves causing water to be ejected from the mantle cavity and an increase in blood pressure producing maximal dilation of the foot. This ensures a firm pedal anchorage (Fig. 3, *pa*) prior to retraction (iv).

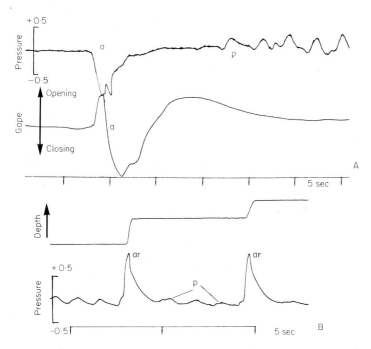

FIG. 5. Recordings of digging cycles of (A) *Margaritifera margaritifera* and (B) *Ensis arcuatus*. In both "pressure" represents the pressure recorded in the sand adjacent to the bivalve. A also shows changes of valve gape (gape, recorded by impedance detection a.c. coupled). Probing (*p*) only recommences when the valves have opened after adduction (*a*). B shows increasing depth of penetration (depth, upward sweep) at successive digging cycles (*ar*).

(iv) Contraction of first the anterior (a) and secondly the posterior (b) retractor muscles resulting in the shell being pulled down into the sand (Figs 3 and 5B). The siphons reopen at or just before the termination of retraction.

(v) Relaxation of adductors, valves open, pedal dilation and anchorage lost.

(vi) Static period equivalent to the plateau in the trace of a whole digging period (Fig. 2A) during which the foot is re-extended with repeated probings (Figs 4 and 5, *p*). Probing movements of the foot occur continually during the static period, recommencing after retraction only when the valves have opened (Fig. 5A), and the number of probes per cycle is related to the duration of the static period (Trueman *et al.*, 1966a).

The downward thrusting pedal movements are brought about by the rhythmically repeated contraction and relaxation of the retractor and transverse muscles in the distal region of the foot. In some genera, e.g. *Donax*, waves of contraction may be observed to spread from the anterior tip of the foot, producing a series of undulatory waves which facilitate forward movement. By contrast with probing, which involves the retractor muscles distally, pedal retraction is carried out principally by the proximal regions of these muscles. Thus in retraction the shell is drawn over the foot which retains its dilated shape and position in the substrate.

Anchorage in the substrate

Digging cycles consist essentially of repeated adduction and opening of the valves integrated with protraction and retraction of the foot. Adduction accomplishes pedal anchorage by the dilation of the foot over a flat broad area as in *Tellina*, into a bulbous swelling as in *Ensis*, or by the outward spreading of a cleft foot as in *Glycymeris*. The strength of pedal anchorage and the factors effecting this have been discussed previously (Trueman, 1967a). Adduction also reduces the width of the shell and the water ejected from the mantle cavity serves to loosen the adjacent sand (Figs 3 and 6) so that at retraction the shell moves down more easily. Loss of pedal anchorage occurs towards the end of retraction when the distal region of the foot may shrink. In many bivalves, e.g. *Mercenaria*, the foot is then pulled up towards the shell leaving a small fluid-filled cavity beneath (Fig. 3, iv b).

When the adductors relax (stage v) the valves open because of the opening moment of the ligament which presses the shell against the sand so forming a shell or secondary anchorage (Fig. 6, *sa*), preventing the animal from being pushed upwards as the foot probes downwards. Drew (1907) and Pohlo (1963) have previously observed the shell of members of the Solenidae gripping the burrow during pedal protraction. The force with which the foot can be pushed downwards is a function of the strength of the shell anchor. When a bivalve lies on the surface of the sand this is limited to its weight for there can be no shell anchorage (Trueman, in press) and when only partially buried too great a probing force causes the shell to be raised in its burrow (Fig. 2B, stage i). This phenomenon has been observed to occur commonly in *Cardium edule* although the ribs on the shell must serve to strengthen the shell anchor (Trueman, Brand and Davis, 1966b).

Whilst discussing the burrowing of worms Clark (1964) suggested that all soft bodied animals use essentially the same method. He observed that some part of the body wall is first dilated to form an

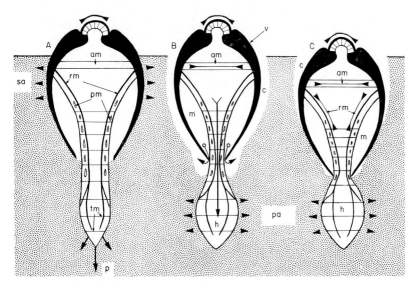

Fig. 6. Diagram of successive stages in the burrowing of a generalized bivalve showing shell (*sa*) and pedal (*pa*) anchorages (arrowheads). A, Represents stages i, ii, or vi of the digging cycle with the valves pressing against the sand by means of the opening thrust of the ligament and the foot extending by probing (*p*); B, stage iii where contraction of the adductor muscles (*am*) ejects water from the mantle cavity (*m*) so loosening the sand (*c*) around the valves (*v*), high pressure simultaneously produced in the haemocoele (*h*) gives rise to pedal dilation; C, stage iv, contraction of retractor muscles (*rm*) pulls the shell down into the loosened sand. *tm*, Transverse pedal muscle; >——< tension in ligament, adductor or retractor muscles; other letters and symbols as in the previous Figures.

anchor while the head is pushed forward into the substrate, and that secondly the anterior end of the worm dilates to form a new anchor while the body is drawn in by the contraction of longitudinal muscles. These two anchorages are respectively the flanging and dilation anchors of *Arenicola* (Trueman, 1966d) and correspond to the shell and pedal anchors of a bivalve. The retractors are equivalent to the longitudinal muscles of a worm and both sets of muscles function in the same manner to draw the animal into the sand.

G

Rate of burrowing

Ansell (1962) has observed that in members of the Veneridae the time/cycle increases and the depth of penetration/cycle decreases as the digging period proceeds. Trueman *et al.* (1966a), while confirming this in respect of certain other bivalves, note that the slowing down of digging is due to a lengthening of the static period. Ansell discussed whether the slowing down is due to some intrinsic nervous mechanism controlling burrowing or to extrinsic environmental factors, and he commented that fatigue is unlikely to be the cause of the retardation or cessation of burrowing since animals will repeatedly burrow after removal from the substrate immediately on completion of a digging period.

The rate of probing during the second phase of the digging period has been determined for a number of bivalves burrowing freely by use of external pressure recordings (Fig. 5). It shows considerable variation between species, e.g. *Ensis arcuatus*, 90 probes/min; *Mercenaria mercenaria*, 16; *Mya arenaria*, 1, although in each the rate of probing slows down very little after many digging cycles. The amount of probing is thus controlled by the duration of the static period rather than by change of probing frequency. Protraction of the static period thus allows more probing, which may be required for pedal extension because of the increasing resistance of the substrate as depth of burial increases (Trueman *et al.*, 1966b).

Although the patterns of both the digging periods and cycles are commonly observed throughout the Bivalvia, there is considerable variation in respect of rate of burrowing in similar substrates. In general, active bivalves with slim shells, e.g. *Tellina* or *Donax*, burrow more rapidly than those with tumid shells, e.g. *Cardium*, *Mercenaria*. A full discussion on the effect of shell shape and substrate on the rate of burrowing is, however, deferred until the results of further experimental work are available.

The advantages of rapid burrowing are most clearly visualized in relation to littoral bivalves. The migratory behaviour of *Donax* in relation to tidal rise and fall is well known (Wade, 1964), and it is certainly to the advantage of any clam near the water's edge to disappear quickly after being uncovered. Ropes and Merrill (1966) observed that repeated dislodgements of *Spisula solidissima* are in a shoreward direction as the tide rises, many being left stranded, and suggested that speed of burrowing is a critical factor in their survival.

The termination of the digging period may be due to different factors in different bivalves. In *Cardium* or *Donax* activity generally ceases when the posterior dorsal margin of the shell is level with the surface of the sand while *Tellina* or *Macoma* burrow more deeply and the end of their digging periods may be related to siphonal extension. The availability of respiratory water currents may control depth in *Glycymeris* (Ansell and Trueman, 1967b) for this bivalve will burrow for several centimetres beneath the surface of coarse shell gravel, whereas as soon as it is covered by fine sand digging ceases and the animal raises itself slightly so as to expose the mantle margins.

THE FLUID DYNAMICS OF BURROWING

Extension of the foot, its probing into the sand and the initial stages of terminal dilation to form an anchor (stage ii, Fig. 2B) are brought about by the intrinsic pedal muscles. Protraction is caused principally by the contraction of the transverse and protractor muscles with the relaxation of the retractors. To allow this antagonistic action the pedal blood must be maintained at constant volume, being prevented from leaving the foot by Keber's valve, and low hydrostatic pressures are generated in the haemocoele (Fig. 7C), although as much as 10 cm of water pressure have been recorded in *Ensis* (Trueman, 1967a). Adduction has been observed to cause rapid pedal extension in *E. arcuatus* with accompanying high pressures but generally, during normal digging, adduction only occurs with the foot fully extended so causing pedal dilation. When the valves *Ensis* are adducted with the foot retracted, the latter is thrust downwards and pushes the shell upwards. This occurs when the animal is returning to its normal location with the siphons at the surface of the sand after being buried more deeply. Recordings of pressure either in the pericardial cavity or in the pedal haemocoele (Fig. 7) show a peak at adduction. The increase in pressure, produced dorsally, affects the whole body, being transmitted to the pedal haemocoele through the circulatory system. The foot, shell and body musculature may be thought of as a hydraulic system in which the force produced by adduction can be transferred to a region of application in the foot where, by causing swelling, it allows a pedal anchorage to be obtained (Figs 3 iii, 6B and C, *pa*). The current investigations confirm Chapman's (1958) observations that the heart is not powerful enough to operate the hydrostatic system of the foot (Trueman, 1966c).

Retraction immediately follows adduction and maintains the pressure in the foot. A multiple pressure peak in which the final peak corresponds to retraction is shown in Fig. 7A. During retraction the volume of blood is retained in the foot by Keber's valve and possibly by the action of the valves of the shell pressing against the sides of the foot (Fig. 6C). This ensures that the foot remains dilated until the end of retraction when the valves open and pedal anchorage is lost.

The fluid-muscle system of a bivalve consists of two fluid-filled chambers, the mantle cavity and the blood system. Adduction affects

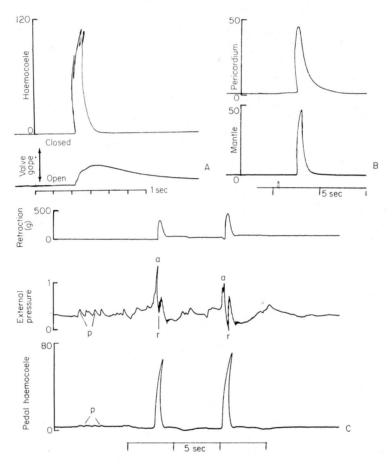

FIG. 7. Internal pressure changes recorded during burrowing activity in *Ensis arcuatus* (A and C) (from Trueman, 1967a) and *Margaritifera margaritifera* (B) (from Trueman, 1966a). A. Pressure peaks in the pedal haemocoele (haemocoele) produced by adduction of the valves (below) and by pedal retraction. B. Pressure peaks of similar amplitude but of different duration in the mantle and pericardial cavities, the mark over the time trace indicates siphonal closure. C. Simultaneous recording of retraction strength, external and internal pressures. The retraction strength represents the pull of the posterior retractor muscles (at stage iv) while the shell remains in almost the same position. The external positive pressure (*a*) occurred at adduction, negative at retraction (*r*) as anchorage is lost. The needle used to record internal pressure (pedal haemocoele) was inserted into the upper part of the foot through the fourth pallial aperture. Probes of the foot (*p*) are indicated in the lower traces.

both simultaneously for pressure peaks of nearly equal amplitude may be recorded during digging both in the pericardial and mantle cavities. In bivalves in which the opposing mantle folds are free and water can escape from around the ventral margin of the mantle cavity, e.g. *Margaritifera*, the duration of the pressure peak in the mantle cavity is markedly less than in the pericardium (Fig. 7B). Keber's valve must function to retain the blood in the foot not only to allow pedal dilation but also so as to prevent a surge of blood into the gills at retraction when the pressure in the mantle cavity is low. In *Ensis*, however, in consequence of extensive mantle fusion (Yonge, 1952; Owen, 1959) the pressure peaks in both cavities are of comparable duration and a pressure gradient does not occur between them (Trueman, 1967a).

Ejection of water from the mantle cavity produces currents which loosen the sand below and adjacent to the valves (Fig. 6B and C) forming a cavity (*c*) into which the shell is pulled at pedal retraction. Ropes and Merrill (1966) compare the use of a stream of water by the surf clam, *Spisula solidissima*, with the modern method of seating pilings through the use of a jet stream of water to displace the substrate. In *Ensis arcuatus* (13 cm in length) the strength of retraction (800 *g*) is approximately 100 times the weight of the animal in water, while in *Mercenaria mercenaria* (6·3 cm in length, retraction strength, 5 *g*) it is only one-quarter (Ansell and Trueman, 1967a). These are extreme examples, but in the latter penetration is principally accomplished by the shell dropping down into the cavity produced, whereas in *Ensis* the strength of the retractor muscles and pedal anchor are most important.

The direction in which the water is ejected is controlled by the mantle margins in bivalves such as *Mercenaria*, *Tellina* or *Cardium*, but the loosening of the sand must not extend so far as to affect the pedal anchor. The quantity of water ejected is largely determined by the angle through which the valves are adducted and the surface area of the shell. Allowing for the increase in volume of the foot, values range from an angular change of 20° in *Ensis arcuatus* (13 cm in length) with approximately 4 ml ejected, to *Tellina tenuis* (1·8 cm in length), 1·25° closure and 0·02 ml ejected. Experiments on the penetration of shells into sand have shown how relatively small quantities of water, injected beneath a bivalve, facilitate penetration (Trueman *et al.*, 1966b).

THE FUNCTION OF THE LIGAMENT AND HINGE TEETH

Recent experimental studies have indicated how the ligament and hinge teeth of bivalves may function during burrowing. Investigation of the strength of the ligament of a variety of genera show that, in general, it is no more powerful in burrowing than attached forms, although the former have additionally the resistance of the substrate to overcome (Trueman, 1964). It has been shown that the function of the ligament is to open the valves at stage v of the digging cycle and to press the valves outward during the static period to produce a shell anchor. The more powerful the ligament the stronger the anchorage and in consequence the more effective the probing. Many bivalves which dig rapidly have powerful opisthodetic parivincular ligaments, e.g. *Tellina, Ensis*, whereas *Nucula*, with a weak internal ligament, burrows sluggishly and superficially. Determination of the rate of opening of the valves with the adductors cut away shows that the ligament always opens the valves more rapidly than occurs during the digging cycles. For example, the respective times in *Tellina tenuis*, for the same angle of gape, are 0·01 and 0·17 sec, in *Macoma balthica* 0·05 and 0·5 sec, the additional time during the digging cycle probably being due to the stretching of the adductors. In some bivalves the opening of the shell occurs rapidly after retraction, e.g. *Tellina, Donax, Ensis*, while the adjacent sand remains loosened, but in others, notably in *Mercenaria* (Ansell and Trueman, 1967a), *Margaritifera* (Fig. 8; Trueman, in press), or *Glycymeris*, the valves open more slowly and the ligament is supplemented by hydraulic forces generated by pedal and siphonal retraction when the shell is more than one-third buried.

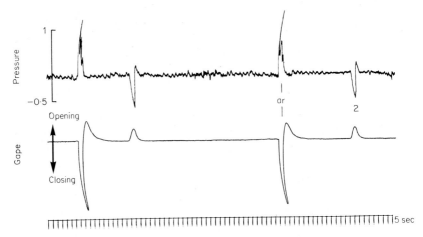

FIG. 8. Simultaneous recording of external pressure (pressure) and valve movements (gape, recorded by impedance detection a.c. coupled) of *Margaritifera margaritifera* when half-buried in sand. During each static period a secondary pushing open of the valves (2) occurs in association with pedal retraction, (negative deflexion of pressure). *ar*, Adduction-retraction.

This is comparable to the hydrodynamic relationship which occurs in *Mya arenaria*, where siphonal retraction causes the stretching of the adductors as the valves gape (Trueman, 1966c). Although this implies that the ligament is not strong enough to cause the valves to open fully in the latter genera, it does not mean that the ligament is unable to press the valves outwards sufficiently to effect a shell anchor. Indeed adequate strength to open the valves against the compacted sand would probably mean the failure of the anchorage. The ligament must be considered to act as a mechanism which effectively stores part of the energy of the adductors for use during the later stages of the digging cycle.

Although much attention has been given to hinge teeth by systematists, there have been few attempts to assess their functional significance. Newell (1954) considered that they do not serve as fulcral points for the valves but rather act to guide the valves so that they will fit correctly at their margins as they close. Evidence from sections, in which epithelium may be seen to lie between the teeth, substantiates his first point whilst the second is obviously of prime importance. During digging, however, the valves are never completely closed but are open more or less widely. The hinge teeth function to maintain contact between the valves dorsally when the valves are gaping ventrally for under these conditions crenulations or teeth along the ventral margin cannot be effective. The continuous contact between the valves dorsally must be particularly important at retraction when the effect of one valve meeting with some greater resistance in the substrate might be to misalign the valves seriously in the absence of any control system.

The possibility of the tissues around the hinge teeth being the site of tactile receptors concerned with the angle of gape or alignment of the valves should not be overlooked, although some preliminary histological investigations have not revealed such structures. Dr. Deforest Mellon Jr. (personal communication) has recently described a sensory function in the hinge region of *Spisula*, stimulation by a probe or closure of the valves causing pedal retraction. The probable importance of postural sense organs for the co-ordination of the digging cycle makes such a function for the hinge teeth realistic.

THE CO-ORDINATION OF DIGGING

Although digging involves the co-ordination of much of the body musculature, apart from the early papers of Drew (1908) and Fraenkel (1927), little has been written on its nervous control. The digging period generally commences shortly after the removal of a bivalve from its substrate. Laid on the surface of the sand it will gradually open the shell and extend the foot but any further disturbance leads to pedal retraction, valve closure, and further delay before the shell is reopened. An exception to this is *Ensis* where rapid burrowing is the normal response, stimulation of the siphons playing a role in this reaction (Fraenkel, 1927). Barnes (1955) has shown in *Anodonta* that gentle manual rotation, or, in sensitive individuals, vibration of the bench, may cause the valves to gape and Ropes and Merrill (1966) have suggested that surf clams (*Spisula solidissima*) are stimulated to dig quickly in nature by the disturbance of wave action. *Donax vittatus* also reacts rapidly to disturbance and recordings of this species, made while *in situ* on the beach, show digging activity only when disturbed by wave action at the edge of the rising or falling tide (Trueman, 1967b). In many bivalves, e.g. *Cardium, Mercenaria, Anodonta, Glycymeris*, however, disturbance is followed by a variable period of quiescence before digging commences.

In all species so far investigated, burrowing continues, if undisturbed, until the animal is beneath the sand. Digging cycles begin when the foot is maximally extended (Trueman, 1967a) and the digging stages (ii–v) follow in a fixed sequence of muscular contractions (Table I).

TABLE I

State of contraction of certain muscles of a bivalve during the different stages of the digging cycle

Muscles	Stages of cycle						
	vi	i	ii	iii	iv	v	vi
Transverse pedal	C R P	C E	C	R	R	R	C R C P
Retractors (distal)	R C	R	R	R			R C R
Siphonal sphincter	R R	R	C	C	C	R	R R R
Adductors	R R	R	R	C	C	R	R R R
Retractors (proximal)	R R	R	R	R	C	R	R R R

C, Contraction; R, relaxation; E, maximum extension of foot; arrows represent a series of events which may affect each other successively. The transverse and retractor muscles give rise to a probing rhythm (P) and in this respect are considered in the distal region of the foot only.

The rhythmic contraction and relaxation of the pedal muscles which occurs during probing is possibly co-ordinated by the pedal ganglia, for Nadort (1943) described complete reflex arcs in the foot after removal of the cerebral ganglia from members of the Unionidae. In *Mytilus*, Woortman (1926) has also shown that the foot could protrude and withdraw after transection of the cerebropedal connectives, although the co-ordinated movements of creeping no longer occurred. The importance of the cerebral ganglia in interganglionic co-ordination is shown in *Ensis* where cutting of the cerebropedal commissures results in the foot becoming flaccid and inability to dig (Bullock and Horridge, 1965). The function of the pedal ganglia and the dependence on the cerebral ganglia possibly vary considerably between genera. The probing rhythm continues throughout the digging period except during stages ii–v of the digging cycles. This may be due to pedal extension affecting stretch receptors and causing both inhibition of the rhythm and the commencement of these stages of the digging cycle. The importance of extension of the foot in the control of digging is indicated both by the long periods of probing which occur during the first phase of the digging period (Fig. 2A at X), when the probing is weak in the absence of a shell anchor, and by the increase in duration of the static period with depth of burial. The substrate becomes more resistant as depth increases and more probing is required for the full extension of the foot.

Little is known of how the successive events of the digging cycle are programmed in a constant time sequence in each animal. Although complex movements involve all three principal pairs of ganglia (Woortman, 1926), the nervous system of the Bivalvia is relatively simple and it is unlikely that the programming of the digging cycle is carried out exclusively in the central nervous system. It is possible that after siphonal closure each stage is brought about by a series of reflex arcs, each being stimulated by the mechanical effect of the previous event as indicated by the arrows in Table I. Thus adduction might affect receptors of either pressure in the foot or of tactile stimuli in the region of the hinge teeth, so as to effect retraction. The discovery of such a mechano-sensory organ in the region of the hinge teeth in *Spisula*, stimulation of which causes pedal retraction, by Mr. G. Mpitsos (Mellon, personal communication) should encourage further researches in this field.

The relaxation of the adductors brings about the opening of the valves (v) and may be related to both the fall of internal pressure and to the termination of retraction. Probing, which has been inhibited while the valves were closed, now recommences (Fig. 5A). The digging cycle appears to be controlled by two principal events, firstly the amount of extension of the foot, which stimulates the changes leading to pedal anchorage, and secondly the termination of retraction after which the shell anchorage is obtained.

DISCUSSION

Bivalves are primitively adapted to shallow burrowing in soft, often unstable, substrates (Morton, 1964). Their most important adaptations in this respect are the form of the foot and the bivalved shell which enables the strength of adduction to be used to anchor the foot. The ability of *Ensis*, a genus highly specialized for rapid burrowing, to change the shape of the foot from a V-shaped tip for penetration to a swollen bulb for anchorage is an outstanding adaptation. Over other molluscs that burrow, e.g. *Dentalium, Natica, Terebra* (Morton, 1964), bivalves have at least three advantages: (1) the adductor muscles are used to cause pedal dilation in addition to the intrinsic pedal musculature; (2) the profile of the shell is reduced before being pulled down into the substrate; (3) the elastic ligament effects a secure shell anchor.

In general high pressures only occur in the haemocoele of bivalves during digging, and they probably increase in amplitude with increasing magnitude of adductions (Fig. 4A) as penetration of the substrate proceeds, in a similar manner to recordings from the coelom of the polychaete, *Arenicola* (Trueman, 1966b). High pressures and pedal dilation are retained for passage over the surface of sand, e.g. *Margaritifera* (Trueman, in press), but bivalves which normally progress over a hard substrate do so without adduction being involved, while retaining the rhythm of extension and retraction of the foot (Morton, 1960, 1964). Pedal anchorage is obtained in a different manner on a hard surface, where there would accordingly seem to be little functional significance in the production of high pressures in the pedal haemocoele. The evolution of the bivalved shell can be satisfactorily explained as an adaptation for active burrowing for it represents a mechanism for generating the high pressures involved in digging into or moving over a soft substrate. By the form of their shell, the Bivalvia have been able to utilize the double fluid muscle system of mantle cavity and haemocoele for burrowing, and this has placed them among the more successful inhabitants of soft substrates.

REFERENCES

Allen, J. A. (1958). The basic form and adaptations to habitat in the Lucinacea (Eulamellibranchia). *Phil. Trans. R. Soc.* (B) **241**: 421–481.

Ansell, A. D. (1962). Observations on burrowing in the Veneridae (Eulamellibranchia). *Biol. Bull. mar. biol. Lab.*, *Woods Hole* **123**: 521–530.

Ansell, A. D. and Trueman, E. R. (1967a). Burrowing in *Mercenaria mercenaria* (L.) (Bivalvia, Veneridae). *J. exp. Biol.* **46**: 105–116.

Ansell, A. D. and Trueman, E. R. (1967b). Observations on burrowing in *Glycymeris glycymeris* (L.) (Arcacea, Bivalvia). *J. exp. Mar. Biol. Ecol.* **1**: 65-75.

Barnes, G. E. (1955). The behaviour of *Anodonta cygnea* L., and its neurophysiological basis. *J. exp. Biol.* **32**: 158–174.

Bullock, T. H. and Horridge, G. A. (1965). *Structure and function in the nervous systems of invertebrates* **2**: 1387–1431. San Francisco and London: Freeman.

Chapman, G. (1958). The hydrostatic skeleton in the invertebrates. *Biol. Rev.* **33**: 338–371.

Clark, R. B. (1964). *Dynamics in metazoan evolution.* Oxford: Clarendon Press.

Drew, G. A. (1907). The habits and movements of the razor-shell clam, *Ensis directus. Biol. Bull. mar. biol. Lab.*, *Woods Hole* **12**: 127–140.

Drew, G. A. (1908). The physiology of the nervous system of the razor-shell clam (*Ensis directus* Con.). *J. exp. Zool.* **5**: 311–326.

Fraenkel, G. V. (1927). Die Grabbewegung der Soleniden. *Z. vergl. Physiol.* **6**: 167–220.

Hoggarth, K. R. and Trueman, E. R. (1967). Techniques for recording the activity of aquatic invertebrates. *Nature, Lond.* **213**: 1050–1051.

Morton, J. E. (1960). The responses and orientation of the bivalve *Lasaea rubra* Montagu. *J. mar. biol. Ass. U.K.* **39**: 5–26.

Morton, J. E. (1964). Locomotion. In *Physiology of Mollusca* **1**: 383–423. Wilbur, K. M. & Yonge, C. M. (eds). New York: Academic Press.

Nadort, W. (1943). Some experiments concerning the nervous system of *Unio pictorum* and *Anodonta cygnea. Archs néerl. Sci. (Physiol.)* **27**: 246–268.

Newell, N. D. (1954). Status of invertebrate paleontology, 1953 V. Mollusca: Pelecypoda. *Bull. Mus. comp. Zool. Harv.* **112**: 161–172.

Owen, G. (1959). Observations on the Solenacea with reasons for excluding the family Glaucomyidae. *Phil. Trans. R. Soc.* (B) **242**: 59–97.

Pohlo, R. H. (1963). Morphology and mode of burrowing in *Siliqua patula* and *Solen rosaceus. Veliger* **6**: 98–104.

Quayle, D. B. (1949). Movements in *Venerupis* (=*Paphia*) *pullastra* (Montagu). *Proc. malac. Soc. Lond.* **28**: 31–37.

Ropes, J. W. and Merrill, A. S. (1966). The burrowing activities of the surf clam. *Underw. Nat.* **3**: 11–17.

Trueman, E. R. (1964). Adaptive morphology in paleoecological interpretation. In *Approaches to paleoecology*: 45–74. Imbrie, J. & Newell, N. D. (eds). New York: Wiley.

Trueman, E. R. (1966a). Bivalve mollusks: fluid dynamics of burrowing. *Science, N.Y.* **152**: 523–525.

Trueman, E. R. (1966b). Observations on the burrowing of *Arenicola marina* (L.). *J. exp. Biol.* **44**: 93–118.

Trueman, E. R. (1966c). The fluid dynamics of the bivalve molluscs, *Mya* and *Margaritifera. J. exp. Biol.* **45**: 369–382.

Trueman, E. R. (1966d). The mechanism of burrowing in the polychaete worm, *Arenicola marina* (L.). *Biol. Bull. mar. biol. Lab.*, *Woods Hole* **131**: 369–377.

Trueman, E. R. (1967a). The dynamics of burrowing in *Ensis* (Bivalvia). *Proc. R. Soc.* (B) **166**: 459–476.

Trueman, E. R. (1967b). The activity and heart rate of bivalve molluscs in their natural habitat. *Nature, Lond.* **214**: 832–833.

Trueman, E. R. (in press). The locomotion of the freshwater clam *Margaritifera margaritifera. Malacologia.*

Trueman, E. R., Brand, A. R. and Davis, P. (1966a). The dynamics of burrowing in some common littoral bivalves. *J. exp. Biol.* **44**: 469–492.

Trueman, E. R., Brand, A. R. and Davis, P. (1966b). The effect of substrate and shell shape on the burrowing of some common bivalves. *Proc. malac. Soc. Lond.* **37**: 97–109.

Wade, B. (1964). Notes on the ecology of *Donax denticulatus* (Linné). *Proc. Gulf Caribb. Fish. Inst.* 17th meeting: 36–41.

Woortman, K.-D. (1926). Beiträge zur Nervenphysiologie von *Mytilus edulis. Z. vergl. Physiol.* **4**: 488–527.

Yonge, C. M. (1952). Studies on Pacific coast mollusks. IV. Observations on *Siliqua patula* Dixon and on evolution within the Solenidae. *Univ. Calif. Publ. Zool.* **55**: 421–438.

Symp. zool. Soc. Lond. (1968) No. 22, 187–192.

ASPECTS OF EXCRETION IN THE MOLLUSCS

W. T. W. POTTS

Department of Biology, University of Lancaster, England

SYNOPSIS

As the author has recently reviewed excretion in the molluscs (Potts, 1967), the opportunity has been taken to discuss some obvious lacunae in our knowledge of molluscan excretion which could be filled by using existing techniques.

Although Picken (1937) thirty years ago estimated the colloid osmotic pressure of the blood of two species of freshwater molluscs, there are few measurements of the colloid osmotic pressure in the literature. It is desirable, now that osmometers are available, that measurements of the colloid osmotic pressure of a wide variety of molluscs should be made.

Although the evidence suggests that urine is first formed by ultrafiltration, the exact sites of the ultrafiltration membranes are not known in any molluscs. It is suggested that the suspected sites of ultrafiltration should be examined with the electron microscope. Suitable colloids such as Thorotrast might help in the localization of the filtration sites.

The high molecular weight of mollusc blood proteins suggest that the filtration membrane may be permeable to small protein molecules. By the use of a variety of compounds of various molecular weights the permeability properties of the filtration membrane could be defined.

With the exception of Bouillon's brief work (1960) in *Helix* the fine structure of molluscan excretory organs has not been examined with the electron microscope. The *Helix* kidney shows some interesting similarities with the vertebrate and crustacean excretory organs but also some striking differences.

In spite of recent work in this field a large number of problems remain. The origin and significance of uric acid in aquatic prosobranchs is not clear. Some terrestrial prosobranchs appear to be uricotelic but the acquisition of uricotelism within the prosobranchs requires investigation.

INTRODUCTION

In 1962 the late Professor Munro Fox asked me to review excretion in molluscs. Since this review has now been published (Potts, 1967), the opportunity will be taken here to draw attention to certain gaps in our knowledge, beginning with a discussion of some aspects of kidney structure and function.

Although some larval molluscs possess protonephridia, adult molluscs always possess kidneys derived from coelomoducts. The evidence available suggests that urine is produced in the molluscs by ultrafiltration and is later modified by resorption and secretion. Ultrafiltration can be recognized by various criteria. Among these are:

(1) The hydrostatic pressure of the blood should exceed colloidal osmotic pressure of the blood at the site of filtration.

(2) At the site of filtration there should be a membrane permeable to water and to small solute molecules but impermeable to proteins.

If these two criteria are fulfilled, a protein-free ultrafiltrate should be produced in Donnan equilibrium with the blood.

COLLOID OSMOTIC PRESSURE OF THE BLOOD

It has been shown in many molluscs that the primary urine consists of a protein-free solution in Donnan equilibrium with the blood. However, direct evidence that the hydrostatic pressure of the blood exceeds the colloidal osmotic pressure is available only in octopus, and the membranes at which ultrafiltration takes place have not yet been demonstrated in a single mollusc.

Thirty years ago Picken (1937) attempted to demonstrate, with the methods then available, that ultrafiltration took place in *Lymnaea* and in *Anodonta*. At that time the direct measurement of low colloidal osmotic pressure in small volumes was not possible. Picken estimated the colloidal osmotic pressure of the blood by the comparison of the refractive indices of the mollusc blood and of mammalian blood plasmas with those of isotonic salines. However, this method involves the assumption that the molecular weight of the blood proteins are the same in mammalian blood and in molluscan blood. This supposition is almost certainly incorrect, at least in the case of *Lymnaea* which contains haemocyanin. It is very desirable, now that sensitive and accurate colloid osmometers are available, that direct measurements should be made of the colloid osmotic pressure of *Anodonta* and *Lymnaea* blood and of a wide variety of other molluscs.

SITE OF ULTRAFILTRATION

The site and properties of the membranes at which ultrafiltration takes place have not yet been defined in any mollusc. There is a great deal of evidence that suggests that the site of filtration was primitively in the heart, and that the urine drained from the pericardium into the kidney ducts. This is probably still the case with most molluscs with the notable exception of the terrestrial pulmonates, where filtration takes place from the renal vein (Martin, Stewart and Harrison, 1965), and in the cephalopods where there is evidence that ultrafiltration takes place through the branchial heart appendage. However, in no case is the exact site of the filtration membrane known. In the vertebrates ultrafiltration takes place through the basement membrane of the glomerulus. Electron micrographs have shown that there are wide channels between the endothelial cells lining the blood capillaries, allowing the blood plasma direct access to the basement membrane, while the other side of the membrane is covered by peculiar podocytes which stand, as it were, on tiptoe, thus leaving large areas of the basement membrane exposed for the exit of the ultrafiltrate. Very similar structures have been demonstrated in the crayfish *Astacus* where the basement membrane of the end sac is also lined with podocytes. Similar structures should be looked for in the molluscs.

The exact site of ultrafiltration might be demonstrated by the injection of suitable colloids such as thorium dioxide (Thorotrast), which should accumulate on the inner surface of the membrane at the site of ultrafiltration. Some experiments have been made with Thorotrast in molluscs, but unfortunately not for the purpose of identifying the filtration membrane. Brown and Brown (1965) injected Thorotrast into the marine gasteropod *Bullia*. Radiographs taken after 4 days show that the heart contained a high concentration of Thorotrast. However, Brown and Brown (1965) were primarily concerned with the means of removal of the colloid and reported that after 4 days the colloid was entirely concentrated in haemocytes. Ten minutes after injection the colloid was uniformly distributed throughout the animal. At some stage between 10 min and 4 days the Thorotrast might have been found lining the filtration membrane but not yet ingested by haemocytes. *Bullia* is a marine gasteropod in which the rate of filtration would be very low. Better results might be expected with freshwater forms such as *Paludina* or *Lymnaea*.

PROPERTIES OF FILTRATION MEMBRANES

The permeability properties of the filtration membranes have not, so far, been defined. Ions and small molecules such as glucose and inulin can filter through the membrane while large molecules such as haemoglobin and haemocyanin can not. It should not be too difficult to measure the pore diameter by the use of a succession of compounds of increasing molecular weight. The very high molecular weight of haemocyanin found in many molluscs and of the haemoglobin found in *Planorbis* suggest that the pore diameter of the filtration membranes is large.

After ultrafiltration, urine is modified by resorption and secretion in the kidney sac and the ureter, if the latter is present. In freshwater and terrestrial species extensive resorption of salt takes place. If the kidney is relatively impermeable to water this results in the formation of a dilute urine, as is produced by freshwater snails, while if the kidney is permeable to water the resorbed ions are followed by the water and a small volume of concentrated urine is formed as in the terrestrial snails. The fine structure of the kidney sac and ureter have been examined only in *Helix* (Bouillon, 1960). It is clear that there is a vast field waiting for someone with an electron microscope and an interest in molluscan excretion. Bouillon has described the fine structure of the cells in the two portions of the kidney sac and in the ureter of *Helix*. All three portions of the excretory system show many similarities. The most common type of cell in all the portions has deeply folded basal membranes. The folds penetrate far into the cells, in the ureter almost to the distal border. Mitochondria lie between the folds. Distally the cells show a brush border. Similar structures have been described in many excretory organs, notably in the proximal convoluted tubules of the mammal and in the nephridial canal of the crayfish (Kümmel, 1964). In the distal convoluted tubules of the mammal, where resorption takes place against a concentration gradient, the mitochondria and membranes are very pronounced but the brush border is absent. Bouillon (1960) suggested that in *Helix* the membranes are a device for the uptake of material from the blood into the kidney, but on the analogy of the vertebrates it is more likely that they are a device for transporting material from the kidney back into the blood. Bouillon's micrographs also show interesting details of the uric acid and other granules so abundant in the kidney cells. Electron micrographs of the fine structure of the kidneys of lamellibranchs, cephalopods and of marine and freshwater gasteropods have not yet been published.

NITROGEN EXCRETION

The kidneys of practically all molluscs are rich in intracellular concretions. These concretions are of varying composition. They include uric acid, guanine, xanthine, calcium carbonate and melanin. The significance of uric acid in terrestrial snails is well known but uric acid also occurs in smaller quantities in the kidneys of all gasteropods and some other molluscs as well. The significance of uric acid in aquatic molluscs is doubtful. Gostan (1965) showed that in the archeogastropods of the genus *Nerita* the marine species *N. albicilla* contains the full sequence of uricotelic enzymes while *N. costata*, which lives in the upper littoral zone contains no detectable urease, allantoinase, nor allantoicase and only a trace of uricase. This suggests that the higher concentration of uric acid found in the more terrestrial *Cyclostoma elegens* (see Needham, 1938) is due to the loss of uricase while the traces of uric acid found in the aquatic forms is presumably due to an incomplete breakdown of uric acid derived from purine metabolism. The advantages to be derived from the loss of uricase is not clear. Full adaptation to terrestrial conditions must include the ability to convert amino nitrogen into uric acid. It is well known that *Helix* can synthesize uric acid from the amino nitrogen, apparently by the same route as occurs in birds and in insects. When this stage has been reached, uricotelic enzymes would be disadvantageous. However, at present, we do not know at what stage the ability to synthesize uric acid from amino nitrogen was acquired. Is the loss of uricotelic enzymes in the more terrestrial prosobranchs associated with the ability to synthesize uric acid, or is the uric acid found in these forms derived only from purine metabolism? In the latter case the advantage to be derived from the loss of uricase must be very small.

The status of urea as an excretory product in the molluscs has been clarified in the last few years. Small quantities of urea are found in a wide variety of molluscs, but no complete ornithine cycle has been found. The relatively small quantities of urea reported in several instances are no doubt formed by the action of arginase on exogenous arginine. However, there is one observation which still requires further explanation. Delaunay (1931) reported that, after "autodigestion", the excretion of the slug, *Arion rufus*, contained between 60 and 80% of the total non-protein nitrogen in the form of urea. If this observation is correct it would be possible that the slug possesses either a high concentration of uricase or an ornithine cycle. It is worth noting that both Spitzer (1937) and Delaunay reported that uric acid accounted for less than 10% of the total nitrogen of *Arion*.

In many other fields of excretory physiology our knowledge of the molluscs is even more limited. For example, nothing is known of the hormonal and nervous regulation of excretion. However, I have confined myself here to those topics in which the experimental approach is self-evident. In some other fields, such as the hormonal control of excretion, the problem is to know where to begin.

REFERENCES

Bouillon, J. (1960). Ultrastructure des cellules rénales des Mollusques. I. Gastéro-podes pulmonés terrestres. *Annls Sci. nat.* (2) **12**: 719–749.

Brown, A. C. and Brown, R. J. (1965). The fate of thorium dioxide injected into the pedal sinus of *Bullia* (Gastropoda: Prosobranchiata). *J. exp. Biol.* **42**: 509–520.

Delaunay, H. (1931). L'excrétion azotée des Invertébrés. *Biol. Rev.* **6**: 265–302.

Gostan, G. (1965). Cytophysiologie de l'excrétion chez les Mollusques pulmonés. *Annls Biol. anim. Biochem. Biophys.* **4**: 481–494.

192   W. T. W. POTTS

Kümmel, G. (1964). Das Cölomsäckchen der Antennendrüse von *Cambarus affinis* Say (Decapoda, Crustacea). *Zool. Beitr.* **10**: 227–252.
Martin, A. W., Stewart, D. M. and Harrison, F. M. (1965). Urine formation in the pulmonate land snail, *Achatina fulica*. *J. exp. Biol.* **42**: 99–124.
Needham, J. (1938). Contributions of chemical physiology to the problem of reversibility in evolution. *Biol. Rev.* **13**: 224–251.
Picken, L. E. R. (1937). Urine formation in freshwater molluscs. *J. exp. Biol.* **14**: 20–34.
Potts, W. T. W. (1967). Excretion in the molluscs. *Biol. Rev.* **42**: 1–41.
Spitzer, J. M. (1937). Physiologisch-ökologische Untersuchungen über den Exkretstoffwechsel der Mollusken. *Zool. Jb.* (Zool.). **57**: 457–496.

Symp. zool. Soc. Lond. (1968) No. 22, 193–211.

THE FINE STRUCTURE OF CARDIAC AND OTHER MOLLUSCAN MUSCLE

R. H. NISBET and JENIFER M. PLUMMER

Department of Physiology (E.M. Unit),
Royal Veterinary College, University of London,
London, England

SYNOPSIS

The fine structure of unstriated muscle from the collar and buccal mass of Achatinidae is compared with that of cardiac muscle in the same animals, with striated fibres from the buccal retractors of Trochidae and with the smooth and striated fibres of the adductor muscle in *Pecten*. The axial periodicities of the thick myofilaments in smooth muscle fibres from *Archachatina* and *Pecten* have been observed to range from 100 to 150 Å units. Such variation may be significant, in view of the mechanism of contraction by means of a dislocation of myosin molecules postulated by Morales (1965).

Absence of a regular pattern of organization of the myofilaments, dense bodies and sarcotubules, with associated mitochondria, is correlated with slow and sustained (tonic) contraction in the smooth muscles. Conversely, phasic contraction is related to the specialized organization of muscle fibres into "sarcomeres" consisting of myofilaments, dense bodies, transverse and longitudinal sarcotubules, and parallel columns of mitochondria. The latter arrangement, together with a dyad association between sarcolemmal invaginations and the sarcotubules, appears to be significantly related to the rapid movement of calcium ions to and from the active sites on the myofilaments, and to excitation-contraction coupling. This organization may, therefore, be of greater significance than the variations of biochemical organization in the myofilaments of tonic and phasic muscles.

INTRODUCTION

Recent work on the fine structure of cardiac muscle in Achatinidae (Nisbet and Plummer, 1966) has led us to questions related to the origin of striated muscle in molluscs. The present communication examines comparative aspects of muscle structure and function in some gastropods, although reference will also be made to the smooth and striated adductor muscle of *Pecten*.

Much of the published work on molluscan muscle has been concerned with the structure and function of the anterior byssal retractor and the adductors of *Mytilus edulis*, and with the adductor muscles of a number of other bivalves (Philpott, Kahlbrock and Szent-György, 1960; Hanson and Lowy, 1961; Rüegg, 1963; Rüegg, Straub and Twarog, 1963; Lowy, Millman and Hanson, 1964; and Rüegg, 1964). In gastropod molluscs, Schlote (1957, 1960) has examined the fine structure of the penial retractor of *Helix pomatia*, Kawaguti (1963) has described fine structure in heart muscle of the snail *Euhadra hickonis* and North (1963) has reported on ventricular muscle fibres in *Helix aspersa*.

From the point of view of the electron microscopist working on muscle, the molluscs present difficult problems of fixation on the one hand and of toughness of their tissues on the other. The tissues are easily damaged during fixation, and the muscle fibres are always associated with large quantities of collagen and frequently with equally tough vesicular connective tissue cells. Nevertheless, the smooth muscle of the collar has been studied in *Archachatina marginata* and *Achatina fulica*, and compared with buccal and cardiac muscle in the same animals. The comparison has been extended to include the striated muscle of the buccal mass in some Trochidae and the smooth and striated muscle of the adductor in *Pecten maximus*.

METHODS

The pulmonate material was fixed in the ways described in an earlier paper (Nisbet and Plummer, 1966). For the marine material a variety of methods was used, with limited success.

(1) After anaesthetization of the animal (*Gibbula cineraria*) with $7\frac{1}{2}\%$ $MgCl_2$ the fine buccal retractor muscles were flooded with cold 10% formalin in sea water, and, after 2 h, post-fixed with osmium tetroxide in veronal acetate buffer (pH 7·8–8·0) made up in molar sucrose.

(2) Without anaesthesia (*Monodonta*; *Pecten*) the muscles were similarly flooded with 10% formol–sea water and post-osmicated.

(3) The tissues were flooded directly with osmium tetroxide in veronal acetate buffer made up in molar sucrose, small pieces being cut out to try to improve fixation.

After 2–4 h, all were processed into Araldite.

Sections were cut with glass knives on Cambridge ultra-microtomes, mounted on grids and stained with either uranyl acetate and potassium permanganate or with lead citrate and uranyl acetate. The grids were then examined in a RCA EMU 3F electron microscope.

UNSTRIATED MUSCLE IN *Archachatina*, *Achatina* AND *Pecten*

Figure 1 is a low magnification print of a section through the collar of *Archachatina* showing one longitudinally sectioned and three transversely sectioned smooth muscle fibres. The thick filaments appear to have a random distribution, the thin filaments forming an ill-defined, fine-textured background. Subsurface cisternae and sarcolemmal invaginations are fairly numerous but very few "dense bodies" (Hanson and Lowy, 1961) are visible. The sarcotubular system is sparse and there are virtually no mitochondria. Higher magnification of part of the section (Fig. 2A) shows subsurface cisternae, sarcolemmal invaginations,

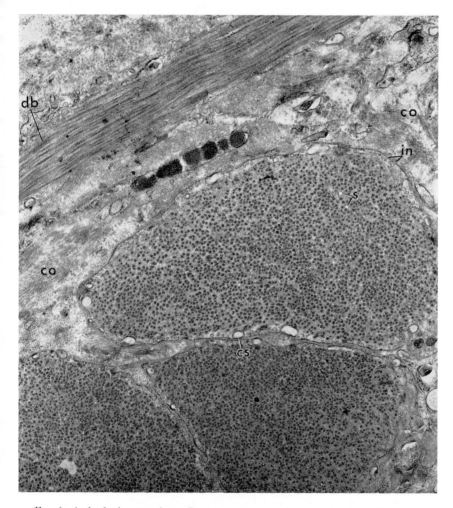

FIG. 1. *Archachatina marginata.* Low magnification photograph of a section through the collar, showing one smooth muscle fibre in longitudinal and three in transverse section. Thin filaments form a fine-textured background to the thick filaments. Few dense bodies (*db*) and sarcotubules (*s*) are visible and no recognizable mitochondria. *co*, Collagen; *cs*, subsurface cisterna; *in*, sarcolemmal invaginations. × 11 500.

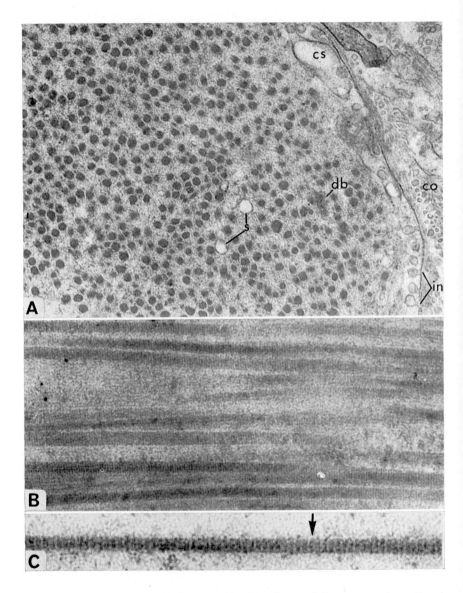

Fig. 2. *A. marginata*. A. Higher magnification of part of the transversely sectioned fibre at centre of Fig. 1. Diameters of thick filaments, 250–700 Å. Note the spiral patterns of thin filaments round the thick filaments and indications of an orbital arrangement of the former. × 36 000. B. High magnification of thick filaments in smooth muscle fibre, showing periods at 120–130 Å. × 53 000. C. *Pecten*. Thick filament from smooth adductor muscle fibre. Note axial periods of 150 Å and (arrowed) the indication of a more darkly staining substructure. × 130 000. *co*, Collagen; *cs*, subsurface cisterna; *db*, dense body; *in*, sarcolemmal invaginati on; *s*, sarcotubule.

a few sarcotubules and one dense body. The thick filaments have diameters ranging from 250 to 700 Å, the overlap of these dimensions in the section suggesting that they are fusiform structures of limited length. The thin filaments form obliquely spiral patterns round the thick filaments, indicating an orbital arrangement that is not clearly defined in this material. The longitudinal section (Fig. 2B) shows thick filaments with maximum diameters of 650 Å and an axial periodicity of 120–130 Å. Comparison with the thick filaments in the smooth adductor muscle of *Pecten* is of interest. The filament shown (Fig. 2C) has a diameter of 260 Å and axial periods at 150 Å. However, in other sections of this muscle, the periodicity may be 100–110 Å. The shorter periods measured in both *Archachatina* and *Pecten* may be significant. Hanson and Lowy (1957) demonstrated similar periods in molluscan smooth muscle. More recently Elliott (1964) has shown the periodicity of paramyosin to be 145 Å, in agreement with Philpott *et al.* (1960). However, Morales (1965) has postulated a mechanism of contraction by means of a dislocation of the myosin molecules that reduces the pitch of the helix by about one-third of its extended length and simultaneously increases the diameter of the filament.

Examination of some, apparently, smooth muscle from the pulmonate buccal mass (Fig. 3A and B) shows marked differences from the foregoing muscle. As already shown (Nisbet and Plummer, 1966) the thick filaments are of limited length and have diameters that are fairly constant at 170–180 Å, their axial periods are approximately 160 Å. Each group of thick filaments is associated with groups of thin (50 Å) filaments that arise from the dense bodies. This arrangement, viz. two dense bodies with their thin filaments forming orbits round a group of thick filaments, will be called a "contractile unit" and will be discussed later (p. 209). Between the contractile units run longitudinal sarcotubules and groups of mitochondria. Transverse sarcotubules are associated with the dense bodies the peripheral members of which are also closely related to sarcolemmal invaginations. Absence of a striated appearance is due to lack of transverse alignment of the dense bodies.

STRIATED MUSCLE IN TROCHIDAE

Most buccal muscles of trochids are striated although in *Monodonta* there is a "herring-bone" appearance to the striae seen in the light microscope (Nisbet, 1954). Examination of electron micrographs of the post-median retractor muscle (Fig. 4) makes clear the reason. Dense bodies and their associated thick filaments form orderly though staggered arrays that may have a spiral arrangement. The lengths of the

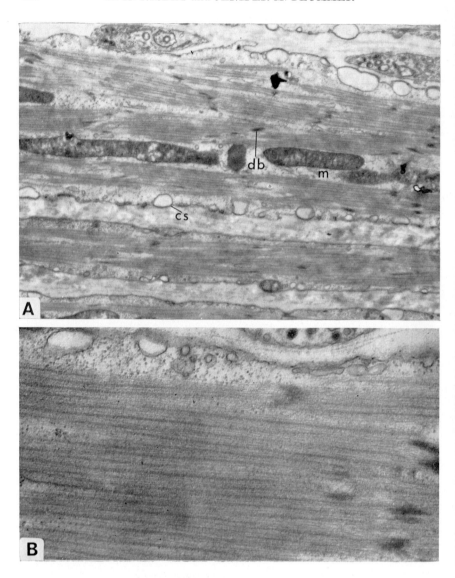

FIG. 3. *Achatina fulica*. A. Portions of two muscle fibres in the matrix of the radular collostyle, the upper near to the middle, the lower near to one end of their respective lengths. Note that the myofilaments are separated into discrete groups ("myofibrils") by columns of mitochondria and longitudinal sarcotubules; and the contractile units are out of register. × 8000. B. Higher magnification of part of (A) showing that thin filaments arise from the dense bodies; that thick filaments may pass, but do not end in, the dense bodies: thick filaments (diameters 170–180 Å) show axial periods at *ca* 160 Å and cross bridges to the thin (50 Å) filaments. × 41 000. *cs*, Subsurface cisterna; *db*, dense body; *m*, mitochondria.

contractile units are approximately 7–8 μ between corresponding dense bodies. The thin filaments and sarcotubules have not been well fixed and few thin filaments can be seen to emerge from the dense bodies.

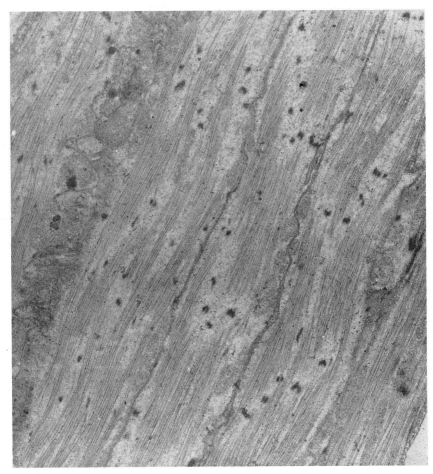

FIG. 4. *Monodonta lineata.* Small portion of the post-median buccal retractor (PMR) muscle, stretched before fixing, showing staggered arrangement of dense bodies and thick filaments that sometimes pass, but do not enter the dense bodies. Thin filaments poorly fixed. Lengths of contractile units are 7–8 μ. × 8000.

Transverse sections through the post-median retractor (Fig. 5A and B) show that the thick filaments have diameters ranging from 170 to 460 Å. They occasionally appear to be double (Fig. 5A, arrowed). Although a few groups give an impression of a hexagonal arrangement,

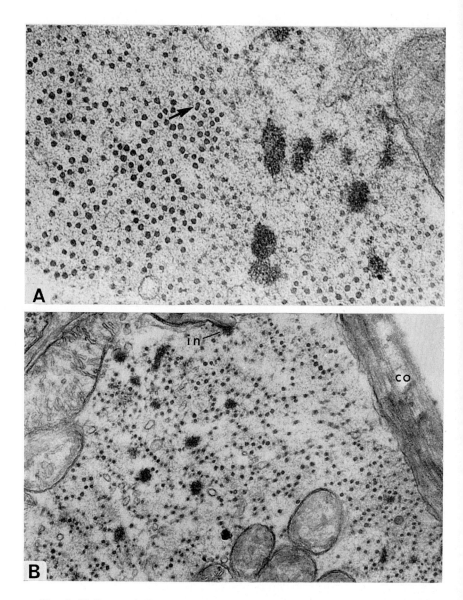

Fig. 5. *M. lineata.* A. Transverse section showing part of a fibre of the PMR. Dense bodies, adjacent "I"-band areas (thin filaments only) and "A"-band areas with thick filaments, can be seen. The range of thick filament diameter is 170–460 Å: occasionally hexagons can be seen. Note double filament (arrow). × 53 000. B. As (A) but at lower magnification, to show peripheral and central mitochondria, dense bodies and longitudinal sarcotubules. Note dense body associated with sarcolemmal invagination (*in*) and linear arrangement of many thick filaments. *co*, Collagen. × 36 000.

linear groups are more frequent than hexagonal ones. Thin filaments are poorly resolved but occasionally an orbital arrangement is apparent. Areas relatively free from thick filaments, and associated with the dense bodies, show the overlapping of "I" band and "A" band zones. A thick layer of collagen fibrils surrounds each group of muscle fibres and penetrates between them (Fig. 5B).

Material obtained from *Gibbula cineraria* is also poorly fixed and transverse sections have not yet been examined, but the longitudinal sections through the post-median retractor (Fig. 6A and B) suggest a more specialized arrangement than that found in *Monodonta*. The low magnification print shows a parallel arrangement of groups of myofilaments with rows of mitochondria and sarcotubules lying between them. Dense bodies and transverse sarcotubules occur with great regularity at intervals of $2 \cdot 6$–$3 \cdot 0$ μ. Fibre diameters range from $4 \cdot 3$ to $6 \cdot 5$ μ and in each fibre the dense body (or "Z") bands, the "A" band and the "I" band zones of the contractile units are aligned with one another, giving rise to the appearance of a series of "sarcomeres", a condition approaching that to be seen in *Pecten* striated muscle (Fig. 10A and B). The higher magnification (Fig. 6B) shows that the dense bodies are closely associated with transverse (or terminal) cisternae of the sarcotubular system but that dense bodies and their associated myofilaments have not formed continuous transverse arrays and may, therefore, still be regarded as discrete contractile units. Longitudinal sarcotubules can be seen between these units.

The mitochondria of this muscle (Fig. 6B) are peculiar. Their tubular and vesiculate cristae are very different from those of *Monodonta* muscle (Fig. 5B) and the muscle of other gastropods studied in the present work.

CARDIAC MUSCLE IN ACHATINIDAE

The muscle fibres of both auricle and ventricle are striated but the degree of symmetry shown in longitudinal sections can vary. In a contracted auricular fibre (Fig. 7A) the "sarcomeres" are symmetrically arranged in the upper region of the fibre. The "A" band zone extends nearly to the dense bodies and each sarcomere is short (*ca* $1 \cdot 14$ μ). Oblique sections show less regularity in the arrangement of contractile units (Fig. 7B) indicating that they may have a spiral arrangement. The lack of transverse alignment becomes more marked in extended fibres and is usually more apparent in ventricular muscle (Fig. 8A). The mitochondria are located mainly in two regions, a peripheral zone and a massive central column.

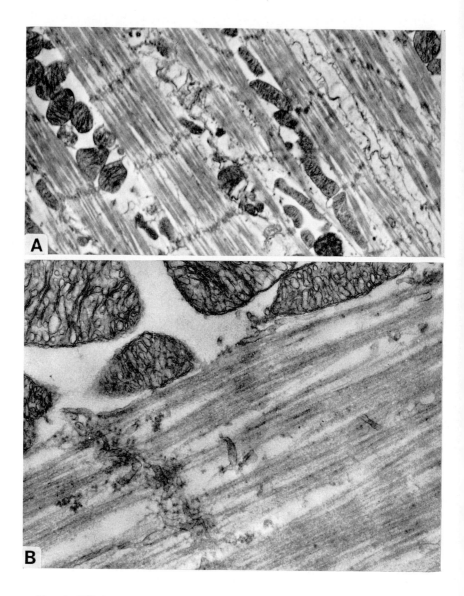

Fig. 6. *Gibbula cineraria*. A. Longitudinal section through three fibres of the contracted PMR, showing parallel arrangement of groups of myofilaments, mitochondria and sarcotubules. Note the regularly repeating pattern of transverse sarcotubules and dense bodies, giving "sarcomere" lengths of 2·6–3·0 μ. "I" band regions very short. × 8000. B. Higher magnification of small portion of (A). Note dense body material (poorly fixed) associated with transverse sarcotubules, evidence of longitudinal sarcotubules, and unusual mitochondria. × 38 000.

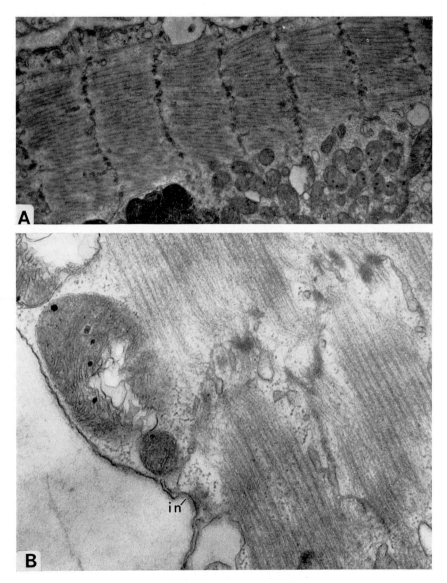

FIG. 7. *A. fulica*. A. Cardiac muscle: part of a longitudinal section through a contracted fibre. Note the organization of contractile units, dense bodies, transverse and longitudinal sarcotubules into parallel arrays, all in register, forming sarcomeres (length *ca* 1·14 μ) with a massive central zone of mitochondria. × 14 000. B. Higher magnification of part of auricular fibre, showing transverse sarcotubules looping round the dense bodies (cf. Fig. 9B), origin of thin filaments from dense bodies and peripheral extension of transverse sarcotubule to a "dyad" association with a sarcolemmal invagination (*in*). × 41 000.

At higher magnifications (Fig. 7B) the thin filaments can be seen to arise from the dense bodies, the latter being closely associated with the transverse sarcotubules. From the transverse cisternae, longitudinal tubules pass between the contractile units. Peripherally, the dense bodies and sarcotubules are in close association with sarcolemmal invaginations (Nisbet and Plummer, 1966). Without an analysis of serial sections it is difficult to determine the extent of the sarcolemmal invaginations and the scale of the dyad association that may be formed (see Page, 1966).

The measured diameters of the thin filaments are very close to 50 Å (range 48–55 Å), the corresponding measurements of the thick filaments giving a range of 120–190 Å.

The longitudinal section of the ventricular muscle shows less regularity in the arrangement of dense bodies (Fig. 8A) and also shows the discrete nature of these bodies and their relations with the thin filaments. Towards the end of the fibre massive half-desmosomes are formed, in which the thin filaments terminate (Fig. 8B).

In transverse sections (Fig. 9A and B) the thick filaments frequently appear to form hexagonal arrays, each with a central filament. Each thick filament is surrounded by an orbit of thin filaments, the number of these being most probably twelve. A hundred counts of these in stretched fibres has given a range from nine to fifteen. Counts in contracted fibres have ranged up to nineteen thin filaments. The ratio of thin to thick filaments is 6 : 1, in agreement with estimates given for the adductor of *Crassostrea angulata* (Hanson and Lowy, 1961) and the ventricular muscle of *Helix aspersa* (North, 1963), although the hexagonal array is not always easy to determine. The relation of the sarcotubules to the dense bodies, to the mitochondria, the sarcolemma and to the sarcolemmal invaginations is similar to that of many other muscle fibres, both vertebrate and invertebrate, the association between the sarcolemma and the sarcotubules being similar to that defined by Porter and Palade (1957) as the "dyad" (see also Page, 1966). Thus excitation-contraction coupling in these fibres may be similar to that demonstrated in isolated frog muscle fibres (Huxley and Taylor, 1958) and in lizard fibres (Huxley and Straub, 1958).

STRIATED MUSCLE IN *Pecten*

Philpott *et al.* (1960) found the "I" band region of this muscle "very sensitive to fixation". We have met with the same difficulty, both the thin filaments and the sarcotubules tending to disappear from the fixed material.

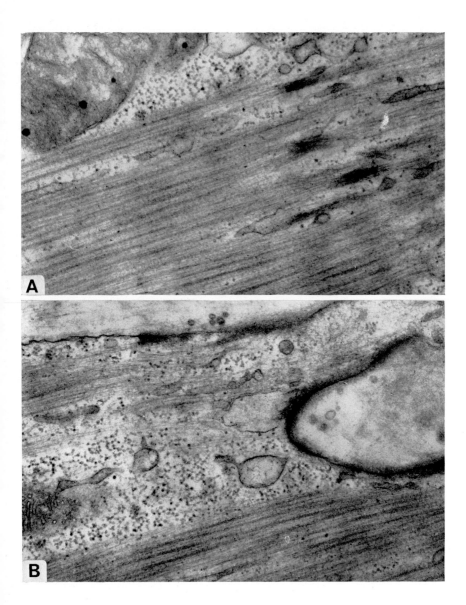

FIG. 8. *A. fulica*. A. Ventricular muscle fibre. Longitudinal section showing "stagger" of dense bodies and the origin of thin filaments from them, i.e. the ends of three clearly defined contractile units. Note sinusoidal passage of thin between thick filaments and the frequent cross-bridges between them. × 42 000. B. As (A) but nearer to the end of the fibre, showing the massive half-desmosomes and the termination in these of the thin filaments. × 42 000.

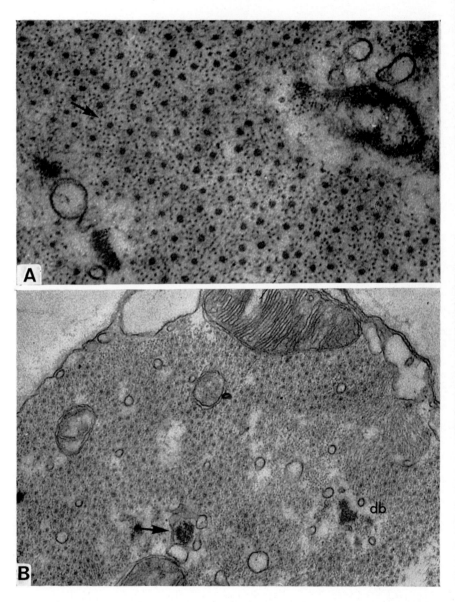

Fig. 9. *A. fulica*. A. High magnification of a transverse section through an auricular fibre. Note (1) dense bodies with associated sarcotubules, that on right showing clearly the origin of the thin filaments, (2) trend to hexagonal arrangement of thick filaments (many showing radial bridges to thin filaments) and (3) orbits of thin filaments with (at arrow) twelve clearly defined. × 80 000. B. Transverse section similar to (A). Note relation of sarcotubules to dense body and to mitochondria (arrow) and the way that the longitudinal sarcotubules surround contractile units (e.g. at *db*). × 41 000.

In longitudinal sections (Fig. 10A) the visible organization of the muscle fibres shows similarities to vertebrate striated muscle. The fibres have diameters of about 1–2 μ. They are divided into series of compartments by "Z" membranes, the lengths of the sarcomeres being about 1·5 μ. "A"-bands (1·2 μ) and "I"-bands (0·3 μ) are clearly defined and at the centre of the "A"-band there is evidence of an "H" zone (Fig. 10B). Differences from vertebrate striated muscle are also apparent. The "myofibril" occupies virtually the whole of the fibre and the thick filaments have diameters considerably greater than those of vertebrate skeletal muscle. The diameters of primary filaments given by Philpott et al. (1960) are 200–250 Å. Our measurements are somewhat less (from 150 to 220 Å). A group of "vesicles" (arrowed in Fig. 10B) that may be either sarcotubules or sarcolemmal invaginations, or both, occurs at the level of many of the "Z" membranes. The latter structures have the appearance of closely organized dense bodies (Fig. 10B). In transverse sections the thick filaments demonstrate a remarkably regular hexagonal pattern (Fig. 10C). Their radial separation is 430 to 480 Å (Fig. 10D). Residues of the thin filaments are visible round the thick filaments but the orbital numbers cannot be estimated.

DISCUSSION

The remarkable ability to maintain a constant level of contraction possessed by the unstriated adductor muscles of many bivalve molluscs has been the subject of investigation and discussion for many years. Those workers who have favoured a "catch" mechanism in such muscles (Nieuwenhoven, 1947; Twarog, 1954; Johnson, 1954, 1958) or a modified form of catch mechanism (Jewell, 1959) have received more recent support for this view from Philpott et al. (1960) who suggest that two separate systems, an actomyosin (phasic) and a paramyosin (tonic) system, may be present in the same muscle. Similar views are held by Rüegg (1964). In opposition to this view Lowy et al. (1964) have argued that the characteristic protein of the thick filaments of "catch" muscles (tropomyosin-A) is present in the thick filaments of the obliquely striated part of the adductor in Crassostrea and that this muscle cannot maintain tonic contraction.

It appears likely, therefore, that the different response characteristics of tonic and phasic muscles may have their origin in aspects of organization other than, or in addition to, the structural characteristics of their myofilaments. An organization that provides for the rapid provision—and removal—of calcium ions at the active sites (Weber,

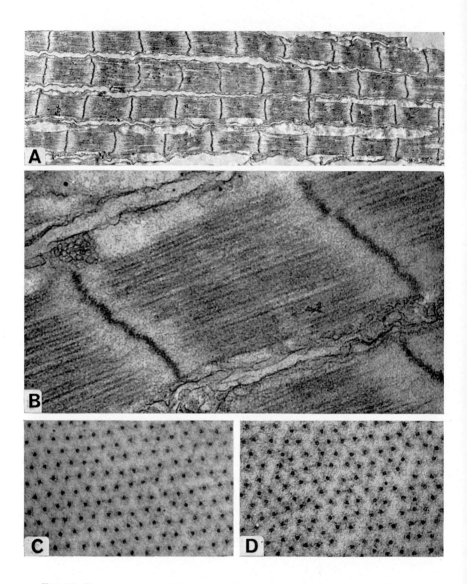

Fig. 10. *Pecten maximus*. A. Low magnification of longitudinal section through four fibres, 1–2 μ diameter, from the striated adductor muscle. Note complete conversion of fibres into series of sarcomeres showing "I"-bands, "A"-bands and "Z" membranes. Sarcomere lengths 1·5 μ ("A"-bands 1·2 μ and "I"-bands 0·3 μ). × 8000. B. Higher magnification of a sarcomere. Note group of vesicles (or tubules) at one "Z" membrane, evidence of tubules in the "Z" membranes and of thin filaments ending in them. × 38 000. C and D. Small portions of transversely cut fibres showing regular hexagonal arrangement of the thick filaments. Thin filaments are poorly defined. × 53 000.

1966) and excitation-contraction coupling (Huxley and Straub, 1958; Huxley and Taylor, 1958; Page, 1966) is one possibility.

Viewed from such a standpoint, the structure of the unstriated muscle fibres in the collar of *Archachatina* (Figs 1 and 2) is interesting. The myofilaments occupy a very large proportion of the fibre area. Few sarcotubules can be seen and even fewer mitochondria. Dense bodies are rare and "contractile units" cannot be distinguished.

The smooth buccal muscle of *Achatina* (Fig. 3) shows a more highly organized structure in which dense bodies and myofilaments are organized into rather long contractile units; sarcolemmal invaginations are plentiful; sarcotubules and mitochondria are associated with the contractile units. It is likely that the lack of striation is due solely to the lack of transverse alignment of these units.

The organization of cardiac and buccal muscles described in the foregoing series shows progressive specialization, in which dense bodies become transversely aligned and the contractile units come to lie in parallel with one another, giving rise to the series of sarcomeres seen most clearly in the longitudinal sections of *Achatina* auricle (Fig. 7) and the post-median retractor muscle of *Gibbula* (Fig. 6). The striated adductor of *Pecten* (Fig. 10) represents the terminal development in this specialization.

The evolution of phasic contractile systems may, therefore, be more dependent upon the development of close associations between myofilaments, tubular conducting systems and mitochondria, than upon the variations of biochemical organization in the myofilaments themselves.

ACKNOWLEDGEMENTS

We gratefully acknowledge the support and encouragement that we have received from Professor E. C. Amoroso, F.R.S., in the prosecution of this work. Our technical staff, Mr. J. T. Gunner, Miss D. G. Lonsdale and Mrs. M. A. Peake have worked hard to help us and for this we are most grateful.

The work was supported by U.S. Public Health Service Grants nos. RG 06489 and HD 01476 from the National Institutes of Health, Bethesda, Maryland, U.S.A.

REFERENCES

Elliott, G. F. (1964). X-ray diffraction studies on striated and smooth muscles. *Proc. R. Soc.* (B) **160**: 467–472.

Hanson, J. and Lowy, J. (1957). Structure of smooth muscles. *Nature, Lond.* **180**: 906–909.

Hanson, J. and Lowy, J. (1961). The structure of the muscle fibres in the translucent part of the adductor of the oyster *Crassostrea angulata*. *Proc. R. Soc.* (B) **154**: 173–196.

Huxley, A. F. and Straub, R. W. (1958). Local activation and interfibrillar structures in striated muscle. *J. Physiol., Lond.* **143**: 40P.

Huxley, A. F. and Taylor, R. E. (1958). Local activation of striated muscle fibres. *J. Physiol., Lond.* **144**: 426–441.

Jewell, B. R. (1959). The nature of the phasic and tonic responses of the anterior byssal retractor muscle of *Mytilus*. *J. Physiol., Lond.* **149**: 154–177.

Johnson, W. H. (1954). A further study of isometric mechanical responses of the anterior byssus retractor muscle of *Mytilus edulis* to electrical stimuli and mechanical stretch. *Biol. Bull. mar. biol. Lab., Woods Hole* **91**: 88–111.

Johnson, W. H. (1958). Further evidence concerning the catch mechanism in molluscan muscles. *J. cell. comp. Physiol.* **52**: 190.

Kawaguti, S. (1963). Electron microscopy on the heart muscle of a snail. *Biol. J. Okayama Univ.* **9**: 140–148.

Lowy, J., Millman, B. M. and Hanson, J. (1964). Structure and function in smooth tonic muscles in lamellibranch molluscs. *Proc. R. Soc.* (B) **160**: 525–536.

Morales, M. F. (1965). On the mechanochemistry of contraction. In *Molecular biophysics*: 397–410. Pullman, B. and Weissbluth, M. (eds). London: Academic Press.

Nieuwenhoven, L. M. van (1947). *An investigation into the structure and function of the anterior byssal retractor muscle of* Mytilus edulis. L. Diss., Nijmegen-Utrecht, Dekker en van de Vegt N.V.

Nisbet, R. H. (1954). *Structure and function of the buccal mass in* Monodonta lineata *(da Costa)*. Thesis for Ph.D. University of London.

Nisbet, R. H. and Plummer, J. M. (1966). Further studies on the fine structure of the heart of Achatinidae. *Proc. malac. Soc. Lond.* **37**: 199–208.

North, R. J. (1963). The fine structure of the myofibers in the heart of the snail *Helix aspersa*. *J. Ultrastruct. Res.* **8**: 206–218.

Page, E. (1966). Tubular systems in purkinje cells of the cat heart. *J. Ultrastruct. Res.* **17**: 72–83.

Philpott, D. E., Kahlbrock, M. and Szent-György, A. (1960). Filamentous organisation of molluscan muscles. *J. Ultrastruct. Res.* **3**: 254–269.

Porter, K. R. and Palade, G. E. (1957). Studies on the endoplasmic reticulum III. Its form and distribution in striated muscle cells. *J. biophys. biochem. Cytol.* **3**: 269–300.

Rüegg, J. C. (1963). Actomyosin inactivation by thiourea and the nature of the viscous tone in a molluscan smooth muscle. *Proc. R. Soc.* (B) **158**: 177–195.

Rüegg, J. C. (1964). Tropomyosin-paramyosin system and "prolonged contraction" in a molluscan smooth muscle. *Proc. R. Soc.* (B) **160**: 536–541.

Rüegg, J. C., Straub, R. W. and Twarog, B. M. (1963). Inhibition of contraction in a molluscan smooth muscle by thiourea, an inhibitor of the actomyosin contractile mechanism. *Proc. R. Soc.* (B) **158**: 156–176.

Schlote, F. W. (1957). Die Myofilamente glatter Muskulatur und ihr Verhältnis zu den Myofilamenten quergestreifter Muskulatur. Eine electron-mikroscopische Studie an Muskelzellen der Weinbergschnecke. *Z. Naturf.* **12**: 647–653.

Schlote, F. W. (1960). Die kontraktion glatter Muskulatur auf grund von Torsionsspannungen in den Myofilamenten. *Z. Zellforsch. mikrosk. Anat.* **52**: 362–395.

Twarog, B. M. (1954). Responses of molluscan smooth muscle to acetylcholine and 5-hydroxy-triptamine. *J. cell. comp. Physiol.* **44**: 141–164.

Weber, A. (1966). Energized calcium transport and relaxing factors. In *Current topics in bioenergetics* **1**: 203–254. Sanadi, D. R. (ed.). London: Academic Press.

Symp. zool. Soc. Lond. (1968) No. 22, 213–235.

GAMETOGENESIS AND OVIPOSITION IN *LYMNAEA STAGNALIS* AS INFLUENCED BY γ-IRRADIATION AND HUNGER

J. JOOSSE, MARIA H. BOER and C. J. CORNELISSE

Department of Zoology, Free University, Amsterdam,
The Netherlands

SYNOPSIS

A study of some aspects of gametogenesis and oviposition in the freshwater snail *Lymnaea stagnalis* (Basommatophora, Pulmonata) reveals the phenomenon of resorption of gametes. In the spermoviduct the sperms are not only stored, but also resorbed by the gland cells of the wall of this duct. Horstmann (1955) has already shown that the majority of the sperms of the ejaculate is hydrolysed in the bursa copulatrix of the copulation partner. The ripe oocytes, which have been available for ovulation during a restricted period, degenerate in the acini of the ovotestis and are resorbed by their nurse cells. These facts suggest that resorption of gametes plays an important role in the metabolism of the snails.

The gametogenesis and the oviposition of adult pond snails are affected by γ-radiation with a cobalt-60 source at a dose rate of 10 000 r. As a result oviposition stops. The developing sex cells in the gonad degenerate within a period of about 30 days. However, gametogenesis starts again and at 78 days after exposure new ripe sex cells can be observed in the acini of the ovotestis.

Although pond snails can survive a starvation period of 6 weeks very well, under these circumstances growth ceases and oviposition stops within 2 weeks. However, gametogenesis does not cease, but appears to be continued at a slower rate: the volume of the ovotestis is reduced. Since the sex cells are not otherwise used they are resorbed. The rapid decrease in weight of the female accessory sex organs (albumen gland, nidamental gland, oviduct) during the first 2 weeks of starvation is mainly due to the production of egg masses.

INTRODUCTION

In a previous study (Joosse, 1964) a first attempt was made to clarify the assumed relations between the endocrine structures situated in the central nervous system and the reproductive phenomena in *Lymnaea stagnalis*. Such relations were supposed on the basis of simultaneous changes observed in spermatogenesis, oviposition, Gomori-positive and phloxinophilic neurosecretory cells in the cerebral ganglia, and dorsal bodies of specimens collected in the field for the study of the annual cycle.

The experimental approach to this problem presents some difficulties. The field specimens of *Lymnaea* show a high frequency of trematode infection which makes them unsuitable for studies concerning reproduction. Hence, for these experiments laboratory bred snails must be used. But snails bred in the laboratory have a more or less constant reproductive activity which needs to be interrupted.

Operative castration is impossible owing to the complicated position of the gonad, which is centrally situated in the visceral coil and surrounded by the digestive gland. Chemical disturbance of the spermatogenesis by injection of cadmium chloride was tried. This is known to cause a rapid, although temporary, sterility in the males of some vertebrates (Setty and Kar, 1964), but it did not affect the male cells in the ovotestes of adult pond snails when injected into the haemocoel or dissolved in the water of the breeding jars. However, two other methods, viz. exposing the snails to *γ-radiation* from a cobalt-60 source or to a hunger period, appeared to influence oviposition as well as gametogenesis. In this paper the results of these experiments are presented and a preface is given on the fate of the sex cells in *Lymnaea stagnalis*. In the pond snails ripe sex cells are resorbed in such quantities that their products may play an important role in the metabolism.

MATERIAL AND METHODS

The experiments were carried out with pond snails bred in the laboratory during several generations; the stock was originally collected from a polder near Amsterdam. From a size of about 10 mm the animals were reared two per glass jar with 500 ml of water. The snails of the irradiation experiment were kept two per jar during the experimental period, those of the starvation experiment were kept individually. The water was renewed daily. The food consisted of lettuce only. The temperature of the water was 22°C during the irradiation experiments and 25°C during the starvation experiment. The experimental animals were chosen at random. Before fixation or dissection, the snails were decapitated (radiation experiment) or narcotized (starvation experiment) following the method of Joosse and Lever (1959). They were fixed in Stieve-sublimate and embedded in paraffin (m.p. 58°C). Sections were cut at 5–15 μ and stained with Gomori's chrome-haematoxylin-phloxin (1941). To assess changes in weight the accessory sex organs were weighed before fixation and the ovotestes after fixation for 24 h in a 5% formalin solution, followed by dissection.

A cobalt-60 source that delivered 100 r/sec was used for the γ-radiation.

GAMETOGENESIS

Morphology of the genital tract

The gonad of the pond snail is composed of a large number of acini. In each acinus male and female sex cells develop simultaneously.

These acini are connected with the spermoviduct by vasa efferentia. The spermoviduct has a number of pouchlike evaginations, the vesiculae seminales, in which sperm cells are stored. It bifurcates at its distal end and from here the female cells are transported into the oviduct and the male cells into the vas deferens. The oviduct has two distinct accessory glands, the albumen gland and the nidamental gland. The sperm are transported through the vas deferens and prostate gland to the penis. During copulation the ejaculate is brought into the bursa copulatrix of the copulation partner. This bursa is a blind sac the duct of which opens into the vagina. For details and schemes see Holm (1946), Bretschneider (1948a,b), and Duncan (1958, 1960).

Gametogenesis

Both male and female gametes develop from the undifferentiated cells of the germinal epithelium (Fig. 1). (For histological data on the ovotestis, see Archie, 1941; Aubry, 1962; Quattrini and Lanza, 1965.)

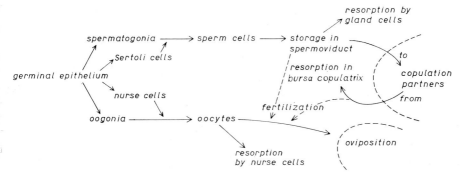

Fig. 1. Scheme illustrating the ultimate fate of the gametes in *Lymnaea stagnalis*. For details see text.

According to the observations of Merton (1930) in *Planorbis* the spermatogonia and oogonia differentiate from the germinal epithelium near the ciliated vas efferens of the acinus. From here they move to the distal end of the acinus. The oogonia migrate to a retro-epithelial position by amoeboid movements (Bretschneider and Raven, 1951). They become sessile when surrounded by the nurse cells during the phase of vitellogenesis. The spermatogonia are displaced in the same direction by their Sertoli cells. As a result of these processes the ripe gametes are found on the floor of the bottle-shaped acini.

In *Lymnaea stagnalis* the shape of the acini is more irregular when compared with that of members of the Planorbidae. However, in general the observations of Merton are also true for the pond snail (Fig. 2A and C).

The ultimate fate of the gametes

It has been suggested that the male gametes of *Lymnaea* can have two destinies. Either they may be transported as yellow coloured balls up the oviduct to the spermoviduct and used for the fertilization of eggs (Bretschneider, 1948a,b), or the ejaculate may disintegrate: "it seems probable that disintegration, the fate of many sperm, is a natural process and is not assisted by the secretions of the bursa copulatrix" (Duncan, 1958).

The opinions on this problem must be changed since the work of Horstmann (1955), who has shown that in *L. stagnalis* the bulk of the cells transported into the bursa during a copulation is actively hydrolysed and resorbed by the epithelium. The products of this process have a stimulating effect on the first oviposition of virgin snails. Only a few thousand sperm are transported to the spermoviduct where they fertilize the eggs.

In the laboratory pond snails are bred often individually. Consequently copulation cannot occur, and self-copulation was not observed. In these specimens self-fertilization is normal. The sperm are stored in the blind sacs of the spermoviduct and, as spermatogenesis seems to be a continuous process, a discharge of sperm has to occur. It was found that sperm cells can be resorbed by the gland cells of the spermoviduct (Fig. 2B). Such a resorption had already been observed in the spermoviduct of *Helix pomatia* by Breucker (1964).

Similar phenomena can be mentioned as regards the oocytes. In field specimens as well as in laboratory bred pond snails degenerating oocytes can be frequently observed in the acini of the ovotestis. These abortive cells are situated in the distal region of the acini. According to the theory of Merton (1930), the ripe oocytes must lie there. Apparently the oocytes can await the ovulation stimulus only during a certain period of time, after which they are resorbed by their nurse cells.

Fig. 2. A. Section through the ovotestis of a pond snail collected in the field during April. Three acini can be seen, in which no ripe male cells are present. × 125. B. Section through the spermoviduct, showing the resorption of sperm by the gland cells. × 600. C. Section through the ovotestis of a pond snail collected in the field during May (cf. A). There can be seen many groups of sperm on their Sertoli cells in the aboral region of the acini. × 125. Gomori stain.

a, Amoeboid oocyte; *dg*, digestive gland; *fo*, full-grown oocyte; *gc*, gland cells; *ge*, germinal epithelium; *s*, sperm; *sc*, spermatocytes; *sg*, spermatogonia; *st*, spermatids; *v*, vacuoles with sperm; *ve*, vas efferens.

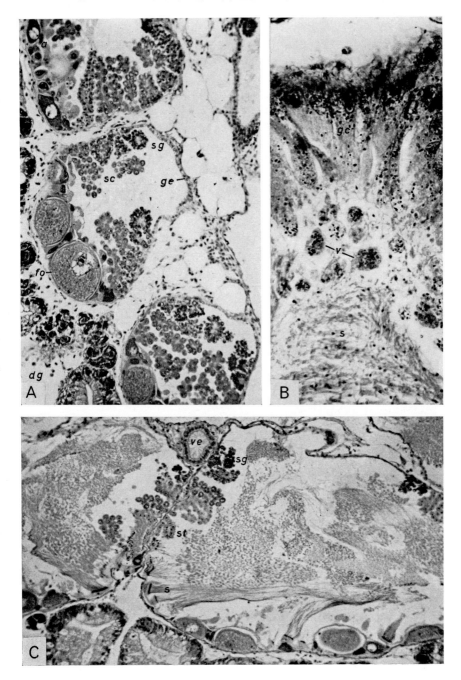

FIG. 2.

This period may be rather short. During the study of the postembryonic development of the ovotestis of pond snails, degenerating oocytes appeared simultaneously with the first ripe oocytes. Four weeks later their number had increased to 10% of the total number of the female cells.

In field specimens the degenerating oocytes appeared to represent about 5% of the female cells, whereas in laboratory bred specimens this figure varied around 30% (Joosse, 1964). These data suggest that the percentage of degenerating cells may be influenced by external conditions. In the laboratory the food is given *ad libitum* and the temperature is constant and rather high. Such favourable conditions are not constantly present in the field. They may have an effect on oviposition (discharge of eggs from the ovotestis) and resorption (degeneration of eggs within the ovotestis) which may cause a variable picture of the ovotestis. Lūsis (1961) has described degenerating oocytes in *Arion ater rufus*. In this species degeneration of ripe oocytes was only observed after the period of oviposition. In the pond snails, however, degenerating oocytes can be observed in the ovotestis of specimens during all phases of the adult life period.

The data presented above are assembled into a scheme (Fig. 1). It is concluded that in the pond snail resorption of gametes is a frequently occurring phenomenon. The products of the resorbed gametes may play a role in the metabolism of this species. As will be shown below, this latter conclusion can be strengthened by the results of the starvation experiments.

γ-IRRADIATION EXPERIMENTS

The resistance of molluscs to irradiation is very high. Bonham and Palumbo (1951) irradiated adult snails of the genera *Radix* and *Thais* with hard X-rays. They found LD_{50} values at 40 days after irradiation of 8000 r for *Radix* and at 160 days of 13 000 r for *Thais*. Laviolette and Cuir (1959) found no raise in mortality as compared to controls in *Arion rufus* after total irradiation with X-rays up to a dose of 18 000 r. The ovotestes of these 2-month-old animals ("phase juvénile"), showed a great radiosensibility. At this age the acini contain only young oocytes and spermatogonia. These degenerated in the animals treated with the high dose rates. After a short phase of sterility the production of gametes was restored.

The growth of 1-week-old *Arion rufus* and *Lymnaea stagnalis* after X-irradiation was stopped by dose rates of 3000 r for *Arion* and at least 18 000 r for *Lymnaea* (Laviolette and Voulot, 1961).

The radiosensibility of *Australorbis glabratus* has been studied in detail by Perlowagora-Szumlewicz (1964a–d). She observed a decrease in radiosensibility with increasing age at exposure. The degree of radiation damage to the reproductive capability was age- and dose-dependent. A maximum sensitivity was found in newborn snails and older adults. The changes in the ovotestes were not studied histologically, but from the results of oviposition the damage appeared to be a temporary one. A sublethal damage to egg production was reparable within 3–5 months after exposure.

In *Physa acuta* 100 000 r produced a permanent sterility, 10 000 r a temporary one (50 days) and 2000 r decreased only the number of viable eggs (Ravera, 1965).

From these data it is clear that gastropods are highly resistant to irradiation. Moreover, their reproduction, when influenced by irradiation, shows a great regenerative capacity. Therefore this method may be successful in changing gametogenesis in pond snails.

Lethality of γ-irradiation

The LD_{50} curve for laboratory bred specimens of pond snails with a shell length of 27–28 mm was determined. These specimens had just entered the period of egg production. There were thirteen groups each consisting of ten specimens. Twelve experimental groups received a range of doses varying from 800–40 000 r (100 r/sec). The control group was handled similarly to the experimental snails, except for entering the Co_{60}-source. An immediate reaction of the animals to a dose rate of $\geqslant 6400$ r was observed. The buccal mass was extruded and violent contraction of the foot occurred. In several cases even blood was extruded via the haemal pore (Lever and Bekius, 1965). Within 1 h after the irradiation these irritated animals recovered to a normal behaviour.

The lethality showed a clear relation to the dose rate of the γ-irradiation (Fig. 3). The lowest doses (800–3200 r) suggest a positive effect of the treatment on the life-span. The higher doses give increasing lethality. The highest dose of 40 000 r resulted in an LD_{50} of 28 days.

Oviposition

The oviposition during the experimental period of 90 days is presented in Fig. 4. During this period the low doses (up to 2400 r) do not seem to influence oviposition. On the other hand, the production of egg masses is restricted to the first 20 days after exposure in all groups above the level of 9600 r. The doses 3200 and 6400 r are transitional between the two extremes. It is of interest to notice here the return of egg production during the last 10 days of the experimental period of the 6400 r group. Apparently, the possibility of a recovery as observed in *Australorbis glabratus* by Perlowagora-Szumlewicz (1964b) and in *Physa acuta* by Ravera (1965) is present in *Lymnaea* too.

General effects and preliminary conclusions

Although the attention is focused in this paper on the reproductive organs, irradiation treatment at high dose rates has a number of effects on the snails, as was observed in this and in other experiments. Thus, no shell growth occurred in the groups with dose rates $\geqslant 6400$ r, except during the first 2 weeks after exposure. The general activity of the irradiated snails is reduced. The respiration frequency and the amount of food eaten is decreased.

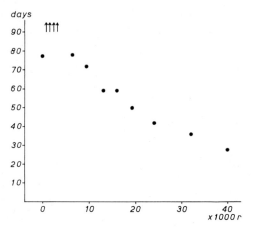

FIG. 3. LD_{50}-values of pond snails exposed to various dose rates of γ-radiation. The values of the four lowest doses exceeded the experimental period of 90 days.

From this introductory experiment it can be concluded that in adult *Lymnaea stagnalis* increasing dose rates of γ-rays cause increasing mortality. However, a dose of 800–3200 r may lengthen the life span. The oviposition is decreased by dose rates $\geqslant 6400$ r and is stopped by dose rates $\geqslant 9600$ r during at least 90 days. From these results the lowest dose rate can be determined at which the oviposition is ceased whilst the mortality is still low. This dose apparently is about 10 000 r.

Histology of the ovotestis after irradiation

For the histological investigations two experiments have been carried out, the first one with 105 snails of 28–32 mm shell height, the second one with 180 snails of 24–30 mm shell height. The dose rates were 9600 and 10 300 r, respectively. Because the results of both series showed no great differences, the data of the second experiment are presented only (Fig. 5). On the days indicated in Fig. 5, four

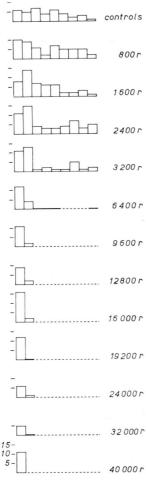

controls

800 r

1 600 r

2 400 r

3 200 r

6 400 r

9 600 r

12 800 r

16 000 r

19 200 r

24 000 r

32 000 r

40 000 r

FIG. 4. The oviposition of pond snails after γ-radiation treatments at various dose rates. Each column represents the mean number of oocytes per snail per day during a period of 10 days.

control and four irradiated animals were fixed. From every ovotestis about ten series of sections were cut, regularly distributed over the entire organ.

The ovotestes of the control animals show the normal pattern of specimens bred in the laboratory. All stages of the spermatogenesis and oogenesis are present in every acinus, including degenerating oocytes (Fig. 2C). In the irradiated snails the damage caused by the radiation treatment is already visible after one day in the *male cells*.

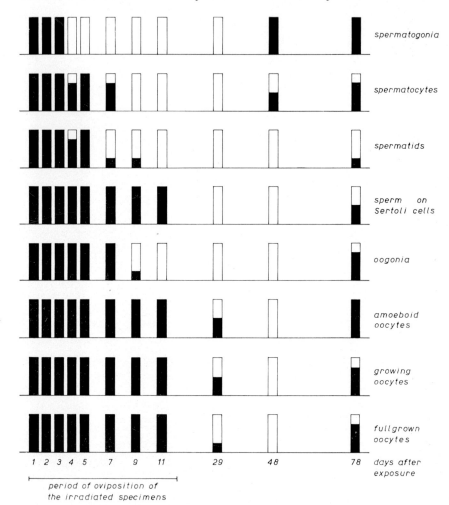

Fɪɢ. 5. Histogram of the ovotestis of pond snails fixed at various days after exposure to a γ-radiation treatment of 10 000 r. Each column represents the data of four irradiated specimens. The ovotestes of the controls appeared to contain all stages of gametogenesis during the entire experimental period.

These cells become globular, the cytoplasm stains more densely, the nuclei appear pycnotic and the cells become loose from the acinar wall. In the lumen of the acinus such cell groups may still remain together.

Beginning with the spermatogonia 4 days after exposure, the other stages of the male cells show the same processes (Fig. 7A). Already after 11 days only sperm cells on Sertoli cells can be observed (Fig. 6A). These had entirely disappeared out of the ovotestis on the 29th day after exposure (Fig. 5). Fragments of sperms, however, persisted in the lumen of the acini. In the acini the Sertoli cells, having lost their spermia, persist near the wall of the acinus. Here they show a hypertrophy (Fig. 7C) comparable to that observed by Lūsis (1961) in the gonad of *Arion* during ageing processes. Apparently the Sertoli cells are less sensitive to irradiation.

In the *female cells* the degeneration can be observed at first in the oogonia. These are absent from the ninth day, and from about 48 days after exposure the gonad contains no female cells (Fig. 6B). However, at this stage degenerating oocytes are still present. Most probably the nurse cells resorb them, although in some cases degenerating oocytes were seen in the spermoviduct.

The role of the gland cells of the spermoviduct may be important in the removal of all degeneration products of the acini. In the irradiated animals the gland cells were filled with sperm heads, especially in the series fixed 78 days after exposure (Fig. 7B). Perhaps in this phase the old irradiated sperm are removed since new ones are produced in the acini again.

It must be emphasized that Fig. 5 does not give a quantitative but a qualitative impression of the presence of the sex cells in the whole ovotestis. In fact the acini are already almost free from normal sex cells after 29 days (Fig. 6B). The spermoviduct, however, remains filled with sperm cells during the entire experimental period.

The degeneration of the sex cells described above, is followed by a phase of renewed gametogenesis. From Fig. 5 it is clear that spermatogenesis and oogenesis are started again at 48 and at least 78 days after exposure respectively. Soon most of the developmental stages of the sex cells are present (Fig. 6C). Thus, adult pond snails show the same regenerative capacity of gametogenesis as has been reported for juvenile *Arion rufus* (see Laviolette and Cuir, 1959), for adult *Australorbis glabratus* (see Perlowagora-Szumlewicz, 1964b) and for *Physa acuta* (see Ravera, 1965).

Conclusions

The lethality of pond snails shows a clear relation to the dose rate of γ-irradiation. The LD_{50} for a dose rate of 40 000 r appeared to be 28 days. The oviposition is not decreased by γ-irradiation up to a dose

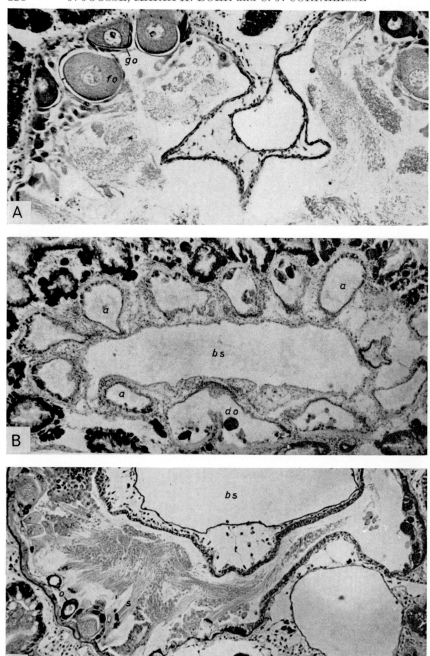

Fig. 6.

of 2400 r. Dose rates above 9600 r stop the oviposition during at least 90 days. As a result of a dose rate of 10 000 r the sex cells degenerate. Beginning with the spermatogonia and the oogonia all stages of sex cells are removed from the acini of the ovotestis. After 29 days the snails are considered to be castrated. This sterile phase persists only a short time, since spermatogenesis and oogonesis appeared to have started again at the 48th or 78th day after exposure.

STARVATION EXPERIMENT

In gastropods starvation experiments have been performed mainly to investigate the identity of the reserve materials. This problem has recently been discussed by Goddard and Martin (1966). The question is still unsolved as to whether fat or glycogen serves as the main deposit of reserve material. The storage centre would be the digestive gland. May (1934) has found that in molluscs apart from glycogen the polysaccharide galactogen can be identified. In *Helix pomatia* it appeared to be present in the albumen gland only. During a starvation period *Helix* first used up the glycogen reserves and later on the galactogen was drawn upon:

"One may speculate that the role of the galactogen is primarily to act as a reserve for reproduction and, because it is protected from easy mobilization, the snail is driven to the continued exertion of finding food and thus carries through the very demanding reproductive process successfully" (Goddard and Martin, 1966, p. 285).

The effect of starvation on the gonad was studied by Neuhaus (1949) in *Bithynia tentaculata*. He found ripe sex cells in animals starved for 85 days, whereas the size of their gonad had not changed very much.

Starvation experiments with *L. stagnalis* have not been published.

The aim of the experiment described below was to find out whether the expected interruption of the oviposition during a hunger period would become manifest in the histology of the ovotestis and the weight of the accessory sex organs.

FIG. 6. Sections through the ovotestes of γ-irradiated pond snails (dose rate 10 000 r). A. Fixed at 9 days after exposure. Normal female cells, male cells only represented by sperm. × 140. B. Fixed at 48 days after exposure. Around a blood space a number of acini can be seen, containing only some degenerating oocytes, and hypertrophied nurse cells (see also Fig. 7C). × 70. C. Fixed at 78 days after exposure. Regeneration of the gametogenesis. × 140.

a, Acini; *bs*, blood space; *do*, degenerating oocyte; *fo*, full-grown oocyte; *go*, growing oocyte; *o*, oogonia; *s*, sperm.

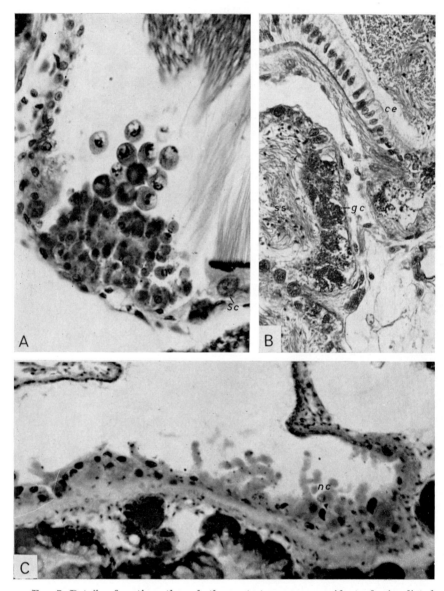

Fig. 7. Details of sections through the ovotestes or spermoviduct of γ-irradiated pond snails (dose rate 10 000 r). A. Degenerating male cells containing pycnotic nuclei in an acinus of a specimen fixed 3 days after exposure. × 320. B. Spermoviduct of a specimen fixed at 78 days after exposure. The gland cells are densely packed with sperm. × 525. C. Hypertrophied nurse cells in an acinus of a pond snail fixed at 48 days after exposure. × 300.

ce, Ciliated epithelium; *gc*, gland cells; *nc*, nurse cells; *Sc*, Sertoli cell; *ss*, sperm storage.

Apart from the controls, ninety animals were starved during a maximum of 7 weeks. The shell height varied from 34 to 36 mm. Every week five snails were fixed for histological studies. For the determination of the weight of the sex organs another group of five snails was used.

General observations

During this experiment the mortality was extremely low. Only five specimens died at the end of the sixth and during the seventh week. Thus, pond snails can survive a starvation period of about 6 weeks. To check the result of feeding after this period, sixteen animals were fed after having been starved for 6 weeks. None of them died, whilst they started the interrupted oviposition within 1 week (see below).

Starved pond snails exhibit a slow radula movement as they move around. Most probably they eat their own slime paths. Gradually the body becomes more transparent. After 6 weeks of starvation the digestive gland has become brown instead of green, and it is so transparent that the ovotestis can be seen *in situ*. The gut is reduced to a thin tube of a brown colour. The production of faeces, although decreased, continues through the whole period of starvation.

Growth

During the starvation period shell growth had ceased completely. The growth process must have stopped immediately at the beginning of the starvation period (Fig. 8).

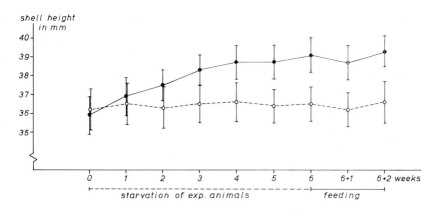

FIG. 8. Growth curves of the control (●) and the starved (○) pond snails.

Oviposition

The data on the egg production of the snails of the starvation experiment are presented in Table I. The controls show a continuous oviposition varying from 20 to 38 eggs per snail per day or 0·2–0·4 egg masses per snail per day. In the first week of the experimental period the starved snails produced a lower number of egg masses and eggs when compared to the control specimens. This means that not only the number but also the size of the egg masses is reduced. From the beginning of the second week and onwards, the oviposition of the starved animals had ceased. After 6 weeks of starvation a number of starved snails were fed. Already on the seventh day the first egg masses were produced. During the second week of feeding these snails deposited a surprising high number of egg masses, although with a small number of eggs.

TABLE I

Egg production of the snails of the starvation experiment

| | No. of weeks | | | | | | | |
	1	2	3	4	5	6	7	8
A. Mean number of oocytes per snail per day during successive periods of 1 week								
Controls	20·6	32·6	34·9	22·3	38·3	28·2	22·7	21·3	27·7
		Starvation						Feeding	
Starved specimens	22·4	13·8	0·1	—	—	—	—	0·6	7·2
B. Mean number of egg masses per snail per day								
Controls	0·40	0·31	0·31	0·23	0·33	0·24	0·19	0·19	0·26
		Starvation						Feeding	
Starved specimens	0·26	0·20	0·006	—	—	—	—	0·02	0·19

From these results it can be concluded that as a result of starvation the oviposition of pond snails is stopped within 1 week. If food is given after 6 weeks of starvation oviposition occurs within 1 week. However, the number of eggs in the egg masses remains reduced, at least during the second week of feeding.

Size and histology of the ovotestis

The size of the ovotestis was determined by weighing the tissues after dissecting them out of the surrounding digestive gland material. The mean weights and the standard deviations are presented in Fig. 9. The volume of the ovotestis decreases regularly during the starvation

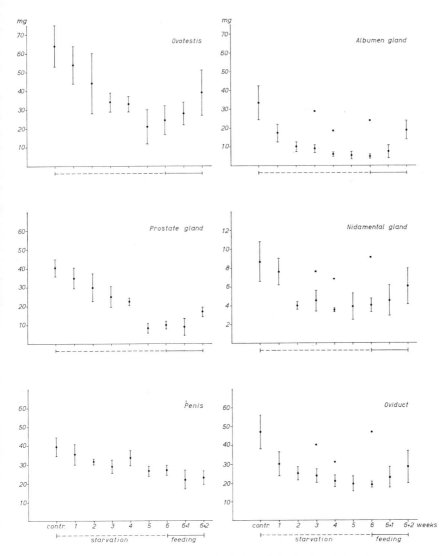

Fig. 9. The mean and the standard deviation of the fresh weights of sex organs of pond snails from the starvation experiment.

period. After 6 weeks the mean weight of the ovotestes is reduced to about 30% of the control values.

For the histological study of the ovotestes of starved pond snails five specimens were fixed weekly. The results appeared to be surprising. In contrast to the data of the irradiated snails, the ovotestes of the starved specimens contained all stages of the spermatogenesis and the oogenesis during the entire experimental period (Fig. 10A). Apparently gametogenesis is not interrupted by the absence of food. However, the number of sex cells in the acini has decreased considerably. Moreover, the decrease of the volume of the ovotestes (Fig. 9) must be considered in this respect. Thus gametogenesis is continued during starvation, but at a much lower level as regards the number of cells. As a result there must occur a removal of sperm and oocytes. Most probably the sperm are transported to the spermoviduct. On the other hand, the oocytes are resorbed by their nurse cells. This becomes apparent from the large number of degenerating oocytes in the acini of starved snails. It is of interest to mention that no degeneration of developing sex cells was observed.

In summarizing these results it can be said that the ovotestes of starving pond snails decrease in size. The number of sex cells is greatly reduced but spermatogenesis and oogenesis are still represented by all stages. The decrease in the number of sex cells is effected by a transport of the sperm out of the acini and by a resorption of the oocytes by their nurse cells. Since the sperm will be resorbed by the gland cells of the spermoviduct, all the materials of the degenerating sex cells become available to the starved snails. In this way two important conclusions can be drawn from the above mentioned results.

(1) Starving snails can mobilize the materials of their ripe sex cells.

(2) Notwithstanding the reduction of the volume of the gonad and the number of sex cells during starvation, the pond snails continue to dispose of ripe sex cells during a starvation period of at least 6 weeks.

Therefore it is not surprising that as a result of food supply after 6 weeks of starvation oviposition starts within 1 week (Table I; Fig. 10B).

FIG. 10. Starvation experiment. A. Section through the ovotestis of a specimen starved for 4 weeks. All the stages of the gametogenesis can be observed, but the number of cells is small. At the aboral side of an acinus the accumulation of degenerating oocytes (do) can be seen. × 170. B. Section through the ovotestis of a specimen starved during 6 weeks, followed by a period of food supply during 2 weeks. The number of male gametes per group has increased considerably (cf. Fig. 10A). × 170.

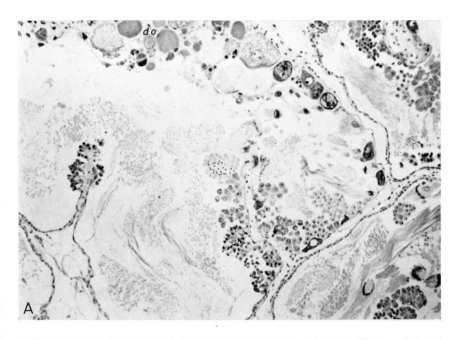

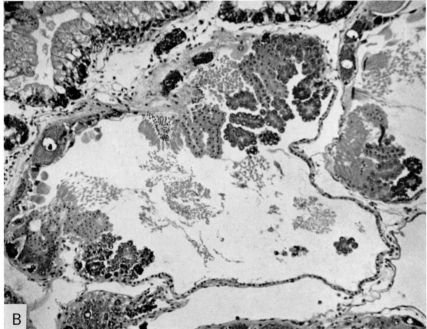

Fig. 10.

It may be of interest to recall the speculation of Goddard and Martin (1966), cited above, that snails should favour their reproductive activities. In fact this speculation was based on the data of the accessory sex glands of snails. The results of the present experiment showed that during starvation the gonad is not protected from the general processes to mobilize the reserve materials. However, the remaining sex cells are kept ready for reproductive activities.

Accessory sex organs

The fresh weight of the albumen gland, the nidamental gland, the oviduct, the prostate gland and the penis (including the penis sheath) of the starvation experiment are presented in Fig. 9.

The decrease in weight of the female organs takes place mainly during the first 2 weeks of the starvation period (Fig. 9). The changes in weight after that period are not significant. This might suggest that the decrease in weight is due to the production of egg masses. This opinion is supported by the data from three snails having been starved for 3, 4 and 6 weeks (Fig. 9). The weights of the albumen gland, the nidamental gland and the oviduct of these specimens are abnormally high. The data of the specimen that was starved for 6 weeks are astonishing. Because the oviposition of the snails has been registered individually, it can be said now that these specimens did not produce egg masses during the entire starvation period. Apparently the materials present in the female organs could not be mobilized even after 6 weeks of starvation. The weight of the prostate gland (Fig. 9) decreases more rapidly during the fifth week of starvation, but the weight of the penis (Fig. 9) did not change to such an extent. After feeding, the weight of the accessory sex organs increases again, except in the case of the penis.

These data suggest that the production of the secretory materials of the cells of the accessory sex organs ceases at the beginning of the starvation period. The decrease in weight of the female organs from that moment is mainly due to a loss of material when oviposition was carried on. It cannot be mobilized for other purposes.

In contrast to this, the decrease of the weight of the prostate gland is more regular. As copulation could not occur, this must be due to a mobilizing of its secretion for general metabolic purposes. In case of the muscular tissues of the penis this is only possible to a lesser extent.

SUMMARY

In this study some aspects of the gametogenesis and oviposition of the freshwater snail *Lymnaea stagnalis* (Basommatophora, Pulmonata) are discussed.

1. Attention is focused to the phenomenon of the resorption of gametes. In the spermoviduct the sperm are not only stored, but also resorbed by the gland cells of the wall of this duct. Moreover, Horstmann (1955) has shown that the majority of the sperm of the ejaculate is hydrolysed in the bursa copulatrix of the copulation partner. Furthermore the ripe oocytes, having been available to ovulation during a restricted period, degenerate in the acini of the ovotestis and are resorbed by their nurse cells. From these facts it is suggested that the resorption of gametes may play an important role in the metabolism of pond snails.

2. The gametogenesis and the oviposition of adult pond snails can be affected by γ-radiation with a cobalt-60 source at a dose rate of 10 000 r. As a result oviposition stops. The developing sex cells in the gonad degenerate within a period of about 30 days. However, gametogenesis starts again and at 78 days after exposure new ripe sex cells can be observed in the acini of the ovotestis.

3. Pond snails can survive a starvation period of 6 weeks very well. Under these circumstances growth ceases, and oviposition stops within 2 weeks. During starvation gametogenesis does not cease, but appears to be continued at a lower scale: the volume of the ovotestis is reduced. Consequently a great number of ripe sex cells has to be resorbed. These products become available to the metabolic processes and can be used also for the restricted gametogenesis. The rapid decrease in weight of the female accessory sex organs (albumen gland, nidamental gland, oviduct) during the first 2 weeks of starvation is mainly due to the production of egg masses. Most probably their products cannot be mobilized during a starvation period of 6 weeks.

REFERENCES

Archie, V. E. (1941). *The histology and developmental history of the ovotestis of* Lymnaea stagnalis lillianae. Thesis Univ. Wisconsin.
Aubry, R. (1962). Étude de l'hermaphrodisme et de l'action pharmacodynamique des hormones de Vertébrés chez les Gastéropodes Pulmonés. *Archs Anat. microsc. Morph. exp.* (Suppl.) **50**: 521–602.
Bonham, K. and Palumbo, R. F. (1951). Effects of X-rays on snails, crustacea and algae. *Growth* **15**: 155–188.
Bretschneider, L. H. (1948a). Insemination in *Limnaea stagnalis. Proc. K. ned. Akad. Wet.* (C) **51**: 358–362.
Bretschneider, L. H. (1948b). The mechanism of oviposition in *Limnaea stagnalis. Proc. K. ned. Akad. Wet.* (C) **51**: 616–626.

Bretschneider, L. H. and Raven, Chr. P. (1951). Structural and topochemical changes in the egg cells of *Limnaea stagnalis* L. during oogenesis. *Archs néerl. Zool.* **10**: 1–31.

Breucker, H. (1964). Cytologische Untersuchungen des Zwitterganges und des Spermoviduktes von *Helix pomatia* L. *Protoplasma* **58**: 1–41.

Duncan, C. J. (1958). The anatomy and physiology of the reproductive system of the freshwater snail *Physa fontinalis* (L.). *Proc. zool. Soc. Lond.* **131**: 55–84.

Duncan, C. J. (1960). The genital systems of the freshwater Basommatophora. *Proc. zool. Soc. Lond.* **135**: 339–356.

Goddard, C. K. and Martin, A. W. (1966). Carbohydrate metabolism. In *Physiology of Mollusca* **2**: 275–308. Wilbur, K. M. and Yonge, C. M. (eds). London: Academic Press.

Gomori, G. (1941). Observations with differential stains on human islets of Langerhans. *Am. J. Path.* **17**: 395–406.

Holm, L. W. (1946). Histological and functional studies on the genital tract of *Lymnaea stagnalis appressa* Say. *Trans. Am. microsc. Soc.* **65**: 45–68.

Horstmann, H.-J. (1955). Untersuchungen zur Physiologie der Begattung und Befruchtung der Schlammschnecke *Lymnaea stagnalis* L. *Z. Morph. Ökol. Tiere* **44**: 222–268.

Joosse, J. (1964). Dorsal bodies and dorsal neurosecretory cells of the cerebral ganglia of *Lymnaea stagnalis* L. *Archs néerl. Zool.* **16**: 1–103.

Joosse, J. and Lever, J. (1959). Techniques of narcotization and operation for experiments with *Limnaea stagnalis* (Gastropoda Pulmonata). *Proc. K. ned. Akad. Wet.* (C) **62**: 145–149.

Laviolette, P. and Cuir, P. (1959). Action des rayons-X sur la gonade d'*Arion rufus* L. (Mollusque Gastéropode). *Archs Anat. microsc. Morph. exp.* **48**: 25–47.

Laviolette, P. and Voulot, C. (1961). Étude de la croissance des mollusques irradiés (*Arion rufus* et *Limnaea stagnalis*). *Bull. biol. Fr. Belg.* **95**: 679–694.

Lever, J. and Bekius, R. (1965). On the presence of an external hemal pore in *Lymnaea stagnalis* L. *Experientia* **21**: 395–396.

Lūsis, O. (1961). Postembryonic changes in the reproductive system of the slug *Arion ater rufus* L. *Proc. zool. Soc. Lond.* **137**: 433–468.

May, F. (1934). Chemische und biologische Untersuchungen über Galaktogen (Biologischer Teil: Galaktogen und Glykogengehalt bei hungernden Weinbergschnecken). *Z. Biol.* **95**: 606–613.

Merton, H. (1930). Die Wanderungen der Geschlechtszellen in der Zwitterdrüse von *Planorbis*. *Z. Zellforsch. mikrosk. Anat.* **10**: 527–551.

Neuhaus, W. (1949). Hungerversuche zur Frage der parasitären Kastration bei *Bithynia tentaculata*. *Z. ParasitKde* **14**: 300–319.

Perlowagora-Szumlewicz, A. (1964a). Effect of radiation on the population kinetics of the snail *Australorbis glabratus*: Age at exposure and immediate and late effects of X-rays. *Radiat. Res.* **23**: 377–391.

Perlowagora-Szumlewicz, A. (1964b). Effects of ionizing radiation on the population kinetics of the snail *Australorbis glabratus*: Age at exposure and the effects on reproduction. *Radiat. Res.* **23**: 392–404.

Perlowagora-Szumlewicz, A. (1964c). Effects of ionizing radiation on *Australorbis glabratus* eggs. *Expl Parasit.* **15**: 226–231.

Perlowagora-Szumlewicz, A. (1964d). Survival, growth, and fecundity of *Australorbis glabratus* snails developed from eggs exposed to ionizing radiation. *Expl Parasit.* **15**: 232–241.

Quattrini, D. and Lanza, B. (1965). Ricerche sulla biologia dei Veronicellidae (Gastropoda Soleolifera), II. Struttura della gonade, ovogenesi e spermato-genesi in *Vaginulus borellianus* (Colosi) e in *Laevicaulis alte* (Férussac). *Monit. zool. ital.* **73**: 1–60.

Ravera, A. (1965). X-rays effect on demographic characteristics of freshwater gastropods. *Second Europ. Malacol. Congr.* Copenhagen.

Setty, B. S. and Kar, A. B. (1964). Chemical sterilization of male frogs. (*Rana tigrina*, Daud). *Gen. comp. Endocr.* **4**: 353–359.

Symp. zool. Soc. Lond. (1968) No. 22, 237–256.

ELECTRON MICROSCOPE STUDY OF NEUROSECRETORY CELLS AND NEUROHAEMAL ORGANS IN THE POND SNAIL *LYMNAEA STAGNALIS*

H. H. BOER, ELISABETH DOUMA and JENNEKE M. A. KOKSMA

Zoological Department, Free University, Amsterdam,
The Netherlands

SYNOPSIS

In the cerebral ganglion of *Lymnaea stagnalis* two groups of "Gomori-positive" neurons (medio- and latero-dorsal cells) and a group of "Gomori-negative" neurons (caudo-dorsal cells) occur. The neurosecretory material (NSM) of the "Gomori-positive" and "Gomori-negative" cells is transported to the periphery of the median lip nerve and of the inter-cerebral commissure, respectively (neurohaemal areas). Ultrastructural investigations of the perikarya and the neurohaemal areas show the NSM to consist of electron-dense elementary granules, measuring 2000 Å in the "Gomori-positive" and 1500 Å in the "Gomori-negative" system. Cells from the anterior lobe of the ganglion (non-neuro-secretory cells), which were studied for comparison, differ in several ultrastructural aspects from the neurosecretory cells.

It is concluded that the medio-, latero-, and caudo-dorsal cells are true neuro-secretory cells, i.e. produce neurohormones, especially as synaptic termination of their axons in the neurohaemal areas was not found. The way in which NSM is discharged from the axon terminals is discussed. It is suggested that through fragmentation of elementary granules the NSM is liberated, and can now reach the blood at the molecular level. The elementary granules probably originate in the Golgi area by fusion of small electron lucent vesicles. It is suggested that part of these vesicles arises from Golgi lamellae and another part from cisternae of the endoplasmic reticulum by a process of budding.

INTRODUCTION

Recently Knowles and Bern (1966) suggested that nerve cells should be regarded as "neurosecretory" only if there is evidence that they are part of the endocrinon (endocrine system) of an animal. Criteria for using the term neurosecretion can be experimental or morphological. In many animal groups, e.g. in molluscs, the number of experimental studies on neurosecretion is limited and only morphological criteria can as yet be used.

One of the most valuable morphological data for judging whether nerve cells are neurosecretory is non-synaptic termination of their axons within so-called neurohaemal organs, because this strongly suggests the release of the neurosecretory material (NSM) from the axons directly into the blood. However, a nerve cell can be part of the endocrinon in a different way. It can terminate—as in certain vertebrates (Knowles, 1965)—on endocrine cells, controlling the function of the endocrine organ by "neurosecretomotor innervation". Further morphological evidence for neurosecretion may be obtained from histochemical tests to indicate the chemical nature of NSM (Boer, 1965).

Most of the work on neurosecretion in the Mollusca is restricted to the use of special histological ("neurosecretory") stains (Simpson, Bern and Nishioka, 1966a). However, caution is needed in interpreting nerve cells as "neurosecretory" solely on this basis, since it has been stated that in addition to NSM other substances, like accumulations of glycogen (Simpson, Bern and Nishioka, 1966b), cytosome-like structures (Boer, 1965; Nolte, Breucker and Kuhlmann, 1965), or still other aggregates of cell organelles (Hagadorn, Bern and Nishioka, 1963) may be stained by the same methods.

Investigations on the ultrastructure of neurosecretory systems of vertebrates and invertebrates revealed NSM to consist of large numbers of small (ϕ 1 000–3 000 Å), membrane limited, electron dense elementary granules. However, it has been found (in molluscs) that granules in this size range occur at synaptic junctions and in axon terminals innervating effector organs (e.g. Baxter and Nisbet, 1963; Gerschenfeld, 1963; Amoroso, Baxter, Chiquoine and Nisbet, 1964). These granules probably contain "ordinary" neurotransmittors.

It therefore seems obvious that, as long as there is no or only little experimental or physiological evidence for neurosecretion, the term may perhaps be applied on a morphological basis only, when a "neurosecretory" system has been studied with a combination of histochemical and electron microscope techniques. Only in a few cases have such studies been conducted in molluscs (Nolte, 1965; Simpson et al., 1966b). In our laboratory the neurosecretory systems in the cerebral ganglion of the basommatophoran snail *Lymnaea stagnalis* have, in the past years, been studied extensively by means of the light microscope as well as by histochemical methods (Joosse, 1964; Boer, 1965). It is the aim of the present study to add ultrastructural information to the evidence already obtained.

MATERIALS AND METHODS

For this investigation cerebral ganglia, median lip nerves and intercerebral commissures of about fifty full grown (shell height > 30 mm) *Lymnaea stagnalis* bred in the laboratory, were used. Furthermore, cerebral ganglia of six snails from the Eempolder, near

Amsterdam, collected on 22 March 1966 were studied. Tissues of the laboratory animals were fixed at times evenly distributed over the year. At the beginning of the investigations fixation was carried out in a 4% buffered (pH 7·4) glutaraldehyde solution prior to OsO_4 fixation (ten ganglia). However, since the neurosecretory elementary granules were badly preserved in these cases, later specimens were not prefixed in glutaraldehyde, but immediately placed into a veronal acetate buffered (pH 7·4) OsO_4 solution (1%) for 2 h at 4°C, then placed in 1% uranylnitrate and embedded in Epon 812.

From the ganglia 2 μ thick sections were cut for identification of the cell types with an anoptral-contrast microscope. Attempts were made to trim the blocks in such a way that as many cell types as possible could be studied by the electron microscope. Ultra-thin sections were cut with glass knives on a Reichert ultramicrotome, picked up on form-var-coated copper grids, usually contrasted with Reynold's lead citrate, and studied with a Zeiss EM 9 electron microscope.

OBSERVATIONS

Light microscopy

The complex cerebral ganglion of *Lymnaea stagnalis* consists of a dorsal and a ventral part (Joosse, 1964; see Fig. 1). The pars dorsalis shows two protruding lobes, the lobus anterior, situated antero-laterally to the cerebral commissure, and the lobus lateralis, lying near the origins of the tentacular and optic nerves. The lobus anterior is merely an extension of the pars dorsalis, whereas the lateral lobe can be regarded as a separate ganglion, since it is joined with the cerebral ganglion by two narrow nervous connexions only. In the lateral lobe some "Gomori-positive" neurosecretory cells—cells stainable with chrome-haematoxylin and paraldehyde-fuchsin—and a vesicular epithelial structure, the "follicle gland", occur (Lever, 1958; Lever, Boer, Duiven, Lammens and Wattel, 1959; Lever and Joosse, 1961). These structures have recently been studied with the electron microscope (Brink and Boer, 1967). There are also two dorsal bodies attached to the cerebral ganglion which lie medio- and latero-dorsally, respectively. They consist of small cells and are considered to be endocrine organs (Joosse, 1964; Boer, Slot and van Andel, 1968).

The perikarya of the neurons are found in a few cell layers at the periphery of the ganglion. Apparently most of them are polyploid. At least three different degrees of polyploidy occur, which have been established by volume measurements (Boer, 1965). Cells with the highest degree of polyploidy take the most peripheral position in the ganglion.

I

In the cerebral ganglion three types of neurosecretory cells have been described (Lever, Kok, Meuleman and Joosse, 1961; Joosse, 1964; Boer, 1965).

1. Two groups, each consisting of about fifty cells, contain NSM intensely stainable with chrome-haematoxylin and paraldehyde-fuchsin ("Gomori-positive" cells). Histochemically, the NSM appears to consist of a lipid and a protein component. It has a high cystine content (Boer, 1965). On the basis of the location of the cell groups, the cells are called medio-dorsal (MDC) and latero-dorsal (LDC), respectively (Joosse, 1964). They show their greatest activity during

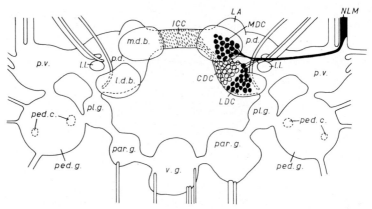

Fig. 1. Schematic drawing (dorsal view) of the central nervous system of *Lymnaea stagnalis*, showing the neurosecretory systems of the cerebral ganglion. The pedal commissures (*ped.c.*) have been cut, and the pedal ganglia (*ped.g.*) and the ventral parts of the cerebral ganglia (*p.v.*) have been put aside in a lateral direction.

CDC, Caudo-dorsal cells; *ICC*, intercerebral commissure; *LA*, lobus anterior; *LDC*, latero-dorsal cells; *MDC*, medio-dorsal cells; *NLM*, nervus labialus medius; *l.d.b.*, latero-dorsal body; *l.l.*, lateral lobe; *m.d.b.*, medio-dorsal body; *p.d.*, pars dorsalis of the cerebral ganglion; *p.v.*, pars ventralis of the cerebral ganglion; *par.g.*, parietal ganglion; *pl.g.*, pleural ganglion; *v.g.*, visceral ganglion.

spring. The NSM of both cell groups is transported via the axons to the lateral peripheral side of the median lip nerve (Fig. 1). The NSM is stored here in bulb-shaped axon endings just underneath the perineurium (neurohaemal area). It has been supposed that at this site the active principles of the NSM are discharged into surrounding blood sinuses (cf. Joosse, 1963, 1964; Nolte, 1964, 1965).

2. In the ganglion a group of neurosecretory cells containing phloxinophilic NSM ("Gomori-negative" cells) is present. It consists of approximately seventy-five cells in the right and of thirty cells in the left cerebral ganglion. The cells are called caudo-dorsal cells (CDC)

according to their position in the ganglion (Fig. 1). Owing to the presence of a large number of concentrically arranged Nissl-disks in their cytoplasm, they have also been called Nissl cells. Like the MDC and LDC these cells show their greatest activity in spring. The phloxinophilic NSM is transported to the peripheral part of the inter-cerebral commissure, which is regarded as the neurohaemal area of the CDC (Joosse, 1964). The NSM, which is stored here in swollen axon endings, is proteinaceous in nature, although the presence of a lipid component has not been excluded beyond doubt. It contains no cystine in large quantities (Boer, 1965).

3. In the pars dorsalis as well as in the pars ventralis, some Sudan black B positive neurosecretory cells occur. Their NSM has been observed to be discharged via the axons (Boer, 1965). These cells are few and relatively small; also it is rather difficult to identify them in the electron microscope. They will not be considered further.

Electron microscopy

In the next section the ultrastructure of the "Gomori-positive" cells (MDC and LDC) and the "Gomori-negative" cells (CDC) will be described, together with their neurohaemal areas (median lip nerve, intercerebral commissure). For comparison attention will be paid to non-neurosecretory ("ordinary") neurons from the lobus anterior (A-cells).

"Gomori-positive" cells (*MDC and LDC*)

Just as with the light microscope, no differences between cells from these two groups were found with the electron microscope. Also no major differences were observed between cells of different sizes. A conspicuous feature of the MDC and LDC lying at the periphery of a group—these are usually the largest cells—is the close contact with the basement membrane of the perineurium. In contrast to most other subcapsular neurons, the MDC and LDC are not separated from this basement membrane by extensions of glial cells. At the site of attach-ment the cell membrane shows many typical infoldings (cf. Brink and Boer, 1967).

The most characteristic elements in the cytoplasm of MDC and LDC are electron-dense elementary granules, which have a mean diameter of 2000 Å. The granules are membrane-limited and their contents seem granulated. Furthermore, in the cytoplasm the usual cell organelles occur: mitochondria (extremely elongated), rough endo-plasmic reticulum (ER), free ribosomes, polyribosomes, Golgi fields, multivesicular bodies, microtubuli (neurotubuli) and cytosomes. In

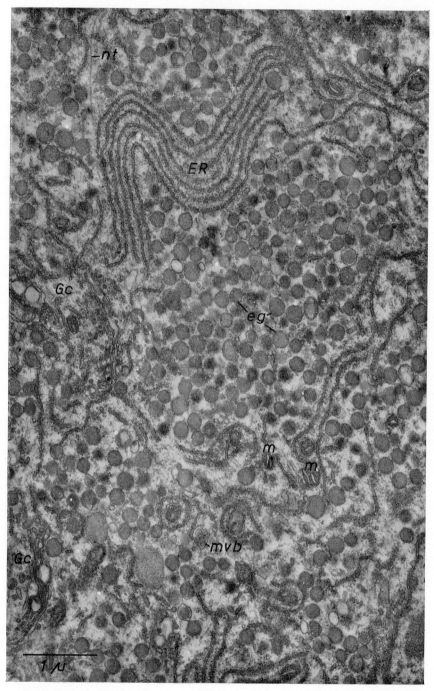

Fig. 2.

contrast to A-cells (see below) MDC and LDC contain only a few complex cytosomes (cf. also Boer, 1965).

The number of elementary granules varies. Some cells are crowded with granules, whereas others contain only a few. In relatively "empty" cells the ER and the Golgi apparatus shows a highly active pattern (active cells, see Fig. 3), whereas these organelles are only poorly developed in cells containing large numbers of elementary granules (inactive cells, see Fig. 2). The ER, which consists of parallel cytomembranes, lined with ribosomes, is mainly situated in the peripheral part of the perikaryon and in the axon-hillock. It can, furthermore, be found around and in contact with the outer nuclear membrane. The cisterns of the ER are relatively wide in active cells, when compared to inactive ones. The ER surrounds numerous Golgi fields, which are located in a zone around the nucleus. In active cells, the Golgi apparatus consists of many large Golgi vacuoles, which are surrounded by hundreds of small Golgi vesicles. In inactive cells the Golgi vacuoles as well as the small vesicles are less prominent. The contents of the Golgi system are of low electron density. In active cells wide cisterns of the ER extend between the Golgi fields. Indications can be found that from these cisterns, which are lined by a few ribosomes only, small vesicles are budded off (Fig. 3).

In inactive cells, all elementary granules are of a similar size and appearance (Fig. 2). They are distributed throughout the cytoplasm. In active cells the granules seem to be associated primarily with the Golgi fields (Figs 3 and 4). Taking into account the form and the size of these granules, many of them may be regarded as immature (Fig. 4). Their distribution in an area containing Golgi fields as well as wide ER cisterns suggests that these cell organelles are involved in granule formation (see Discussion, p. 253).

The median lip nerve

From ultra-thin, cross-sections of the median lip nerve it appears that the main part of the nerve consists of bundles of relatively thin $(0 \cdot 2–1 \ \mu)$ axons. The bundles are surrounded by extensions of glial cells. The total number of axons in the nerve was estimated to exceed 40 000. This large number suggests that the axons branch repeatedly after having left the perikaryon, since the number of nerve cells in the ganglion is limited. However, at the lateral peripheral side of the nerve, large swollen axon terminals $(\phi \ 3–6 \ \mu)$ can be observed (Fig. 5). They

FIG. 2. Electron micrograph of part of an inactive LDC. *ER*, Endoplasmic reticulum; *Gc*, Golgi complex; *eg*, elementary granules; *m*, mitochondrion; *mvb*, multivesicular body; *nt*, neurotubulus.

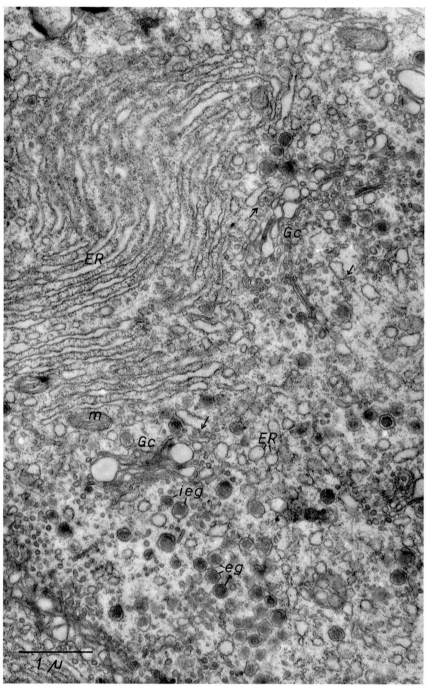

FIG. 3.

lie attached to the basement membrane of the perineurium, in which fibroblasts and muscle cells occur. While in the thin central axons only neurotubuli are prominent, in the peripheral axons great numbers of elementary granules and some mitochondria can be found. The

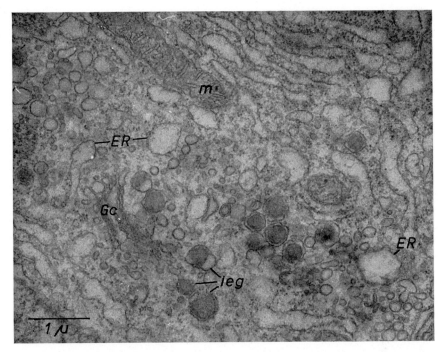

FIG. 4. Part of an active LDC. In the Golgi area (*Gc*), immature elementary granules (*ieg*) of an irregular shape can be noted. In the Golgi area large *ER* cisterns lined by a few ribosomes only, are present. *m*, Mitochondrion.

granules are similar in size (mean diameter 2000 Å) and appearance to those of the MDC and LDC. Furthermore, in some of the axonal bulbs quite a number of small clear vesicles (ϕ 650–850 Å) may occur.

"Gomori-negative" cells (CDC)

The ultrastructure of these neurosecretory cells is different from that of the MDC and LDC in many respects (Figs 6 and 7). They contain

FIG. 3. Electron micrograph of part of an active MDC. In the area of the Golgi complex (*Gc*) large numbers of small vesicles are present, which may partly have arisen by budding from the endoplasmic reticulum (*ER*). Indications for this budding process are seen at arrows. *eg*, Elementary granules; *ieg*, immature elementary granules; *m*, mitochondrion. See also text.

elementary granules with a mean diameter of 1500 Å. The content of
the granules is homogeneous and more electron dense than that of the
MDC and LDC granules. In the present study cells crowded with
elementary granules were not encountered. (However, CDC present in
specimens caught in autumn (Joosse, 1964) have been observed to
contain large quantities of "Gomori-negative" NSM.) Moreover large

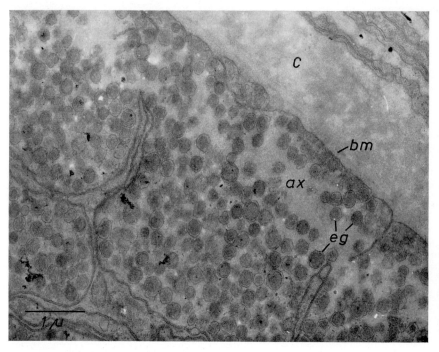

Fig. 5. Peripheral part of the median lip nerve, showing bulb-shaped axon endings
(*ax*) filled with elementary granules (*eg*). *C*, Nerve capsule; *bm*, basement membrane.

($0\cdot5\ \mu$) irregular granules with a yet more electron-dense content occur
in the cells (Fig. 6) and in some electron-dense material was found to be
closely surrounded by small vesicles (Fig. 7), indicating perhaps the
mechanism of formation of the electron-dense granules (see Discussion).

The Nissl disks of the light microscope image were resolved by the
electron microscope into stacks of parallel ER cytomembranes, which

Fig. 6. Electron micrograph of part of a CDC. *NU*, Nucleus; *Gc*, Golgi complex;
ER, endoplasmic reticulum; *eg*, elementary granules; *lg*, large electron dense granules;
m, mitochondrion; *nt*, neurotubulus.

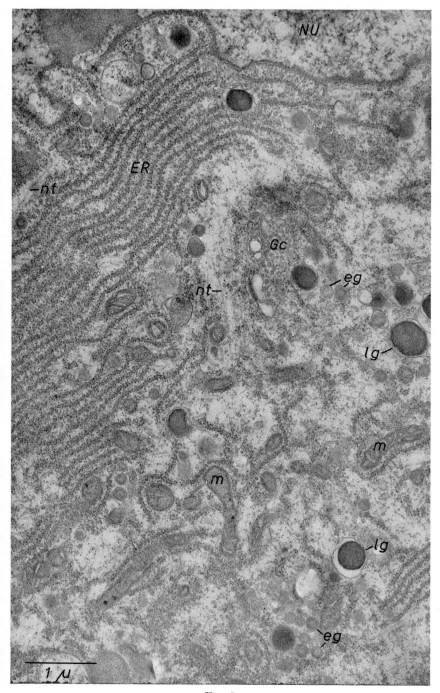

Fig. 6.

are evenly distributed in the cytoplasm. Great numbers of ribosomes line the ER membranes. Moreover, numerous free ribosomes and polyribosomes occur in the cytoplasmic matrix. The great number of ribosomes present in the cytoplasm accounts for the high RNA content in the CDC (Boer, 1965). Although the cells studied are considered as relatively active—storage of large NSM quantities was not found— wide ER cisterns, as in the active MDC and LDC, were not observed.

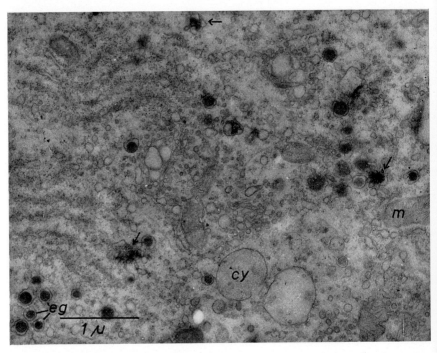

FIG. 7. Part of a CDC. At arrows electron dense material is surrounded by small electron lucent vesicles. *eg*, Elementary granules; *m*, mitochondrion; *cy*, cytosome-like structure.

Golgi fields were not numerous in the CDC. They were usually small and showed an inactive pattern (cf. Boer, 1965). In contrast to the MDC and LDC the contents of the Golgi lamellae were frequently found to be fairly electron dense.

The intercerebral commissure

The general structure of the neurohaemal organ of the CDC is similar to that of the MDC and LDC of the median lip nerves. In the periphery of the intercerebral commissure swollen axon endings containing electron dense elementary granules, some mitochondria and small clear vesicles (φ 550–850 Å), were observed (Fig. 8). The elementary granules had the same appearance and size (1500 Å) as those in the perikarya of the CDC.

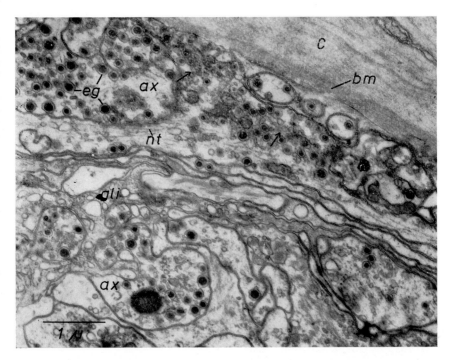

FIG. 8. Electron micrograph of the peripheral part of the intercerebral commissure. In the capsule (*C*) of the nerve collagen fibrils are present. In the axon endings (*ax*) elementary granules (*eg*) and small electron-lucent vesicles (arrows) can be noted. *bm*, Basement membrane; *gli*, extension of glial cell; *nt*, neurotubulus.

Lobus anterior cells

In the lobus anterior, A-cells ("ordinary" neurons) of different size occur (Boer, 1965). With the electron microscope no essential differences between cells of different sizes were found. Although A-cells are considered as non-neurosecretory neurons, in most of them a small number of electron-dense elementary granules (φ 1000–2300 Å) was observed. Usually these granules were associated with the Golgi complex (Fig. 9). Indications of formation of the granules by budding from the Golgi lamellae, which frequently contained electron-dense material, were obtained (Fig. 9). Highly active Golgi fields, showing large vacuoles, were not observed. The ER of the lobus anterior cells is usually poorly developed. Stacks of parallel ER lamellae were not seen, an occasional ER cistern being present only. However, free ribosomes and polyribosomes were numerous. So the lobus anterior cells differ in various aspects from the MDC, LDC and CDC. A further

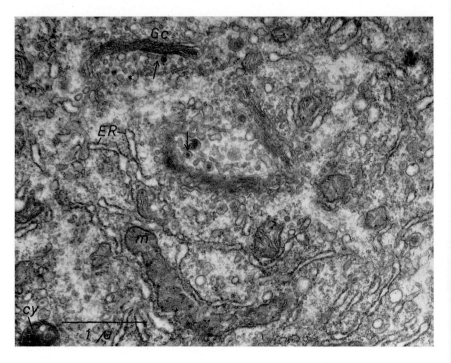

Fig. 9. Electron micrograph of part of a lobus anterior cell. Arrows indicate origination of elementary granules from the Golgi complex (*Gc*) by budding. *ER*, Endoplasmic reticulum; *m*, mitochondrion; *cy*, cytosome-like structure.

difference with these cell types is the presence in the lobus anterior cells of numerous cytosomes, of which two different types may be distinguished: complex cytosomes and lamellar cytosomes (Fig. 10). Moreover, forms which can be regarded as transitional stages between lamellar and complex cytosomes were also observed. In particular the complex cytosomes are usually rather numerous. Furthermore, in the A-cells lipid droplets occur.

DISCUSSION

The NSM of the "Gomori-positive" (MDC, LDC) and "Gomori-negative" (CDC) neurosecretory cells in the cerebral ganglion of *L. stagnalis* is transported via the axons to peripheral areas of the median lip nerve and the intercerebral commissure. Here it is stored in axon endings bounded by the perineurium (Joosse, 1963, 1964).

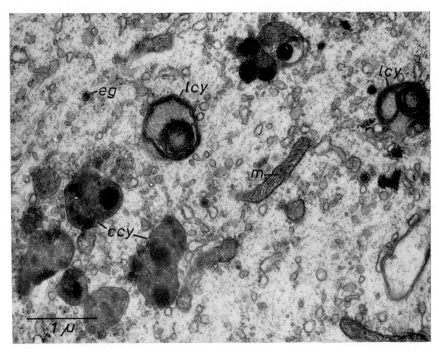

FIG. 10. Part of a lobus anterior cell containing lamellar (*lcy*) and complex (*ccy*) cytosomes. *eg*, Elementary granule; *m*, mitochondrion.

Chemically, the neurosecretory materials are of a complex nature (Boer, 1965). This evidence is in favour of the concept that the cells produce neurohormones. Ultrastructural observations support this view. The NSM of both systems appeared to consist of elementary granules, measuring in the mean 2000 Å in the "Gomori-positive" and 1500 Å in the "Gomori-negative" system. Electron-dense elementary granules (1000–3000 Å) containing neurohormones have been described in "Gomori-positive" systems, as in the hypothalamo-hypophysial system of vertebrates (Lederis, 1964), as well as in "Gomori-negative" systems, as in the urophysis of fishes (Fridberg, Bern and

Nishioka, 1966). However, elementary granules extending into this size range have (in pulmonates) been observed at synaptic junctions, e.g. in *Vaginula solea, Helix pomatia* and *Cryptomphallus aspersa* (see Gerschenfeld, 1963) and also in fibres innervating effector organs, e.g. in *Archachatina* (Baxter and Nisbet, 1963), which suggest that those granules contain "ordinary" neurotransmitters.

There are several reasons for suggesting that the elementary granules in the neurohaemal areas of *L. stagnalis* do not contain neurotransmitters. First, the close contact between neurosecretory axon terminals and perineurium suggests the discharge of the NSM into the perineurium and hence into surrounding blood sinuses. Secondly, structural evidence characteristic for synapses, like axonal membrane thickenings, was not obtained. The presence of synapses in those areas is unlikely because the neurosecretory axon terminals contact only axons of identical structure, i.e. axons containing similar elementary granules. On the other hand, in the axon terminals small clear vesicles (ϕ 650–850 Å) were observed. Vesicles of this type are usually found at synaptic junctions (van der Loos, 1963). However, it has been suggested that such vesicles in neurosecretory axons may result from the fragmentation of elementary granules (Bern, 1963). This fragmentation process is thought to be connected with the release of NSM, in that it liberates the NSM from the elementary granules, so that the active principles of the NSM can readily reach the blood by passing through the cell membrane at the molecule level. It is attractive to assume that a similar process occurs in the neuro-haemal areas of *L. stagnalis*, the more so because no indications were found for other ways of NSM discharge. Thus, no elementary granules were observed outside the axons to suggest the release of intact granules. Furthermore, there were no signs of fusion of the membranes of axons and elementary granules to suggest a sluicing process. Thirdly, pore-like structures in the axon membranes, which have been described in the median lip nerve and the cerebral commissure of *Planorbarius corneus* (see Nolte, 1964, 1965), and which might serve as channels for NSM discharge, were not found in the neurohaemal areas of *L. stagnalis*.

It can be concluded that MDC, LDC and CDC are true neuro-secretory cells. This does not hold for the lobus anterior cells, in spite of the fact that they appeared to contain elementary granules. However, although cells have been called neurosecretory solely on the basis of the presence of elementary granules in *Helix aspersa* and *Planorbis trivolvis* (see Lane, 1966), the observation of such granules in the A-cells of *Lymnaea* is considered to emphasize the statement that their presence is, by itself, not diagnostic for neurosecretion (Knowles and

Bern, 1966), especially as earlier light microscope studies have revealed
no secretory activities in these cells (Boer, 1965). Recent investigations
of the three types of "Gomori-positive" cells in the lateral lobe of *L.
stagnalis* (Brink and Boer, 1967) show that nerve cells, although equally
stainable with neurosecretory stains, may contain different types of
elementary granules. Electron microscopical work (unpublished) on the
"Gomori-positive" MDC of *Australorbis glabratus* suggests that their
elementary granules are different in size (ϕ 1400 Å), and appearance
(fairly electron dense) from those in the MDC and LDC of *Lymnaea*.
However, they might be similar to those in the fuchsinophilic MDC of
Helisoma tenue (ϕ 1 500 Å; see Simpson *et al.*, 1966b). These results
show that caution is needed in interpreting neurosecretory cells in
different species as homologous on the basis of equivalent position in
the ganglion and staining properties.

The association of immature elementary granules with the Golgi
complex suggests this organelle to be involved in granule formation.
This has been stated for many animal groups and species (Bern,
Nishioka and Hagadorn, 1961; Scharrer and Brown, 1961; Follenius,
1963; Lane, 1964). It has been suggested that the Golgi lamellae concentrate material synthesized in the ER: the contents of the lamellae
get an electron-dense appearance. Then the elementary granules are
produced by budding. However, because in the Golgi area small
vesicles have frequently been observed to increase in size and density,
an alternative scheme has been suggested, taking into account the
possible synthesizing role of the small Golgi vesicles (Hagadorn *et al.*,
1963; Bern, 1963; Lane, 1966).

Although in active MDC and LDC of *Lymnaea* the Golgi complex is
extremely large (see also Boer, 1965), Golgi lamellae containing electron-
dense material, or signs of elementary granule formation by budding,
were not found. It therefore seems unlikely that in these cells elementary
granules originate by a budding process. On the other hand, immature
elementary granules and numerous small, electron-lucent Golgi vesicles
occur in the Golgi area. It seems not improbable that part of the small
vesicles arises from the Golgi lamellae. However, another part of them
may well be budded off from the wide ribosome free ER cisternae in the
Golgi area: budding of small vesicles from ER cisterns has been observed
in many other protein-secreting cells (Zeigel and Dalton, 1962). The
immature elementary granules frequently have an irregular shape
(Figs 3 and 4). They may well have been formed by fusion of small
Golgi vesicles with small protein-carrying ER vesicles.

The Golgi complex is, in the CDC, not as prominent as in the MDC
and LDC (see also Boer, 1965). This may indicate that this organelle in

the CDC plays a minor role in elementary granule formation. On the other hand, large electron-dense granules (Fig. 6) and also electron-dense material surrounded by small electron lucent vesicles (Fig. 7) were present in the cells. It would not seem improbable that these structures are in one way or another involved in elementary granule formation. Perhaps the large granules represent sites of NSM synthesis and/or storage, whereas the small electron-lucent vesicles might be considered as carriers of NSM precursors, in the same sense as was suggested for the MDC and LDC. However, in the CDC also Golgi lamellae with electron-dense contents were observed, so that the possibility cannot be excluded that elementary granules can originate by budding from these lamellae. Finally, in the lobus anterior cells granule formation by budding from Golgi lamellae obviously occurs.

In *L. stagnalis* the A-cells are, when compared to neurosecretory cells, characterized by great numbers of cytosomes. Apparently two types, linked by transitional stages, occur: complex and lamellar cytosomes. Furthermore lipid droplets are present. Similar structures have been described in neurons of various molluscs, e.g. in *Helix aspersa* (see Chou and Meek, 1958; Meek and Lane, 1964), in *Planorbis trivolvis* (see Lane, 1966), in *Aplysia californica* (see Simpson, Bern and Nishioka, 1963) and in *Crepidula fornicata* (see Nolte et al., 1965). The cytosomes appeared to contain enzymes. The complex cytosomes are rich in acid phosphatase, and have therefore been considered as lysosomes (Lane, 1963, 1966; Meek and Lane, 1964). Apparently these complex cytosomes can be identified with the "yellow" globules of the light microscope image (Chou, 1957; Chou and Meek, 1958). In *L. stagnalis* these "yellow" globules have been shown to be paraldehyde-fuchsin positive (Boer, 1965). The ultrastructure of the globules supports the view that they do not represent NSM (Boer, 1965; Nolte et al., 1965), although this has been stated for members of the Helicidae (Krause, 1960; Kuhlmann, 1963) on the basis of their staining capacity with the neurosecretory stains.

ACKNOWLEDGEMENT

The authors wish to express their gratitude to Mrs. A. M. van Aardt-Ferreira for her technical assistance.

REFERENCES

Amoroso, E. C., Baxter, M. I., Chiquoine, A. O. and Nisbet, R. H. (1964). The fine structure of neurons and other elements in the nervous system of the giant African land snail, *Archachatina marginata*. Proc. R. Soc. (B) **160**: 167–180.

Baxter, M. I. and Nisbet, R. H. (1963). Features of the nervous system and heart of *Archachatina* revealed by the electron microscope and by electro-physiological recording. *Proc. malac. Soc. Lond.* **35**: 167–177.

Bern, H. A. (1963). The secretory neuron as a doubly specialized cell. In *The general physiology of cell specialisation*: 349–366. Mazia, D. & Tyler, A. (eds). New York: McGraw-Hill.

Bern, H. A., Nishioka, R. S. and Hagadorn, I. R. (1961). Association of elementary neurosecretory granules with the Golgi complex. *J. Ultrastruct. Res.* **5**: 311–320.

Boer, H. H. (1965). A cytological and cytochemical study of neurosecretory cells in Basommatophora, with particular reference to *Lymnaea stagnalis* L. *Archs néerl. Zool.* **16**: 313–386.

Boer, H. H., Slot, J. W. and van Andel, J. (1968). Electron microscopical and histochemical observations on the relation between medio-dorsal bodies and neurosecretory cells in the basommatophoran snails *Lymnaea stagnalis*, *Ancylus fluviatilis*, *Australorcis glabratus* and *Planorbarius corneus*. *Z. Zellforsch. mikrosk. Anat.* **87**: 435–450.

Brink, M. and Boer, H. H. (1967). An electron microscopical investigation of the follicle gland (cerebral gland) and of some neurosecretory cells in the lateral lobe of the cerebral ganglion of the pulmonate gastropod *Lymnaea stagnalis* L. *Z. Zellforsch. mikrosk. Anat.* **79**: 230–243.

Chou, J. T. Y. (1957). The cytoplasmic inclusions of the neurones of *Helix aspersa* and *Limnaea stagnalis*. *Q. Jl microsc. Sci.* **98**: 59–64.

Chou, J. T. Y. and Meek, G. A. (1958). The ultra-fine structure of lipid globules in neurons of *Helix aspersa*. *Q. Jl microsc. Sci.* **99**: 279–284.

Follenius, E. (1963). Etude comparative de la cytologie fine du noyau préoptique (NPO) et du noyau latéral du tuber (NLT) chez la truite (*Salmo irideus* Gibb.) et chez la perche (*Perca fluviatilis*). Comparaison des deux types de neuro-sécrétion. *Gen. comp. Endocr.* **3**: 66–85.

Fridberg, G., Bern, H. A. and Nishioka, R. S. (1966). The caudal neurosecretory system of the isospondylous teleost, *Albula vulpes*, from different habitats. *Gen. comp. Endocr.* **6**: 195–212.

Gerschenfeld, H. M. (1963). Observations in the ultrastructure of synapses in some pulmonate molluscs. *Z. Zellforsch. mikrosk. Anat.* **60**: 258–275.

Hagadorn, I. R., Bern, H. A. and Nishioka, R. S. (1963). The fine structure of the supraesophagial ganglion of the rhynchobdellid leech, *Theromyzon rude*, with special reference to neurosecretion. *Z. Zellforsch. mikrosk. Anat.* **58**: 714–758.

Joosse, J. (1963). The dorsal bodies and neurosecretory cells of the cerebral ganglia of *Lymnaea stagnalis* L. (a preliminary note). *Acta physiol. pharmac. néerl.* **12**: 99–100.

Joosse, J. (1964). Dorsal bodies and dorsal neurosecretory cells of the cerebral ganglia of *Lymnaea stagnalis* L. *Archs néerl. Zool.* **16**: 1–103.

Knowles, F. G. W. (1965). Neuroendocrine correlations at the level of ultra-structure. *Archs Anat. microsc. Morph. exp.* **54**: 343–358.

Knowles, F. G. W. and Bern, H. A. (1966). The function of neurosecretion in endocrine regulation. *Nature, Lond.* **210**: 271.

Krause, E. (1960). Untersuchungen über die Neurosekretion im Schlundring von *Helix pomatia* L. *Z. Zellforsch. mikrosk. Anat.* **51**: 748–776.

Kuhlmann, D. (1963). Neurosekretion bei Heliciden (Gastropoda). *Z. Zellforsch. mikrosk. Anat.* **60**: 909–932.

Lane, N. J. (1963). Thiamine pyrophosphatase, acid phosphatase, and alkaline phosphatase in the neurones of *Helix aspersa. Q. Jl microsc. Sci.* **104**: 401–421.

Lane, N. J. (1964). Elementary neurosecretory granules in the neurones of the snail, *Helix aspersa. Q. Jl microsc. Sci.* **105**: 31–34.

Lane, N. J. (1966). The fine structural localization of phosphatases in neuro-secretory cells within the ganglia of certain gastropod snails. *Am. Zool.* **6**: 139–157.

Lederis, K. (1964). Fine structure and hormone content of the hypothalamo-neurohypophysial system of the rainbow trout (*Salmo irideus*) exposed to sea water. *Gen. comp. Endocr.* **4**: 638–661.

Lever, J. (1958). On the occurrence of a paired follicle gland in the lateral lobes of the cerebral ganglia of some Ancylidae. *Proc. K. ned. Akad. Wet.* (C) **61**: 235–242.

Lever, J., Boer, H. H., Duiven, R. J. Th., Lammens, J. J. and Wattel, J. (1959). Some observations on follicle glands in pulmonates. *Proc. K. ned. Akad. Wet.* (C) **62**: 139–144.

Lever, J. and Joosse, J. (1961). On the influence of the salt content of the medium on some special neurosecretory cells in the lateral lobes of the cerebral ganglia of *Lymnaea stagnalis. Proc. K. ned. Akad. Wet.* (C) **64**: 630–639.

Lever, J., Kok, M., Meuleman, E. A. and Joosse, J. (1961). On the location of Gomori-positive neurosecretory cells in the central ganglia of *Lymnaea stagnalis. Proc. K. ned. Akad. Wet.* (C) **64**: 640–647.

Loos, H. van der (1963). Fine structure of synapses in the cerebral cortex. *Z. Zellforsch. mikrosk. Anat.* **60**: 815–825.

Meek, G. A. and Lane, N. J. (1964). The ultrastructural localization of phos-phatases in the neurones of the snail *Helix aspersa. Jl R. microsc. Soc.* **82**: 193–204.

Nolte, A. (1964). Ultrastruktur des "Neurosekretmantels" des Nervus labialis medius von *Planorbarius corneus* L. (Basommatophora). *Naturwissenschaften* **51**: 148.

Nolte, A. (1965). Neurohämal-"Organe" bei Pulmonaten (Gastropoda). *Zool. Jb.* (Anat.) **82**: 365–380.

Nolte, A., Breucker, H. and Kuhlmann, D. (1965). Cytosomale Einschlüsse und Neurosekert im Nervengewebe von Gastropoden. *Z. Zellforsch. mikrosk. Anat.* **68**: 1–27.

Scharrer, E. and Brown, S. (1961). Neurosecretion XII. The formation of neuro-secretory granules in the earth worm *Lumbricus terrestris* L. *Z. Zellforsch. mikrosk. Anat.* **54**: 530–540.

Simpson, L., Bern, H. A. and Nishioka, R. S. (1963). Inclusions in the neurons of *Aplysia californica* (Cooper, 1863) (Gastropoda, Opisthobranchia). *J. comp. Neurol.* **121**: 237–257.

Simpson, L., Bern, H. A. and Nishioka, R. (1966a). Survey of evidence for neurosecretion in Gastropod molluscs. *Am. Zool.* **6**: 123–138.

Simpson, L., Bern, H. A. and Nishioka, R. (1966b). Examination of the evidence for neurosecretion in the nervous system of *Helisoma tenue* (Gastropoda, Pulmonata). *Gen. comp. Endocr.* **7**: 525–549.

Zeigel, R. F. and Dalton, A. J. (1962). Speculations based on morphology of Golgi-systems in several types of protein secreting cells. *J. Cell. Biol.* **15**: 45–54.

Symp. zool. Soc. Lond. (1968) No. 22, 257–258.

SPONTANEOUS ACTIVITY AND TACTILE PATHWAYS IN THE CENTRAL NERVOUS SYSTEM OF *LYMNAEA STAGNALIS**

T. A. DE VLIEGER

Department of Zoology, Free University, Amsterdam, The Netherlands

ABSTRACT

Investigations of tactile pathways in the central nervous system (CNS) of *Lymnaea stagnalis* were carried out with the aid of stimulus response techniques. One or both lips were mechanically stimulated by means of manipulation of a hair or by using a needle attached to a vibrator. Each lip was connected with the CNS by the median lip nerve only. The fronto-labial nerve was cut (see below). The electrical effects were studied in various peripheral nerves of the CNS. The proximal stumps of the nerves were sucked into glass capillary electrodes.

It was found that even without stimulation, potentials, signalling activity in motor nerve fibres, could be recorded in all nerves investigated. A quantitative study of this spontaneous activity, carried out by means of a level selector, showed the existence of a time pattern. This pattern is called "burst activity", since it was characterized by an alternation of periods (2–3 min) of high activity and periods (3–6 min) of low activity. Abrupt changes in temperature temporarily destroy this pattern.

Responses to mechanical stimulation had to be discriminated from this spontaneous activity. In some experiments this was done with the aid of an averaged response technique. It was found that, upon tactile stimulation of the lip, in all nerves investigated a response could be evoked. Of all nerves the threshold appeared to be lowest in the fronto-labial nerve innervating the stimulated lip.

Cutting of connectives or commissures resulted in quantitative alterations of the stimulus response relations, i.e. these cuts heighten the thresholds. An exception to this rule was found with the inter-cerebral commissure, which transmits only fast adapting tactile signals.

---

* A detailed study on this subject will be found in de Vleiger, T. A. (1968). An experimental study of the tactile system of *Lymnaea stagnalis*. *Neth. J. Zool.* **18:** 105–154.

From these results it could be concluded that pathways for the transmission of tactile signals through the CNS are determined essentially by the stimulus strength.

Study of the behaviour of the animal upon tactile stimulation revealed that, with increasing stimulus strength, withdrawal reactions spread from the stimulated lip over the entire body of the snail. This seems to be in accordance with the electrophysiological observations. However, there is another type of integrated behaviour upon mechanical stimuli. When the foot of the snail does not make contact with a substrate, weak tactile stimulations (of the lip, but also of other body parts) are followed by movements of the foot in the direction of the stimulus.

Experiments with isolated lip-nerve preparations indicated that two types of responses exist, which differ in the length of the adaptation periods (seconds and minutes respectively). The possibility is discussed that these responses are associated with the two types of behaviour mentioned above.

Symp. zool. Soc. Lond. (1968) No. 22, 259–271.

SORTING PHENOMENA DURING THE TRANSPORT OF SHELL VALVES ON SANDY BEACHES STUDIED WITH THE USE OF ARTIFICIAL VALVES*

J. LEVER and R. THIJSSEN

Department of Zoology, Free University, Amsterdam,
The Netherlands

SYNOPSIS

On a sandy beach of the Dutch North Sea coast experiments were carried out with 58 000 brightly coloured artificial shell valves of *Donax vittatus*. Unperforated and perforated (artificial *Natica*-holes of 1, 2 or 3 mm inner diameter, located in the anterior, central or posterior part of the valves), left and right valves of three sizes (19·7, 24·2, and 28·7 mm shell length) and of three specific weights (1·8, 2·3, 2·8) were used.

It appeared that under the influence of waves and water currents a clear selection occurred according to the size, to the symmetry (left and right valves), to the presence or absence of one hole or two holes, to the diameter of the hole, to the location of the hole, and to the specific weight of the valves.

INTRODUCTION

Sands and other bottom particles are selectively transported by water currents, depending on size, shape and specific gravity. Comparable phenomena have been observed occasionally with shell valves. For instance, a selective transportation of left and right valves of the pelecypods *Pitar dione* and *Arca incongrua* along the coast of Trinidad is described by Martin-Kaye (1951), and Boucot, Brace and de Mar (1958) observed an unequal distribution of pedicle and brachial valves of brachiopods in American strata of Devonian age. (See also Müller (1950) and Lever (1958).) Such selection phenomena have rarely been investigated experimentally (Richter, 1922, 1924; Kornicker and Armstrong, 1959).

For some years we have paid special attention to these selection problems, studying them under natural conditions on sandy beaches of the Dutch North Sea coast. Initially, by using stained natural valves, an analysis could be made of the unequal washing ashore of left and right valves of several species (Lever, 1958). Later on, a difference in transport between intact valves and valves which had been perforated by naticid snails was observed (Lever, Kessler, van Overbeeke and Thijssen, 1961).

* This study was supported by the Royal Dutch/Shell Prize 1959.

As these investigations had shown that small differences between valves can cause considerable sorting effects, it was concluded that more precise experiments could be carried out only by using sets of identical valves. Therefore, artificial copies from natural specimens were needed. From the left and right valves of three complete shells of *Donax vittatus*—lengths 19·7, 24·2, 28·7 mm (intervals 4·5 mm)— moulds were made by the London and Scandinavian Metallurgical Co. Ltd. in London (England). The Royal Dutch/Shell Plastics Laboratory in Delft (Netherlands) prepared three synthetic compounds and produced artificial valves with specific weights of 2·8 (the normal s.w. of wet valves of *Donax vittatus*), 2·3, and 1·8 respectively. By adding dyes to the compounds the valves could be brightly coloured red, blue, yellow, green, or violet. This allowed an easy detection on the beach, and also increased the experimental possibilities. In our Institute in Amsterdam conical holes (inner diameter of 1, 2 or 3 mm) were drilled in part of the valves. The shapes of these holes were identical to those bored by *Natica* in valves of *Donax vittatus* of the same size as the artificial valves. (For more technical details, see Lever, van der Bosch, Cook, van Dijk, Thiadens and Thijssen, 1964.)

METHODS

The investigations described below were carried out on the sandy beach of the North Sea island of Schiermonnikoog, which runs approximately west to east, with the sea on the northern side. The experiments started when the tide was nearly at its lowest. The artificial valves were laid out on the lower part of the off-shore slope of a sand bar or of the beach proper, where, generally, the majority of the natural valves are also accumulated. A thin iron rod was used as a marker, and around

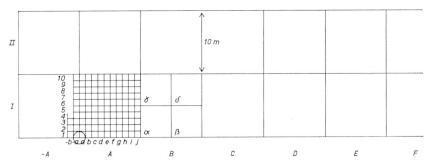

Fig. 1. Division of the slope during the ebb-tide(s) following that at the beginning of an experiment. The valves started from the circle.

this, in an even layer (one specimen thick), the valves were spread out in a circle with a radius of 1 m. During the next ebb-tide the area in which artificial valves appeared to be present was divided into squares with the aid of stainless steel pegs. The border line of these squares ran parallel to the lower margin of the slope and through the centre of the circle. Squares 10 × 10 m were always used, but were often, especially near the circle, subdivided into smaller squares of 5 × 5 m, or even 1 × 1 m (see Fig. 1). During this ebb-tide all valves seen higher on the slope than the line mentioned were collected, with the exception of those that were still situated within the original circle. Below this line very few valves were observed. Often, valves were also collected during later ebb-tides, and at the end of some experiments their occurrence in the sand under the original circle was studied.

EXPERIMENTS

With the artificial valves available three experiments were carried out (see Fig. 2).

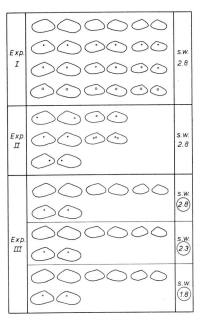

FIG. 2. Scheme of the categories of artificial valves used in the three experiments.

Expt I (20 August 1962). Valves with the normal specific weight of 2·8 were used. The experiment was done to compare the transport of *left and of right* valves, that of valves of *different sizes* (lengths: 28·7, 24·2, 19·7 mm), that of *intact and of perforated* valves, and that of valves with *holes of different diameters* (1, 2, 3 mm). All holes had been drilled in the centre of the valves. The total number of categories was twenty-four.

Expt II (7 August 1963). Valves of the same s.w. (2·8) as those of expt. I were used. Special attention was paid to the transport of valves with *one hole bored at different locations* (in the anterior, central, or posterior part of the valve; all holes were 1 mm in diameter, and all valves had a length of 28·7 mm), and to that of *valves with one or two holes* (valve length 24·2 mm, hole diameter 2 mm). The total number of categories was ten.

Expt III (30 June 1965). The transport of valves with *different specific weights* was studied (2·8, 2·3, 1·8). Moreover, part of the valves of the largest size class were *perforated* with a 1 mm hole in the centre of the valve. The total number of categories was twenty-four.

In all three experiments each category was represented by 1000 valves, so a total of 58 000 artificial valves were used.

RESULTS

Experiment I

During this experiment* the wind came mainly from the north and east, and the waves were moderate in height and approached the coast from an easterly direction. During the first ebb-tide following that of the initiation of the experiment it appeared that artificial valves had been transported over a large area west of the marker pole (maximum distance 70–80 m), and therefore squares were made as in the scheme of Fig. 1 (the sea lies below, and squares A, B, etc., are made in a westerly direction).

The total number of valves of each category found outside the circle is given in Table I.

It is clear that the quantitative transport of the categories had differed considerably, varying from 236 left unperforated valves of the smallest category to twice as many (473) right valves of the largest type which had a 3 mm hole.

A statistical study (analysis of variance with a $3 \times 2 \times 4$ design after arc-sin. transformation) showed that highly significant differences ($P \ll 0·001$) exist between the groups of different size, between those with or without a hole, and between those with holes of different

---

* For a detailed description of this experiment, see Lever *et al.* (1964).

diameter. On the other hand, no statistically significant differences were found between the numbers of left and right valves. Figure 3 illustrates the three sorting phenomena demonstrated so far.

TABLE I

Total numbers of valves collected outside the circle during the first ebb-tide of expt. I.

L, left valves; R, right valves; L_1, left valves with 1 mm hole; L_2, left valves with 2 mm hole; etc.

Size class	L	R	L_1	R_1	L_2	R_2	L_3	R_3
I (28·7 mm)	330	369	372	405	459	427	456	473
II (24·2 mm)	321	321	386	367	404	387	407	426
III (19·7 mm)	236	247	334	314	351	363	380	377

From a detailed analysis of the distribution of the valves, especially in the area that had been divided in 1×1 m squares (see Fig. 1), it could be concluded (see Lever *et al.*, 1964) that: (1) large valves of *Donax vittatus* are better transported than smaller valves, against the slope as well as parallel along the bar; (2) perforated valves are moved more easily onto the slopes than unperforated specimens, whereas in the direction along the bar intact valves move farther than perforated valves; (3) in both directions valves with a 1 mm hole are better transported than those with larger holes; (4) although the *total* numbers of recovered left and right valves of comparable categories were approximately identical (Table I), their distribution appeared to be remarkably different: right valves move much farther in both directions; at the larger distances right valves were found almost exclusively.

Experiment II

During this experiment the wind came from northerly directions and the waves, which were much higher than those in Expt I, approached the slope from the north-west, which resulted in a transportation of the artificial valves east of the marker during the ebb-tides following the start of the experiment. During three successive ebb-tides transported valves were collected and their distribution was noted. Finally, a sample of the sand under the original circle was sieved, as it appeared that many valves had been covered by sand near the marker.

The total results are given in Table II, and the distribution of the eastward transported valves collected during the first ebb-tide is shown in Fig. 4.

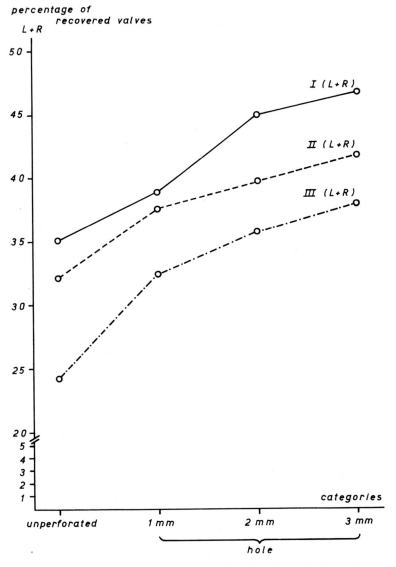

FIG. 3. Expt I. The numbers of unperforated and perforated valves of the three size-classes—expressed as percentages of the initial quantities—collected during the first ebb-tide.

From this Table and this Figure it can be concluded, that, with
respect to the valves collected outside the circle: (1) left valves are
better transported than right ones in both directions (against and
along the slope); (2) the total numbers collected from the categories
which had a 1 mm hole (drilled at different locations) do not differ

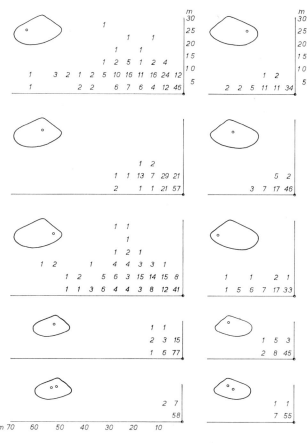

FIG. 4. Expt II. The distribution of the valves collected during the first ebb-tide.
Squares 10 × 10 m.

(Table II), but the valves with an anterior or a posterior hole covered
larger distances in both directions than those with a central hole. This
was especially the case with the left valves (Fig. 4); (3) valves with
one 2 mm hole are better transported in both directions than valves
with two 2 mm holes.

It is interesting that, in accordance .with these conclusions, in the sand under the original circle more right than left valves, more valves with a central than with an eccentric 1 mm hole, and more valves with two than with one 2 mm hole(s), were found.

TABLE II

Total numbers of valves collected during Expt II

A, B, C, one 1 mm hole in the anterior, central, or posterior part of the valve, respectively; L_{22}, left valves with two 2 mm holes. For other symbols, see Table 1.

| Size class | I (28·7 mm) | | | | | | II (24·2 mm) | | | |
| | L_1 | | | R_1 | | | L | | R | |
Hole(s)	A	B	C	A	B	C	L_2	L_{22}	R_2	R_{22}
1st ebb-tide	209	157	178	68	80	74	108	67	64	64
2nd ebb-tide	260	252	247	137	142	140	179	165	161	161
3rd ebb-tide	29	47	51	7	10	10	60	46	21	24
Total	498	456	476	212	232	224	347	278	246	249
In sand, under circle	112	163	112	172	192	179	207	221	262	269

Experiment III

This experiment was carried out under rather unfavourable conditions. The waves were fairly high and their direction changed, and the speed and the direction of the wind also varied considerably. During two subsequent ebb-tides only a small proportion of the valves was found transported outside the circle, and therefore at the end of the second ebb-tide special attention was paid to the valves buried in the sand under the original circle. As exactly as possible the circle was divided into four quadrants, and their valve contents were studied separately. The total numbers of each category recovered, upon the slope and in the sand, are presented in Table III.

From this table it is clear that, under the extremely variable conditions of this experiment, a sorting of left and right valves during transportation could not be found. It is still shown that large valves are better transported than smaller specimens: compare I (unperforated) with II and III. It seems also that, in this case, unperforated valves dominated perforated valves. This is an unexplainable and exceptional observation, as all our earlier and later studies of the distribution of natural valves of various species confirm the results of Expt I.

The most evident result of Expt III, however, is the clear sorting of the valves according to specific weight: in each size class (left as well as right) light valves are much better transported than heavier specimens.

* TABLE III

Total numbers of valves collected during Expt III

s.w., specific weight. For other symbols used, see Table I.

Size class	s.w.	I (28·7 mm)				II (24·2 mm)		III (19·7 mm)		Total
		L	R	L$_1$	R$_1$	L	R	L	R	
Transported	2·8	7	6	—	17	10	10	9	8	67
(2 ebb-	2·3	14	20	13	27	13	15	12	10	124
tides)	1·8	84	94	42	67	73	67	44	37	508
	Total	105	120	55	111	96	92	65	55	699
In sand,	2·8	672	676	755	752	747	707	747	731	5787
under	2·3	390	437	530	484	494	517	513	494	3859
circle	1·8	96	90	126	160	164	129	204	169	1138
	Total	1158	1203	1411	1396	1405	1353	1464	1394	10 784

The results obtained with the valves that were found outside the circle are clearly complemented by the conclusions than can be drawn from the numbers of valves sieved out of the sand under the original circle. Here, the numbers of small valves are larger than those of bigger specimens (III > II > I), and more perforated valves than intact valves are present. Finally, the heavier the valves are, the more have remained in the circle. This is illustrated in Fig. 5, in which the percentages of the original numbers irrespective of size, symmetry (L, R) or perforation, that were found still present in the quadrants of the circle are given (for each of the categories separately, comparable results were obtained). Apparently, the general transport direction had been the same in each weight class, but the quantitative effect differed considerably.

CONCLUSIONS AND DISCUSSION

From the results of the three experiments some conclusions can be drawn about the transport of valves of *Donax vittatus* in tidal areas of sand beaches.

1. It is clear that when a water current runs along a slope, while waves with a beachward direction move over this current, a sorting of *left and right* valves occurs, depending on the direction of current and waves, as shown in Fig. 6A.

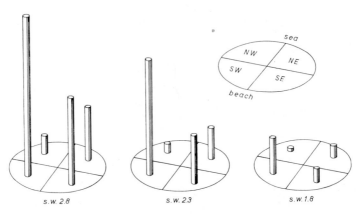

FIG. 5. Expt III. Total numbers of valves of the three specific weights (expressed as percentages of the original quantities) found in quadrants of the sand under the original circle.

An explanation of this phenomenon is given in Fig. 7. The upsurge orients the left and the right valves in the positions I and II, respectively. The backwash has little effect on the left valve as it lies with its sloping side turned towards the waterstream, while the steep sides of the heaviest part of the valve are resisted by the sand. For the right valve the reverse is true. The water of the backwash presses against the steep sides of this valve, and if the pressure is sufficient the valve will start turning. If the valve moves it is probable that it will be transported in a seaward direction. If the valve comes to rest it will lie oriented in position III. This means that the following upsurge passes relatively easily over it as it now has the same position with regard to the oncoming wave as the left valve had to the backwash. As the total resistance of the sand against the steep sides is without doubt higher than that against the point of the left valve (still in position I), the latter will be transported towards the beach more easily than the former. Repeated many times this will result in a better beachward transport of left than of right specimens. It is clear that the result will be reversed when the upsurge comes from the upper right corner of Fig. 7, and the backwash from the lower right. This explanation is based upon field observations, published in detail by Lever (1958).

2. It can be concluded that, with *Donax vittatus*, a difference in transport occurs with respect to the *size* of the valves, large specimens being transported better than small valves, onto the slopes as well as along the slopes, irrespective of the direction of the currents along the beach. (See Fig. 6B.)

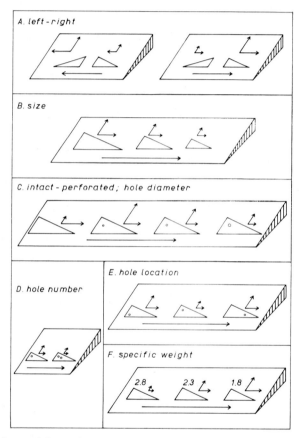

Fig. 6. Scheme of the sorting phenomena demonstrated during the three experiments.

3. A clear difference was observed between *perforated* and *unper-forated* valves. (See Fig. 6C.)

The former are transported more easily towards the beach than the latter, but along the slopes the unperforated valves cover greater distances. This has considerable consequences, because it means also that with alternating current directions (as often occur in tidal areas) the unperforated valves move back and forth over great distances along the slopes. The perforated valves, on the other hand, are less mobile in the currents, but they are transported better onto the slopes, no matter which way the currents move. For a preliminary explanation of this difference, see Lever *et al.* (1961).

4. Differences in transport were found also with respect to the *diameter of the hole* in perforated valves. The mobility along the slopes as well as in beachward direction decreases as the diameter of the hole increases (see Fig. 6C).

5. The same holds for valves with *one hole* compared with valves with *two holes* (Fig. 6D). Apparently, the total perforated area is more important than the number of perforations.

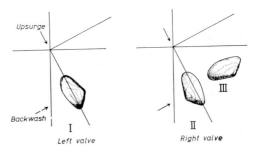

FIG. 7. Explanation of the unequal transport of left and right valves of *Donax vittatus*.

6. Valves with an *eccentric hole* are better transported in both directions than valves with a *central hole* (Fig. 6E).

7. Finally, a clear sorting of valves according to *specific weight* was found: the lighter the valves the better they are transported (Fig. 6F).

The experiments have demonstrated that, under natural conditions, each small difference between the artificial valves resulted in unequal transportations. Therefore, some general assumptions are warranted. Most probably, valves of species which have different shapes, smooth valves and valves with projections or longitudinal or transversal ridges, and also flat, rounded, circular or projected valves, are all unequally distributed. This will also be true for gastropod shells: round shells will be transported otherwise than oval, conical, or slender shells. The same will hold for valves that normally possess one or more holes compared with unperforated valves of approximately the same shape, and the manner of transport of valves that show one or more openings when they lie on the sand will probably differ from that of those which have a closed line of contact with the bottom. The results also permit the expectation that in the transport to the beach or along the sea bottom of objects other than shells (live molluscs, seaweeds, dead fish, crabs, jellyfish, sea urchins, etc.) selection phenomena will play a role. This means that experiments of the kind described are needed to elucidate many biological, sedimentological, palaeontological and palaeo-ecological problems related to recent or fossilized marine sediments that contain remains of organisms.

REFERENCES

Boucot, A. J., Brace, W. and Mar, R. de (1958). Distribution of brachiopod and pelecypod shells by currents. *J. sedim. Petrol.* **28**: 321–332.

Kornicker, L. S. and Armstrong, N. (1959). Mobility of partially submerged shells. *Publs Inst. mar. Sci. Univ. Texas* **6**: 171–185.

Lever, J. (1958). Quantitative beach research. I. The "left-right phenomenon": sorting of lamellibranch valves on sandy beaches. *Basteria* **22**: 21–51.

Lever, J., Bosch, M. van den, Cook, H., Dijk, T. van, Thiadens, A. J. H. and Thijssen, R. (1964). Quantitative beach research. III. An experiment with artificial valves of *Donax vittatus*. *Neth. J. Sea Res.* **2**: 458–492.

Lever, J., Kessler, A., Overbeeke, A. P. van, and Thijssen, R. (1961). Quantitative beach research. II. The "hole effect": a second mode of sorting of lamellibranch valves on sandy beaches. *Neth. J. Sea Res.* **1**: 339–358.

Martin-Kaye, P. (1951). Sorting of lamellibranch valves on beaches in Trinidad. *Geol. Mag.* **88**: 432–434.

Müller, A. H. (1950). Grundlagen der Biostratonomie. *Abh. dt. Akad. Wiss. Berl.* (Kl. Math. Naturw.) **1950** (3): 1–147.

Richter, R. (1922). Flachseebeobachtungen zur Paläontologie und Geologie. III–IV. *Senckenbergiana* **4**: 103–141.

Richter, R. (1924). Flachseebeobachtungen zur Paläontologie und Geologie. VII–XI. *Senckenbergiana* **6**: 119–165.

K

Symp. zool. Soc. Lond. (1968) No. 22, 273–291.

BRITAIN'S FAUNA OF LAND MOLLUSCA AND ITS RELATION TO THE POST-GLACIAL THERMAL OPTIMUM

M. P. KERNEY

Department of Geology, Imperial College, London, England

SYNOPSIS

The possible effect of the Thermal Optimum of Post-glacial times on the land Mollusca of Britain is considered. Changes in the distributions of three species are discussed: *Pomatias elegans* (Müller), *Lauria cylindracea* (da Costa) and *Ena montana* (Drapernaud). Although precise experimental evidence is lacking, it is suggested that the first two are intolerant of winter cold, whereas the third requires fairly high summer temperatures. The fossil record of these species in Britain is in keeping with palaeobotanical evidence for the nature of climatic change during and since the Thermal Optimum.

In this paper I propose to discuss some of the changes in Britain's fauna of land Mollusca caused by the climatic fluctuations of the Post-glacial period. In particular, I shall consider some possible effects of the Thermal Optimum and of the decline in temperature which followed it.

The broad pattern of Post-glacial climatic change in north-west Europe is now well established, principally through the work of palaeobotanists. In the last century it was observed that many peat bogs in Scandinavia, which must have begun to form subsequent to the last glacial retreat, yielded from low levels in the peat remains of birch or pine trees, but from levels nearer the present surface remains or more warmth-demanding trees, such as oak or hazel. These changes were interpreted as a reflexion of the climatic improvement following the disappearance of the ice, and, on evidence such as this, the botanists Blytt and Sernander established a series of climatic periods (Pre-boreal, Boreal, Atlantic, Sub-boreal, Sub-atlantic), each characterized by a particular kind of forest vegetation. The development of the technique of pollen analysis by von Post and others gave much greater precision to the recognition of these vegetational changes. In particular, it was possible to suggest a threefold division of Post-glacial time, the validity of which has since been amply confirmed: (a) a period of increasing warmth; (b) a period of maximum warmth (the Thermal Optimum); and (c) a period of decreasing warmth in which we are now living. Considerably finer subdivisions (pollen zones) can indeed be recognized

over wide areas (Godwin, 1956). However, the precise climatic significance of the zones is by no means always clear. There are numerous
reasons for this. Information on the climatic requirements of particular
plants may not be available. Changes in pollen representation at different levels within a deposit may not reflect true changes in forest composition, or, even if this can be assumed, the changes observed may be
due to the natural ecological successions inherent in all plant communities, or to a progressive leaching of the soil, or to interference by
man, or to other such causes which are not directly dependent on climate
(Iversen, 1960; Frenzel, 1966). The recognition of changes in humidity
has proved particularly difficult, and the evidence for these is usually
stratigraphical rather than palaeobotanical (Godwin, 1954, 1960). In
spite of these uncertainties, the approximate synchroneity across Europe
of many pollen zone boundaries, as has been revealed by radiocarbon
dating, suggests that they have an essentially climatic basis.

The Post-glacial period also witnessed the great rise in ocean levels
consequent on the melting of the ice sheets, and which had important
biological repercussions. At first, the British Isles were united with the
European mainland, and animals and plants could therefore enter
freely from the south as rising temperatures enabled them to do so.
England was isolated from France probably by about 5500 B.C.
(early Atlantic, zone VIIa); Ireland from Britain somewhat earlier.
The presence or absence of particular species at the present day in
Britain or Ireland may therefore give some clue as to their time of
immigration. For example, the absence of *Pomatias elegans* from
Ireland (Fig. 3) accords with our knowledge that this species probably
did not enter southern England until after the time when the Irish
Sea had been refilled. Much has been made of evidence of this kind in
controversies concerning the origin or survival of certain elements in
the British fauna and flora, but the dangers are obvious, and such
biogeographical arguments can carry little weight in the absence of
supporting palaeontological evidence.

Palaeontological work on our Post-glacial molluscan faunas is still
in its early stages. But already it is clear that the story is broadly
analogous to that of the flowering plants. Climatically tolerant forms
appear and spread early, whereas the more warmth-demanding species
arrive later. The land Mollusca also reflect the Post-glacial vegetational
changes. Species adapted to open environments are commonest in the
open landscape of the Late-glacial and earliest Post-glacial periods,
whereas shade-loving species increasingly dominate as the forest cover
spreads.

This kind of sequence is demonstrated very clearly in Fig. 1, which shows the changing percentage frequency of land Mollusca through a series of calcareous deposits filling the bottom of a chalk valley near Brook in east Kent (Kerney, Brown and Chandler, 1964). The earliest deposits were formed by frost-shattering and solifluxion near the end of the Late-glacial period (pollen zone III). The commonest species in this and other Late-glacial deposits which have been studied in southeast England have a palaearctic or holarctic range and a considerable tolerance of cold [*Cochlicopa lubrica* (Müller), *Pupilla muscorum* (L.), *Vallonia* spp., *Arianta arbustorum* (L.), *Vitrina pellucida* (Müller), etc.]. We also find species which at the present day have Boreal-Alpine distributions and have since become extinct in Britain (*Columella columella* (Martens), *Vertigo genesii genesii* (Gredler)). But also present are several snails with rather southern distributions [*Abida secale* (Draparnaud), *Helicella itala* (L.)], which are presumably more thermophilous, and whose presence reflects the first stages of climatic improvement; these species first appear in strength in the rather mild Allerød Interstadial (zone II) (Kerney, 1963). At higher levels in the Brook section (Fig. 1) we find a great increase in Mollusca which prefer shaded environments, reflecting the spread of Post-glacial forests over the area. Conversely, open-ground species (*Pupilla muscorum*, *Abida secale*, *Vallonia* spp.) decline. The most striking decrease is shown by *Hygromia hispida* (L.), a species which often thrives in damp situations in freshly broken ground, but which is much less common in mature, well vegetated habitats. As temperatures rose, increasingly thermophilous snails appeared. The behaviour of two closely related species, *Discus ruderatus* (Férussac) and *Discus rotundatus* (Müller) is of particular interest. The former has at the present day an Alpine/Scandinavian/east European/Asiatic distribution, suggesting an intolerance of mild oceanic climates (Ant, 1963: 102). At Brook *D. ruderatus* appears for a while in the earliest part of the Post-glacial, and then becomes extinct, to be replaced by the rather southern thermophile *D. rotundatus*, the only species of the genus which now exists in Britain. Between about 190 cm and 130 cm in the deposits the Mollusca reflect a densely wooded, humid environment, with an abundance of *Carychium tridentatum* (Risso), Zonitidae, *Acanthinula aculeata* (Müller), and other shade-loving forms. The climate was now fully temperate. These layers are assigned to the Late Boreal and Atlantic periods (zones VI and VIIa).

The upper half of the diagram shows a very striking change. The environment became much more open, and the molluscan fauna is dominated by the grassland genus *Vallonia*. Late-glacial species, such as *Pupilla muscorum*, return. However, this dramatic change is not to

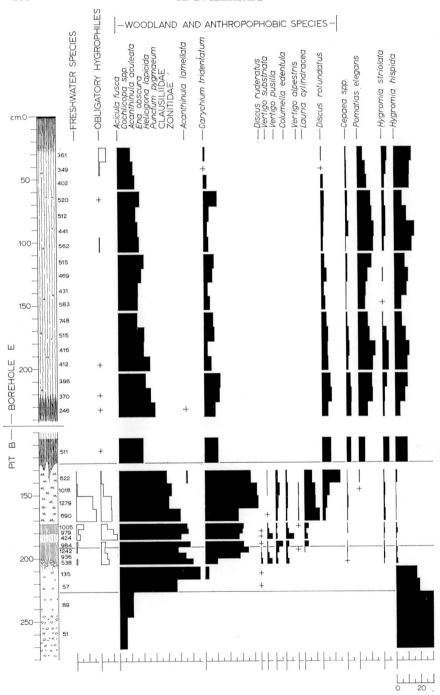

FIG. 1. See

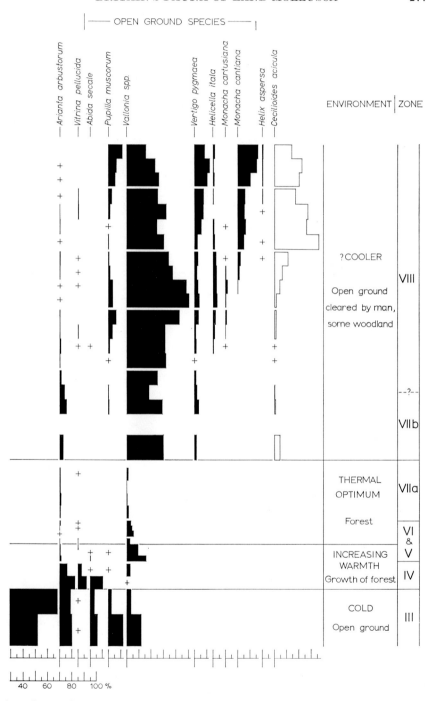

legend on p. 278.

be thought of as necessarily climatic, but can be explained entirely in terms of human interference; pollen-analytic studies reveal that from about 3000 B.C. onwards the chalk landscape of southern England began to be cleared of forest by Neolithic and Bronze Age farmers (Godwin, 1962). Many molluscan species disappeared or became very scarce, not because the climate worsened, but because man had altered the environment and destroyed the critical conditions which these species required. This kind of effect makes it very hard to detect faunal changes which may be due to climate. For example, in the past the fossil land Mollusca obtained from Neolithic and Bronze Age sites on the chalk of southern England have often been considered to reflect a much more humid climate than that which now exists. This view was strongly argued in numerous papers by the late A. S. Kennard (e.g. *in* Cunnington, 1933; *in* Curwen, 1934; *in* Stone, 1933), who noted the presence of hygrophilic species such as *Acicula fusca* and *Arianta arbustorum* in places which are now very dry. But such discrepancies can equally have been caused by a stripping of the forest cover. I have elsewhere considered some of the effects of human interference on Britain's molluscan fauna (Kerney, 1966), and J. G. Evans has shown by his careful studies of the fossil Mollusca of archaeological sites on the chalk how nearly all the observed changes can convincingly be explained in such terms, without the need to invoke any climatic alteration (Evans, 1967). Man has indeed produced a great drying of the landscape, favouring xerophiles at the expense of hygrophiles. The present restricted geographical distributions within the British Isles of a considerable number of woodland and marsh species which were much more widespread in the Boreal and Atlantic periods of the Post-glacial have probably been brought about largely in this way (e.g. *Acicula fusca* (Montagu), *Vertigo pusilla* Müller, *V. substriata* (Jeffreys), *V. moulinsiana* (Dupuy), *V. alpestris* Alder, *V. angustior* Jeffreys, *Lauria anglica* (Wood), *Acanthinula lamellata* (Jeffreys)).

However, leaving aside the question of possible changes in precipitation, the reality of a Post-glacial thermal decline is incontrovertible, and the question therefore arises whether it is possible among the Mollusca to recognize and isolate any effects which can be ascribed to this cause. Most of the evidence for the Thermal Optimum is indeed of

FIG. 1. Molluscan histogram through Late-glacial and Post-glacial subaerial deposits, Brook, Kent. Freshwater species (*Pisidium* spp.), obligatory hygrophiles (*Carychium minimum, Lymnaea truncatula, Vertigo genesii geyeri, V. moulinsiana, V. angustior, Zonitoides nitidus*), and the burrowing species *Cecilioides acicula*, are calculated as percentages over and above 100 and are shown as open histograms. The ecological groupings into "woodland and anthropophobic species" and "open ground species" should be regarded only as approximations (for fuller details, see Kerney *et al.*, 1964).

a biological nature. For example, at the present day the European pond tortoise (*Emys orbicularis* L.) is not found in northern Europe, as the species requires fairly high summer temperatures for breeding and the development of the eggs. Yet in earlier Post-glacial times *Emys* was common in Denmark and southernmost Sweden (Degerbøl and Krog, 1951), and there is also a record for East Anglia (Newton, 1862). Degerbøl and Krog suggest that in Denmark mean summer temperatures in the Late Boreal, Atlantic and Sub-boreal periods were at least 2°C higher than at the present day, and that the species was exterminated by the onset of cooler and more oceanic conditions at the beginning of the Sub-atlantic period (*ca* 600 B.C.). Even more revealing is the work of Iversen (1944), who chose three plants (*Hedera helix* L., *Ilex aquifolium* L. and *Viscum album* L.) which are today in Denmark close to their climatic limits. First Iversen made a careful analysis of the precise meteorological conditions existing at places in Denmark where these plants lived and where they did not live. Broadly, he was able to show that *Hedera* and *Ilex* are markedly intolerant of winter frosts, and require minimum mean summer temperatures of about + 14°C. As might be expected, both plants have Atlantic-Mediterranean distributions in Europe. On the other hand, *Viscum* will stand winter temperatures of at least − 7°C, but needs rather higher mean summer temperatures, of the order of + 17°C. From exact observations such as these, Godwin (1956) and Iversen (1960) offer convincing conclusions, based on the relative frequencies of pollen of these plants in Post-glacial deposits in Denmark and in England, for the precise character of the Climatic Optimum: during the Late Boreal (6500–5500 B.C.), Atlantic (5500–3000 B.C.) and Sub-boreal (3000–600 B.C.) periods, summer temperatures appear to have been somewhat higher than those of today; winter temperatures, on the other hand, declined at the Atlantic/Sub-boreal transition.

Unfortunately, very little comparable work has yet been done on the thermal tolerances of Mollusca. In order to demonstrate past climatic changes, species which are today in Britain near their apparent climatic limits would obviously be the most illuminating. Also, they should be relatively common within the area of their range, adapted to a wide spectrum of habitats, and reasonably independent of human interference. But even granted these conditions, we must not expect to find any single thermal factor controlling distribution. The annual range in temperatures may be as important as the annual extremes. In effect, this may mean that a southern European species may be limited to the east by winter cold, but to the north by lack of summer warmth.

There is no case, comparable to *Emys orbicularis*, of a land mollusc becoming extinct in Britain since the Climatic Optimum. Perhaps this is not surprising, since land snails, being small, have a great capacity for surviving in places with favourable microclimates. Also, some degree of adaption may conceivably occur if the climatic deterioration is not too sudden. Of great interest in this connexion is the work of Sparks (1964a) on the non-marine Mollusca of the Last (Eemian) Interglacial. He shows that a number of thermophilous species survived in south-east England quite unexpectedly late into the latter part of the interglacial. For example, *Belgrandia marginata* (Michaud), *Planorbis vorticulus* Troschel, *Segmentina nitida* (Müller), *Azeca goodalli* (Férus-sac), *Helix nemoralis* L., *Discus rotundatus* (Müller) and *Corbicula fluminalis* (Müller) are all found in deposits of pollen zones h and i (*Pinus* forest).

I wish to discuss three species, all still living in Britain, whose behaviour may reflect the thermal decline since the Post-glacial Optimum: *Pomatias elegans* (Müller), *Lauria cylindracea* (da Costa) and *Ena montana* (Drapernaud).

Pomatias elegans is a "southern" species. The northern limit in Europe of its main area of distribution approximately follows the alignment of the January mean isotherm for about $+2°C$ (Fig. 2). This suggests an intolerance of prolonged cold, although the species can undoubtedly survive an occasional hard frost in hibernation (Kilian, 1951). Such a distribution is very similar to that of *Helix aspersa* Müller, which is frequently killed by severe winters in Britain. *Pomatias elegans* also lives in a number of isolated colonies further to the north-east, as far north as the Danish islands. Schlesch (1961) has shown that in these places, in Zealand and Fünen, the colonies occur on sheltered, south-west facing limestone slopes, from which cold air might drain at night and which probably possess much more favourable microclimates than the surrounding region. The pattern of such isolated localities across Europe (Fig. 2) becomes comprehensible if they are regarded as relicts of a wider distribution to the north-east during the Climatic Optimum, attained when winter temperatures were less severe than those of today. Post-glacial fossil records extend the range further still, to north Jutland (Schlesch, 1961) and to Czechoslovakia (Ložek, 1964).

The British distribution reinforces this hypothesis. On Fig. 3, I have plotted, using the 10 km square grid adopted by the Nature Conservancy Biological Recording Unit, all living and Post-glacial fossil records of *Pomatias elegans* available to me. Surface dead shells are shown by a separate symbol; although some may be quite modern, others have probably been washed out of Post-glacial deposits of considerable age. The species is strongly calcicole. It favours a certain amount of scrubby cover, but not dense woodland, and needs a loose soil in which to burrow. Given these requirements, it is fairly tolerant of human interference. It is abundant on most chalk areas in the south,

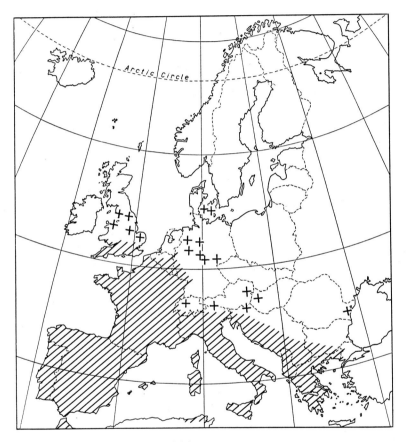

FIG. 2. *Pomatias elegans* (Müller). ///, approximate main area of distribution; +, relict colonies.

on Carboniferous limestone in south and north Wales, and on calcareous sand-dunes in Cornwall, but is represented only by widely separated colonies on the chalk and limestone areas of west Norfolk, Northamptonshire, Lincolnshire and Yorkshire; it is of interest that the localities at Burwell in Lincolnshire (NGR 53/38) and Oglethorpe in Yorkshire (NGR 44/44) were known already to Martin Lister (1678:

120) nearly three centuries ago. The most northerly occurrence, at
Forge Valley near Scarborough (NGR 44/98), is on a wooded limestone
slope facing south-west (A. Norris, personal communication). Judging

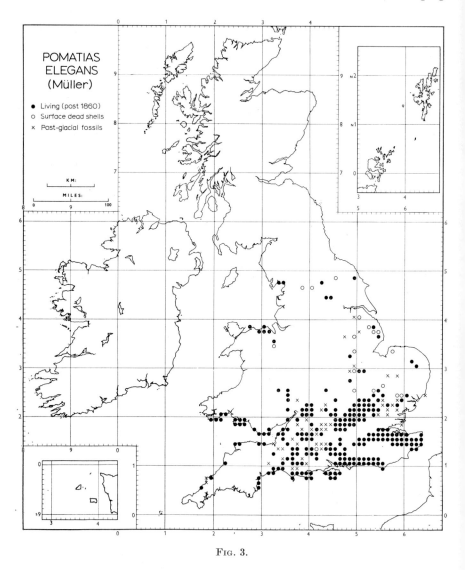

FIG. 3.

from the frequency of dead shells, these northern and eastern colonies
are slowly being eroded. Not many suitable deposits yielding Mollusca
have yet been examined in this region, but it is already clear that the

species formerly possessed a wider distribution in Norfolk and Lincoln-shire. Since *P. elegans* is an obligatory calcicole (Boycott, 1934), the question arises how it was able to reach certain isolated areas of limestone, such as the Carboniferous limestone of north Wales, where it was present already in the Atlantic period (McMillan, 1947; Anon., 1956), and the similar limestones around Warton and Grange-over-Sands in north Lancashire (Kendall, Dean and Rankin, 1909; Jackson, 1912). Presumably this was achieved by some method of passive dispersal, as by birds or large flying insects, and remarkably enough living examples of this species have been observed actually in transport on at least two occasions, nipped onto the legs of bees by the closure of the operculum (Tomlin, 1910; Stalley, 1911). Yet at the present day such means of dispersal must be very ineffective, even on a small scale. Boycott (1921) points out how in Hertfordshire the species has failed to colonize old chalk pits made by clearing away the overlying non-calcareous clays and sands. And Evans (1966) has recently shown how the species was entirely absent from a medieval deposit filling a chalk pit near Pitstone (Bucks), although abundant in earlier Post-glacial deposits in the vicinity, and still surviving locally in small numbers at the present day. These facts suggest that effective dispersal has become much more difficult. Furthermore, it is noticeable that the decline of *P. elegans* is most marked in East Anglia, Lincolnshire and the east Midlands, an area of rather continental climate much of which has minimum mean February temperatures of below $+1°C$ (Perring, 1961). *P. elegans* is recorded as going into hibernation in Essex by the end of September (French, 1890: 129), and at Meathop Fell, north Lancashire, remaining in hibernation between mid-October and the first week in May (Kendall *et al.*, 1909). This period seems unduly protracted, and it would be interesting to have comparable data from southern Europe. Kilian (1951) gives the period of hibernation in the Mainz-Heidelberg area, also near the limits of distribution but in a region of warmer summers than Britain, as November to March, the snails emerging when soil temperatures rise above about $+10–12°C$.

An interesting shrinkage in the range of *Pomatias elegans* is revealed in Fig. 3 by the many fossil records for Essex (Kennard and Woodward, 1897) and Cambridgeshire. In Essex, this may have been partly brought about by a progressive leaching during the Post-glacial of the surface of the chalky boulder clay, on which *P. elegans* locally lives in the absence of chalk or limestone (French, 1890). Alternatively, a climatic deterioration would have the effect of raising the calcium carbonate requirements for this species, a well-known phenomenon often observed among land snails at the northern edges of their geographical ranges. But these explanations can hardly account for the similar extinction of *P. elegans* in the adjacent areas of Cambridgeshire, where it is now also very rare in a living state. Perhaps atmospheric dryness may have been a contributory factor; for Essex and south Cambridgeshire show the lowest humidity values in the British Isles (Perring, 1961: Fig. 4),

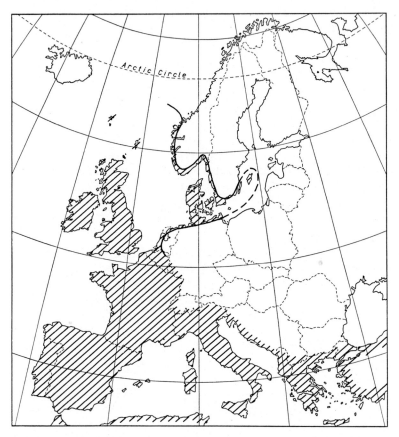

FIG. 4. *Lauria cylindracea* (da Costa) ⧄, Approximate area of distribution.

and it has been observed that *P. elegans* is only active when the relative humidity rises above 95% (Kilian, 1951). The apparent recession of *P. elegans* in parts of Wiltshire and the west country (Fig. 3) is, however, probably not analogous to the situation in Essex; partly this appearance is due to collection-failure, the region being more poorly studied, and partly to the very open nature of the Wiltshire chalk downs, which now provide scant cover for the species. Such complications warn us against "explaining" distributions in an over-simple manner, since all animal populations are subject to a great variety of limiting pressures, the relative importance of which will vary from place to place (Elton, 1966).

Lauria cylindracea has a European distribution which also suggests a susceptibility to winter cold (Fig. 4). This distribution is strikingly similar to that of holly (*Ilex aquifolium*) (Godwin, 1956: Fig. 40). *Lauria cylindracea* is recorded living from every vice-county in the British Isles (Ellis, 1951), but this is misleading: although an exceedingly abundant species in Ireland and in the more oceanic western parts of Britain, it becomes distinctly sporadic in eastern England. This is unlikely to be the result of human interference, as it stands disturbance well, and may be found in gardens (Boycott, 1934: 22). Yet in deposits ascribed to the Atlantic period the species is frequently abundant in areas where it is now rare or absent, for example at Brook (Fig. 1). Again, I suggest that this may have been brought about by a decline in winter temperatures since the Thermal Optimum of the Post-glacial.

Geological evidence about the precise history of these two species in the British Post-glacial is as yet rather scanty. *Lauria cylindracea* occurs rarely in zone VI (Late Boreal) at West Hartlepool (Kennard *in* Trechmann, 1936; Barker and Mackey, 1961: 41). *Pomatias elegans* first appears in a securely dated context in the tufa at Blashenwell, Dorset, in zone VIIa (Atlantic) (Kennard *in* Bury, 1950; Barker and Mackey, 1961: 40). Both species are, however, common in deposits which are ascribed on fairly good general grounds to the Atlantic period, particularly in calcareous tufas whose formation is held to require a moister and warmer climate than now exists (e.g. Broughton, Lincolnshire (Kennard and Musham, 1937), Takeley, Essex (Kennard *in* Warren, 1945), Caerwys and Prestatyn, Flintshire (McMillan, 1947; Anon., 1956), Wateringbury, Kent (Kennard *in* Brown, 1939; Kerney, 1956) and Brook, Kent (Kerney *et al.*, 1964)).

The third species I wish to discuss, *Ena montana*, has a very different European distribution (Fig. 5). It extends far to the north in the Baltic States and European Russia, and can clearly withstand very cold winters. It is evidently waning in Britain, where it lives in scattered colonies in old woodland in the south (Boycott, 1939); the Irish record for Co. Cork (Phillips, 1914; Stelfox, 1929) is too doubtful to be accepted. Its northerly range was formerly greater, being found fossil in Northamptonshire (Shaw, 1908) and west Norfolk (Kerney, unpublished). *Ena montana* is usually classified as an anthropophobe,

"quite intolerant of cultivation or disturbance" (Boycott, 1939: 159).
But shells of this species are surprisingly common in Neolithic and
Bronze Age (Sub-boreal) layers in archaeological sites in the Wiltshire
area, in contexts where much human interference was clearly already
present (Kennard *in* Stone, 1933; Kennard *in* Drew and Piggott, 1936;

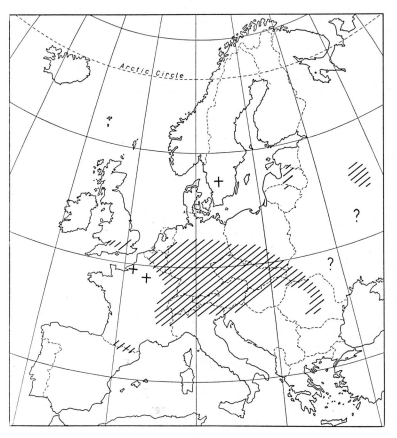

Fig. 5. *Ena montana* (Draparnaud).///, Approximate main area of distribution;
+, relict colonies.

Kennard *in* Clifford, 1938; Kennard *in* Stone and Hill, 1940; Kerney
in Rhatz, 1963; Evans, 1967). At Waylands Smithy, near Uffington,
Berkshire, *E. montana* succeeded in colonizing a ditch between two
phases in the construction of a Neolithic long barrow, after intense disturbance of a large area of ground (M. P. Kerney, unpublished). Within
its main area of distribution, as for example in the Jura of eastern

France, *E. montana* is certainly not anthropophobe, but is a common hedgerow species. Therefore its present dislike of disturbance and rather critical ecological requirements in Britain may be an expression of a climate adverse to its well-being, possibly a lack of summer warmth. It is noteworthy that the surviving British sites are in the part of the country with the highest summer temperatures, nearly all lying within the July isotherm for about $+16\cdot5°C$. Unlike *Pomatias elegans* and *Lauria cylindracea*, nowhere does it occur near the sea. It is nearly always confined to old, undisturbed woodland, and lives in small numbers, but a striking exception is provided by a site at Buntingford, Hertfordshire (NGR 52/33), where it is exceedingly abundant on an artificial bank by the A10 road, among rubbish and nettles and in association with anthropophiles such as *Milax budapestensis* (Hazay), *M. sowerbyi* (Férussac) and *Monacha cantiana* (Montagu). Yet a little over a mile away *Ena montana* lives in its usual low numbers in an ancient mixed oak wood. Clearly some compensatory factor is operating, and it would be interesting if an investigation of the microclimate of the roadbank site were to reveal especially favourable thermal conditions, perhaps caused by summer radiation from the adjacent road surface (Lamb, 1964: 90).

Ena montana had reached south-east England already by the Boreal period, being found in pollen-dated deposits of zones V and VIa at Nazeing, Essex (Allison, Godwin and Warren, 1952: 190, 195–196, 205). It remained frequent both in the Atlantic and the Sub-boreal, dated deposits from both of these periods having provided a number of fossil records outside the present range of the species. If, as I have suggested, *E. montana* was indeed more common in England during these periods, and better able to withstand human pressure than is now the case, this would accord with Iversen's and Godwin's conclusions as to the nature of the Climatic Optimum, with rather warmer summers from at least the Late Boreal until the Sub-atlantic deterioration, about 600 B.C.

Finally, it is interesting to consider the occurrence of *Pomatias elegans*, *Lauria cylindracea* and *Ena montana* in the Last Interglacial in south-east England (Sparks, 1964a). Palaeobotanical evidence, such as the presence of abundant *Carpinus* (hornbeam) pollen a little after the middle of the interglacial (zone g), suggests a climate of continental rather than oceanic type, with warm summers and probably rather cold winters. As Sparks points out (1964a,b), the land Mollusca lend some support to this view. For example, species of continental range occur which failed to reach Britain during the Post-glacial (*Clausilia pumila* C. Pfeiffer, *Fruticicola fruticum* (Müller)). Again, the continental

Discus ruderatus was fairly frequent in the warmest part of the inter-glacial (zones f and g), whereas the oceanic *Discus rotundatus*, abundant in the Post-glacial, was relatively scarce. When we come to the three species whose Post-glacial history we have already discussed, we find that *Ena montana* was common, whereas *Pomatias elegans* is unknown with certainty, and *Lauria cylindracea* "was astonishingly rare in view of its present wide distribution" (Sparks, 1964a: 18).

This paper scarcely consists of more than suggestions and indications as to the past behaviour of land Mollusca. It is clear that a great deal of further research is needed before firmer conclusions can be drawn. Three lines of work suggest themselves. First, precise information on modern geographical range is required. Even within the British Isles, this knowledge is not always available. A vice-comital census exists (Ellis, 1951), but this may be highly misleading with regard to the frequency of certain species in different parts of the country; an example already given is that of *Lauria cylindracea*, recorded as occurring in every vice-county of the British Isles, but which in fact is distinctly local in the continental east as against the oceanic west. A fresh survey of Britain's non-marine Mollusca is now being made, based on distri-bution in the 10 km squares of the National Grid, on the method used by the Botanical Society and the Nature Conservancy and demonstrated for *Pomatias elegans* in Fig. 3 (Perring and Walters, 1962; Kerney, 1967), but it will be some years before the whole fauna can be adequately dealt with in this way. In other European countries similar studies are even more badly needed, and it would be most valuable if through international co-operation an atlas of the European Mollusca could eventually be produced, comparable to Hultén's magnificent atlas for the flowering plants of northern Europe (1950).

Secondly, direct experimental and observational work is required on the present-day thermal tolerances of land snails, and the way in which life histories may be affected by climate. Such work should not be difficult to organize, both in the field and in the laboratory. In particular, studies should be made of the precise microclimatic condi-tions in isolated stations of species beyond their main geographical ranges, as in the case of the Yorkshire colonies of *Pomatias elegans*.

Finally, a great deal of further precise palaeontological work is necessary. Many more deposits yielding non-marine Mollusca must be studied, using modern stratigraphical methods, and a corpus of information thus built up on the ranges in time and space of the different species. Only when this is done can conclusions about the origins of the fauna of today be raised above the level of intelligent conjecture.

BRITAIN'S FAUNA OF LAND MOLLUSCA 289

REFERENCES

Allison, J., Godwin, H. and Warren, S. H. (1952). Late-glacial deposits at Nazeing in the Lea Valley, North London. *Phil. Trans. R. Soc.* (B) **236**: 169–240.

Anon. (1956). The Caerwys tufa. *Lpool Manchr geol. J.* **1**: xxiv–xxviii.

Ant, H. (1963). Faunistische, ökologische und tiergeographische Untersuchungen zur Verbreitung der Landschnecken in Nordwestdeutschland. *Abh. Landesmus. Naturk. Münster* **25**: 1–125.

Barker, H. and Mackey, J. (1961). British Museum natural radiocarbon measurements III. *Radiocarbon* **3**: 39–45.

Boycott, A. E. (1921). The land Mollusca of the parish of Aldenham. *Trans. Herts. nat. Hist. Soc.* **17**: 220–245.

Boycott, A. E. (1934). The habitats of land Mollusca in Britain. *J. Ecol.* **22**: 1–38.

Boycott, A. E. (1939). Distribution and habitats of *Ena montana* in England. *J. Conch., Lond.* **21**: 153–159.

Brown, E. E. S. (1939). A tufaceous deposit near Wateringbury, Kent. *Proc. Geol. Ass.* **50**: 77–82.

Bury, H. (1950). Blashenwell tufa. *Proc. Bournemouth nat. Sci. Soc.* **39**: 48–51.

Clifford, E. M. (1938). The excavation of Nympsfield long barrow, Gloucestershire. *Proc. prehist. Soc.* **4**: 188–213.

Cunnington, M. E. (1933). Evidence of climate derived from snail shells and its bearing on the date of Stonehenge. *Wilts. archaeol. nat. Hist. Mag.* **46**: 350–355.

Curwen, E. C. (1934). Excavations in Whitehawk Neolithic camp, Brighton, 1932–33. *Antiq. J.* **14**: 99–133.

Degerbøl, M. and Krog, H. (1951). Den europaeiske Sumpskildpadde (*Emys orbicularis* L.) i Danmark. *Danm. geol. Unders.* (2) No. 78: 1–130.

Drew, C. D. and Piggott, D. (1936). The excavation of long barrow 163a on Thickthorn Down, Dorset. *Proc. prehist. Soc.* **2**: 77–96.

Ellis, A. E. (1951). Census of the distribution of British non-marine Mollusca. *J. Conch., Lond.* **23**: 171–244.

Elton, C. S. (1966). *The pattern of animal communities.* London: Methuen.

Evans, J. G. (1966). Late-glacial and Post-glacial deposits at Pitstone, Buckinghamshire. *Proc. Geol. Ass.* **77**: 347–364.

Evans, J. G. (1967). *The stratification of Mollusca in chalk soils and their relevance to archaeology.* Unpublished Ph.D. thesis, University of London.

French, J. (1890). On the occurrence of *Cyclostoma elegans* in a living state at Felstead. *Essex Nat.* **4**: 92–94; 129.

Frenzel, B. (1966). Climatic change in the Atlantic/sub-Boreal transition on the Northern Hemisphere: botanical evidence. In *World Climate from 8000 to 0 B.C.*: 99–123. London: Royal Meteorological Society.

Godwin, H. (1954). Recurrence surfaces. *Danm. geol. Unders.* (2) No. 80: 22–30.

Godwin, H. (1956). *The history of British flora.* Cambridge: University Press.

Godwin, H. (1960). Prehistoric wooden trackways of the Somerset Levels: their construction, age and relation to climatic change. *Proc. prehist. Soc.* **26**: 1–36.

Godwin, H. (1962). Vegetational history of the Kentish Chalk Downs as seen at Wingham and Frogholt. *Veröff. geobot. Inst., Zürich (Festschrift Franz Firbas.)* **37**: 83–99.

Hultén, E. (1950). *Atlas över Kärlväxterna i Norden.* Stockholm.

Iversen, J. (1944). *Viscum, Hedera* and *Ilex* as climatic indicators. *Geol. För. Stockh. Förh.* **66**: 463.

Iversen, J. (1960). Problems of the early Post-glacial forest development in Denmark. *Danm. geol. Unders.* (4) No. 3: 1–32.

Jackson, J. W. (1912). On the former range of *Pomatias elegans* in the Warton district. *Lancs. Nat.* **5**: 170–171.

Kendall, C. E. Y., Dean, J. D. and Rankin, W. M. (1909). On the geographical distribution of Mollusca in S. Lonsdale. *Naturalist, Hull* **1909** (635): 435–437.

Kennard, A. S. and Musham, J. F. (1937). On the Mollusca from a Holocene tufaceous deposit at Broughton-Brigg, Lincolnshire. *Proc. malac. Soc. Lond.* **22**: 374–379.

Kennard, A. S. and Woodward, B. B. (1897). The Post-Pliocene non-marine Mollusca of Essex. *Essex Nat.* **10**: 87–109.

Kerney, M. P. (1956). Note on the fauna of an early Holocene tufa at Wateringbury, Kent. *Proc. Geol. Ass.* **66**: 293–296.

Kerney, M. P. (1963). Late-glacial deposits on the chalk of south-east England. *Phil. Trans. R. Soc.* (B) **246**: 203–254.

Kerney, M. P. (1966). Snails and man in Britain. *J. Conch., Lond.* **26**: 3–14.

Kerney, M. P. (1967). Distribution mapping of land and freshwater Mollusca in the British Isles. *J. Conch., Lond.* **26**: 152–160.

Kerney, M. P., Brown, E. H. and Chandler, T. J. (1964). The Late-glacial and Post-glacial history of the Chalk escarpment near Brook, Kent. *Phil. Trans. R. Soc.* (B) **248**: 135–204.

Kilian, E. F. (1951). Untersuchungen zur Biologie von *Pomatias elegans* (Müller) und ihrer 'Kronkrementdrüse'. *Arch. Molluskenk.* **80**: 1–16.

Lamb, H. H. (1964). *The English climate.* London: English Universities Press.

Lister, M. (1678). *Historiae Animalium Angliae tres tractatus.* London.

Ložek, V. (1964). Quartärmollusken der Tschechoslowakei. *Rozpr. ústřed. Úst. geol.* **31**: 1–374.

McMillan, N. F. (1947). The molluscan faunas of some tufas in Cheshire and Flintshire. *Proc. Lpool geol. Soc.* **19**: 240–248.

Newton, A. (1862). On the discovery of ancient remains of *Emys lutaria* in Norfolk. *Ann. Mag. nat. Hist.* (3) **10**: 224–228.

Perring, F. H. (1961). Mapping the distribution of flowering plants. *New Scient.* **7**: 1522–1525.

Perring, F. H. and Walters, S. M. (1962). *Atlas of the British flora.* London: Nelson.

Phillips, R. A. (1914). *Helicigona lapicida* in Ireland. *Ir. Nat.* **23**: 37–38.

Rhatz, P. A. (1963). Farncombe Down barrow, Berkshire. *Berksh. archaeol. J.* **60**: 1–24.

Schlesch, H. (1961). Zwei neue rezente Vorkommen von *Pomatias elegans* (O. F. Müller) in Südseeland und die nördliche Verbreitung dieser Art sowie Bemerkungen über die Verbreitung verschiedener Landschnecken. *Arch. Molluskenk.* **90**: 215–226.

Shaw, W. A. (1908). *Ena montana* in Northamptonshire. *J. Conch., Lond.* **12**: 106.

Sparks, B. W. (1964a). The distribution of non-marine Mollusca in the Last Interglacial in south-east England. *Proc. malac. Soc. Lond.* **36**: 7–25.

Sparks, B. W. (1964b). A note on the Pleistocene deposit at Grantchester, Cambridgeshire. *Geol. Mag.* **101**: 334–339.

Stalley, H. J. (1911). The dispersal of shells by insects. *J. Conch., Lond.* **13**: 163.

Stelfox, A. W. (1929). Report on recent additions to the Irish fauna and flora. Land and freshwater Mollusca. *Proc. R. Ir. Acad.* (B) **39**: 6–10.

Stone, J. F. S. (1933). Excavations at Easton Down, Winterslow, 1931–1932. *Wilts. archaeol. nat. Hist. Mag.* **46**: 225–242.

Stone, J. F. S. and Hill, N. G. (1940). A round barrow on Stockbridge Down, Hampshire. *Antiq. J.* **20**: 39–51.

Tomlin, J. R. le B. (1910). The dispersal of shells by insects. *J. Conch., Lond.* **13**: 108.

Trechmann, C. T. (1936). Mesolithic flints from the submerged forest at West Hartlepool. *Proc. prehist. Soc.* **2**: 161–168.

Warren, S. H. (1945). Some geological and prehistoric records on the north-west border of Essex *Essex Nat.* **27**: 273–280.

Symp. zool. Soc. Lond. (1968) No. 22, 293–317.

CHANGES IN THE COMPOSITION OF LAND MOLLUSCAN POPULATIONS IN NORTH WILTSHIRE DURING THE LAST 5000 YEARS

J. G. EVANS

Institute of Archaeology, London, England

SYNOPSIS

Faunas of land Mollusca in archaeological deposits from north Wiltshire have been used to deduce something of the overall faunal changes in the area during the past 5000 years. The effects of climate in bringing about such changes are treated only in the most general of terms; major faunal changes are attributed solely to the activities of man. Two Late-glacial faunas and one of Climatic Optimum age are discussed to give some indication of the factors which govern the composition of faunas uninfluenced by man. In general, the variety of ecological niches occupied by the species of snail in a fauna is responsible for the relative abundance of each species. With increasing severity of the environment, however, restriction of the fauna becomes apparent, and instead of a fairly uniform distribution of numbers among the various species present, a certain few species occur in enormous numbers, while the majority are either poorly represented or entirely absent. In addition, while apparently negligible in faunas existing under equable conditions, in a severe environment, competition appears to play some part in determining the composition of molluscan populations.

The pattern of faunal change since the beginnings of agriculture in the Neolithic period has been towards a restriction of the fauna in favour of the grassland and xerophilous species. However, it is shown that the creation of a variety of habitats and the introduction of a number of species during the Roman and medieval periods has resulted in a fauna, which in its overall composition, is as diverse as that which occurred during the Climatic Optimum.

The succession of climatic changes, especially with respect to temperature, which has taken place through the Late- and Post-glacial periods has had its effect on the land molluscan populations of the British Isles. This has recently been well demonstrated in detailed studies of the faunal changes during these periods in south-east England (Kerney, 1963; Kerney, Brown and Chandler, 1964). From the beginning of the Neolithic period onwards, the activities of man come to have an increasingly important influence on land Mollusca, enhancing or overriding the effects of climate by such operations as continuous agriculture or the creation of microhabitats such as ditches and hedgerows, and it becomes correspondingly more difficult to disentangle the two influences, especially when, as at the zone VIIa/VIIb transition and at the climatic deterioration of *ca* 550 B.C., they seem to produce similar results. In this paper, the sites to be considered are, apart from that at

Box, all less than 4 miles distant from the village of Avebury in north
Wiltshire in an area which was one of the centres of human activity in
the later prehistoric period. In what follows, therefore, the effects of
climate are treated only in the most general of terms; major changes
are attributed solely to the activities of man.

PRE-NEOLITHIC FAUNAS

As a preliminary to the main discussion, two Late-glacial faunas
and a Climatic Optimum fauna will be considered, to give some indication of the factors which govern the composition of faunas uninfluenced
by the presence of man.

From two sites, Waden Hill and Bishops Cannings, Late-glacial
faunas have been recognized (Fig. 1). These are composed of a restricted
number of species, only one or two of which attain any degree of
abundance. This is more evident at Bishops Cannings where the dominance of *Vallonia pulchella* (Müller) (64%) is virtually absolute; at
Waden Hill both *Pupilla muscorum* (Linné) and *V. pulchella* are well
represented. The lithology of the deposits from which these faunas
derive—chalk muds at Bishops Cannings and well-bedded chalk and
flint gravels at Waden Hill—suggests an environmental background of
open-country with a bare surface colonized here and there by vegetation
but never stable for long enough to allow a full cover to develop.
Kerney (1963) discusses the formation of such deposits. The climate at
this time was colder than at present.

The faunas represent a locally damp environment as shown by the
abundance of the relatively hygrophilous species *Vallonia pulchella*
and the paucity of the more rupestral species *Vallonia costata* (Müller).
At Waden Hill this is supported by the presence of true marsh species
Lymnaea truncatula (Müller), cf. *Succinea oblonga* Draparnaud and
Pisidium spp. (The one or two juveniles of cf. *S. oblonga* could equally
well be ascribed to the closely similar species *Catinella arenaria* (Bou-
chard-Chantereaux).) In addition, it should be remembered that
Pupilla muscorum could apparently tolerate a damper environment
during the Late-glacial period than that in which it is found today
(Kerney *et al.*, 1964).

At Bishops Cannings the fine nature of the deposits and the
severely restricted nature of the fauna would seem to indicate a zone I
date; only one species, *Hygromia hispida* (Linné), at all diagnostic of
zones II and III, being present. At Waden Hill, however, the coarse and
well-bedded nature of the deposits and the presence of both *H. hispida*
and *Helicella itala* (Linné) may suggest a zone III date.

The restricted number of species in these faunas and their dominance by only one or two of these, others occurring in extremely low abundance, appears to be a reaction to the rather severe environment. It is a feature which will be met with later in connexion with certain Post-glacial faunas to be described.

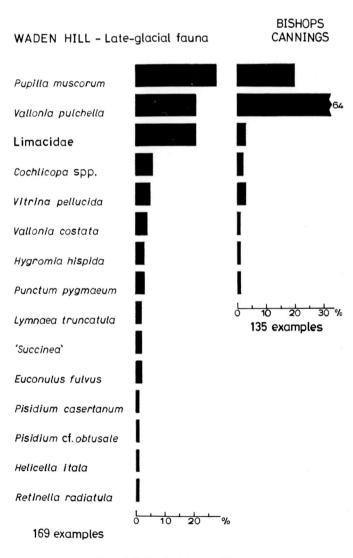

FIG. 1. Late glacial assemblages.

With the amelioration of the climate marking the beginning of the Post-glacial period and the development of a forest cover culminating in the climatic climax forest of the Atlantic period (Godwin, 1956), the molluscan fauna became greatly enriched and by about 5000 B.C. the native fauna was probably complete (Kerney, 1965). Kennard (Bury and Kennard, 1940) lists a fauna from marls and tufas at Box in north Wiltshire, belonging to the Atlantic period. As with many of these tufaceous faunas a rather curious environment is suggested for which it would be hard to find modern parallels. The paucity or absence of the true grassland and xerophile species *Vertigo pygmaea* (Draparnaud), *Pupilla muscorum*, *Vallonia excentrica* Sterki and *Helicella itala* and the dominance of shade-lovers notably the Zonitidae, *Discus rotundatus* (Müller) and *Carychium tridentatum* (Risso), would suggest an environment of closed woodland, with the presence of marsh species such as *Carychium minimum* Müller, *Lymnaea truncatula*, *Zonitoides nitidus* (Müller) and *Pisidium casertanum* (Poli) indicating numerous swampy pools. The fauna is a rich one of which three species, namely *Lauria anglica* (Wood), *Vertigo pusilla* Müller and *Acanthinula lamellata* (Jeffreys) are not known living in Wiltshire today.

It is a curious feature of this and certain other faunas from such very amenable habitats that though most species are more or less equally represented, *Carychium tridentatum* (50%) and, to a lesser extent, *Discus rotundatus* (22%) are extremely abundant, while a number of species such as *Columella edentula* (Draparnaud) and *Euconulus fulvus* (Müller) are sparse even though they may be well within their limits of climatic tolerance. The same phenomenon occurs in the case of Climatic Optimum faunas from Tile Kiln Green in Essex (Kennard, *in* Warren, 1945) and from Brook in Kent (Kerney *et al.*, 1964). In view of the rather equable conditions presented by these environments it is difficult to know how to explain this situation, and with the apparent absence of competition among land Mollusca (Boycott, 1934) one must look to some sort of internal control on the rate of reproduction and the numbers of eggs laid rather than to any form of environmental control. Such physiological control would presumably be tied into the niche occupied by the species in question. Elton (1927) points out that "Each species has certain hereditary powers of increase, which are more or less fixed in amount for any particular conditions." The conditions under consideration in this case are optimum as far as most species of Mollusca are concerned, and it follows therefore from the differences in abundance of certain species under such conditions that a number of different niches are involved. Just what these are is far from clear.

In some instances one can point to specific environmental factors which may act in a limiting capacity as with the relatively rupestral species *Acanthinula aculeata* (Müller) and *Columella edentula* whose abundance may depend on the availability of a firm surface as provided by fallen branches, and this applies especially to the Clausiliidae. However, this in no way accounts for the paucity of other species such as *Retinella pura* (Alder), *Punctum pygmaeum* (Draparnaud) and *Euconulus fulvus* which find their best conditions for life in leaf litter in apparently identical environments to *Carychium* and *Discus*.

Nor is the question of size differences between the various species involved to any extent, although this is probably responsible for the low numbers of the larger helicids such as *Hygromia striolata* (C. Pfeiffer) and *Helix hortensis* Müller. *Carychium tridentatum* and *Discus rotundatus* occur in an abundance out of all proportion to their size.

It should be mentioned too that the absence of grassland and xerophile species may reflect some form of physiological control, for it is difficult to understand otherwise, again in view of the apparent absence of competition, why such species should not occur in a shaded and moist environment.

It appears, therefore, that two groups of factors, the one environmental and the other physiological, govern the composition of molluscan populations, usually acting in conjunction. The environment acts in a restricting capacity as in the case of the Late-glacial faunas in which only two species seem completely adapted to the environment in question and are able to attain the degree of abundance which their hereditary powers of increase permit. In the case of the Box fauna the environment is an optimum one and only restricting with respect to certain rupestral, xerophilous and grassland species. The degree of abundance attained by most species is therefore their hereditary maximum and physiologically controlled. With departure from the optimum, that is with increasing disturbance, drying-out and loss of suitable substratum, a number of the more conservative species will be restricted and eventually excluded. Others will become reduced but may yet maintain themselves. Certain others however, notably *Vallonia costata*, may be able to increase in abundance and even others to migrate into the area from elsewhere. The important point is that each species of snail appears to have an environmental range over which its physiology, behaviour patterns and reproductive capacity enable it to operate, and beyond the limits of which it cannot occur. Thus those species which live in the most severe environments are not necessarily to be found in all kinds of habitat and may in fact be closely restricted to one particular environment. In that environment the hereditary powers of increase will permit of a certain degree of abundance to be attained.

In addition, there is evidence that with increasing severity of the environment a third factor, namely competition for living space, food and shelter, does in fact come to play an important part in determining the composition of molluscan populations.

THE NATURAL FAUNA OF CHALK DOWNLAND
PRIOR TO INTERFERENCE BY MAN

Faunas from downland sites fairly representative of a shaded environment have been recognized from the Neolithic camps on Knap Hill (soil beneath the bank; Fig. 2; data from Sparks, *in* Connah, 1965) and Windmill Hill (soil beneath the bank; Fig. 2; Evans, 1966a) and from the Early Iron Age lynchet on Fyfield Down (Fig. 6, Fyfield Down I, roothole fill and ploughed pre-lynchet soil). No one species dominates these faunas, although at Windmill Hill, *Hygromia hispida* is certainly abundant; *Carychium tridentatum* and *Discus rotundatus* are well represented but not to the extent as in the fauna from Box. In fact, as Sparks points out in the case of Knap Hill, these faunas show "the most even distribution of individuals among the various species present."

The presence of open-country species *Pupilla muscorum*, *Vertigo pygmaea*, *Vallonia excentrica* and *Helicella itala* may be ascribed to contamination from above or perhaps to the presence of some open ground in the environment; it is difficult to be certain about this. *Vallonia costata* however, a species generally considered to be exclusive of open habitats (e.g. Boycott, 1934; Ellis, 1926), is almost certainly an integral part of these faunas. At Knap Hill and Windmill Hill, at any rate, it occurs in much greater abundance than the other open-country species and there is evidence from sites in the Chilterns (Evans, 1966b) that this species can occur in closed woodland and likewise at Box, Brook and Tile Kiln Green. Ant (1963) puts forward evidence from modern ecological studies showing *V. costata* to be tolerant of rather more shaded conditions than is *Vallonia excentrica*, while Waldén (1955) quotes *V. costata* as not rare in ". . . deciduous woods and groves, in shady as well as open places, but does not like dry localities as much as *V. excentrica*."

The presence of *Arianta arbustorum* (Linné) at Knap Hill, Windmill Hill and Fyfield Down and of *Helix hortensis* and *Helix nemoralis* Linné at the latter two sites should be noted. It is a feature of prehistoric sites that these species occur together especially in the Neolithic period (e.g. Kennard, *in* Curwen, 1936). At the present day it is un-common to find *H. hortensis* and *H. nemoralis* in the same place; the latter seems to be confined to areas of high downland, while the former occurs only in more low-lying places, often in association with *Arianta*.

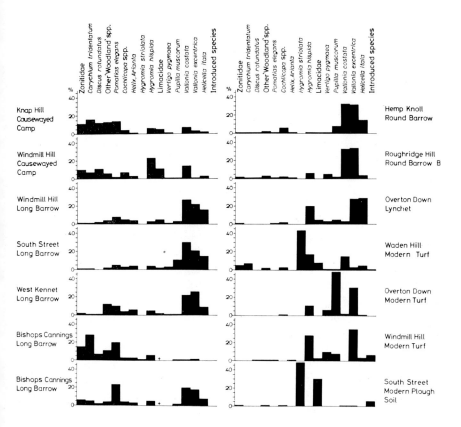

Fig. 2. Molluscan assemblages from ancient and modern soils.

Cain and Currey (1963) suggest this to be due to a temperature effect correlated with the more northerly distribution of *H. hortensis*; cold air drains down on still, clear nights from the higher land and forms cold pools in the valley bottoms. They note that when *H. hortensis* occurs at higher altitudes it does so in strict association with trees, suggesting that there is competition between *H. hortensis* and *H. nemoralis* which is somewhat relaxed in or near trees and woods.

The occurrence of the three species on the downs therefore, during the Neolithic period, may be attributed perhaps to the higher summer temperatures at that time, perhaps to the more shaded environment, or to a combination of these factors.

As regards the environment which these faunas represent, two points need mentioning. Comparison with modern faunas suggests a woodland environment, more or less closed. Taking into account, however, the more amenable climatic conditions at the end of the Atlantic period and the absence of any serious disturbance by man, it is possible that a fauna with an apparently shaded facies might yet be capable of living in an open-country environment. But if this were the case, in the absence of competition under such circumstances, one would expect a higher proportion of open-country species, so it is felt that this idea is untenable.

A more serious objection to these faunas reflecting a woodland environment is the possibility of their being synanthropic assemblages reflecting just the opposite, namely abandoned occupation. Synanthropy and anthropophoby, although from man's point of view extremes, may be manifestations of a single phenomenon, certain species of Mollusca being incapable of living in an average environment and in the former case remedying this by seeking-out man-made habitats such as rubbish dumps, gardens and waste ground, and in the latter by being confined to primeval woodland and other wild places. Thus species which behave in Britain as anthropophobes such as *Helicodonta obvoluta* (Müller) and *Ena montana* (Draparnaud) are, in France, quite common in man-made habitats as hedgerows and copses. In Britain too the presence of *Acicula fusca* (Montagu) in a rubbish dump below Box Hill and of *Ena montana* in a roadside ditch close to a building site at Buntingford, Herts. may be quoted. *Vallonia costata* too is often an important element in rubbish dump faunas, occurring in association with such species as *Oxychilus draparnaldi* (Beck) and *Hygromia striolata*; Boycott (1934) notes its presence in gardens.

To resolve this question further, the archaeological context of the faunas must be considered. At Knap Hill, no occupation of the area prior to the erection of the camp is believed to have occurred (Connah, 1965) and it is reasonable to assume that here the fauna is indeed a reflection of a woodland environment. At Windmill Hill, however, there is considerable evidence of pre-camp occupation (Smith, 1965a) and the buried soil seems to be more of an occupation horizon than a soil *sensu stricto*. So too at Fyfield Down there is evidence of Neolithic occupation (Bowen and Fowler, 1962), though here the time span between this period and the Iron Age cultivation is great enough to have allowed a full forest cover to regenerate.

Direct evidence from plant remains (charcoal and pollen) for the status of the vegetation cover of the downland at the beginning of the Neolithic period is scanty and inconclusive (e.g. Godwin and Tansley, 1941). It must be remembered that from Neolithic times onwards the downs have been under cultivation and that the thin, dry rendzina soil which now mantles much of the area and which must surely be incapable of supporting a woodland cover, may be a direct result of this. The original soil may have been more moisture retentive, rather thicker and with a more or less decalcified A-horizon, approaching a brown-earth and quite able to support yew and ash or hazel and mixed oak woodland in even the most exposed situations. Charcoals from Neolithic levels at Avebury (Maby, *in* Gray, 1935), Knap Hill (Dimbleby, *in* Connah, 1965), Windmill Hill (Dimbleby, *in* Smith, 1965a) and the Windmill Hill Long Barrow (G. W. Dimbleby, personal communication) support this contention, demonstrating the presence of ash, yew, oak and birch with shrub species blackthorn, buckthorn, hazel, hawthorn and elder often in abundance. Pollen spectra from Neolithic soils, Windmill Hill Camp (Dimbleby, *in* Smith, 1965a), Windmill Hill Long Barrow (Dimbleby, personal communication) and South Street Long Barrow, although demonstrating a vegetational background of open-country and cultivation, show some influence of forest species such as oak, lime, elm and elder. In addition, the dominance of these assemblages by the calciphobe, bracken, the small numbers of fern species *Polypodium* and *Drypoteris* and the presence of birch and pine favours the idea that the soil was less calcareous in its natural state than during the period of cultivation.

A Neolithic soil from a dry valley (Fig. 2; Bishops Cannings) provides an interesting contrast to those from the downland sites. The soil was preserved beneath a Neolithic Long barrow and showed a profile compatible with an environment of closed woodland. This is reflected by the fauna which at one point is virtually devoid of open-country species (Fig. 2; Bishops Cannings Long Barrow, upper graph). *Carychium tridentatum* occurs in abundance reflecting the low-lying situation and cover of clayey drift which has given rise to a soil more moisture retentive than that on the downs. However, in places an occupation horizon rather than a soil *sensu stricto* is present and the fauna at one such point is of open-country type indicating perhaps a clearing in an otherwise shaded environment. *Pomatias elegans* (Müller), a species favoured by the broken surface which forest clearance engenders, is abundant in both assemblages; it is usually much less well represented in true woodland environments.

THE NEOLITHIC PERIOD

Open-country faunas

Evidence of human occupation in the area prior to the Neolithic period is restricted to two possible Mesolithic occupation sites, one on Windmill Hill and the other on Hackpen Hill. It is uncertain how much forest clearance can be attributed to the Mesolithic people but the paucity of sites suggests them to have had little effect.

The onset of the Neolithic period is associated, not with an evolution from the indigenous Mesolithic population, but with the invasion of Neolithic farmers from the continent, and changes in the landscape may, therefore, have been marked and rapid and faunal changes correspondingly so.

Open-country faunas representative of a completely cleared environment have been identified from Neolithic soils beneath the Windmill Hill, West Kennet and South Street Long Barrows (Fig. 2). The prominent feature of these faunas, as with the open-country fauna from Bishops Cannings, is their dominance by *Vallonia costata* and *V. excentrica* each in more or less equal abundance. *Helicella itala* is reasonably abundant but *Pupilla muscorum* and *Vertigo pygmaea* are sparse and the latter never seems to constitute an important element of the fauna at any time. *Abida secale* (Draparnaud) is represented by two fragments at Windmill Hill but is otherwise absent. The few examples of the more shade-loving species are referable in part to the contemporary fauna, certain species such as *Punctum pygmaeum*, *Discus rotundatus*, *Retinella radiatula* (Alder) and *Vitrina pellucida* (Müller) being able to exist in low numbers in the most open situations, and in part to the rather resistant apices of certain species belonging to more ancient faunas representative of a shaded environment. Most notable in this latter category are *Pomatias elegans* and the Clausiliidae; at West Kennet for instance the abundance of Other "Woodland" Species is due almost entirely to worn apices of *Clausilia bidentata* (Ström).

As with the woodland faunas, so too with the grassland faunas, each species seems to maintain a certain density of numbers in the population. But while in the former instance, regulation was maintained by the reproductive capacity of the animal and in certain cases by the environment, here, in the exposed and dry downland conditions, competition must surely play a part, so that only the hardiest species can occur in any great abundance. In the Neolithic period these seem to have been the two species of *Vallonia*, to a lesser extent *Helicella itala*, and with *Pupilla muscorum* a poor fourth.

LAND MOLLUSCA IN WILTSHIRE 303

Similar faunas have been identified from the soil beneath the chambered tomb of Wayland's Smithy in Berkshire (Kerney, personal communication) and from the Thickthorn Down Long Barrow in Dorset (Kennard, *in* Drew and Piggott, 1936) so that it is likely that such a fauna was widespread during the Neolithic period in open landscapes.

Refuges for shade-loving species

During the later part of the Neolithic period regeneration and shading-over took place in a number of artificially created habitats such as ditches and pits, and the molluscan faunas of these became enriched. At Knap Hill (Sparks, *in* Connah, 1965) the fauna in the lower levels of the ditch is dominated by *Carychium tridentatum* (43%) and *Discus rotundatus* (23%) and approaches the Climatic Optimum fauna from Box, a reaction to an exceptionally favourable environment of dense scrub or woodland with a well-developed litter layer. The presence of *Clausilia rolphi* Turton, *Lauria cylindracea* (da Costa) and *Acicula fusca*, species absent from the buried soil beneath the bank, show that refuges for these were present. At South Street (Fig. 3), the reversion to woodland was not so complete but here too, species not present in the pre-barrow soil appear namely *Vertigo pusilla*, *Clausilia rolphi*, *Ena montana* and *Euconulus fulvus*. The latter was recorded from the Sanctuary (late Neolithic) on Overton Hill as occurring in a fairly shaded context (Kennard, *in* Cunnington, 1931). From the ditch of the camp on Windmill Hill (Kennard, unpublished; Howard, *in* Smith, 1965a) a similar fauna was recorded, virtually devoid of open-country species, and here a single example of *Columella edentula* was recovered. At Avebury too, the fauna from the ditch was fairly representative of shaded conditions (Kennard and Woodward, *in* Gray, 1935); and likewise at Wayland's Smithy where a single example of *Vertigo pusilla* was recovered. *Arianta arbustorum*, *Helix hortensis* and *H. nemoralis* are present in association at a number of these sites.

The important point is that while clearance and agriculture must have reduced much of the area to open-country it was not so complete as to remove all trace of habitats in which the more conservative species might seek refuge. These refuges too, may need to be of a certain minimum size, for, as Elton (1927) points out, a habitat although otherwise quite suitable must be able to support a population large enough to combat any external checks on its numbers.

Note on the sequence at South Street (Fig. 3)

The Neolithic long barrow at South Street near Avebury has been

L

worked on in some detail and it may be worth recording some of the
results at this stage.

Faunal changes in the buried soil beneath the mound (Figs 2 and
3B, 25–40 cm) suggest some opening-up of the environment though
these are so slight as to have little significance. A turf in the
lower levels of the ditch (Fig. 3A, 215–225 cm) derived from this soil
has a virtually identical fauna. The notable absence of *Cecilioides
acicula* (Müller) from this turf demonstrates that the presence of this
species in the soil beneath the mound is due entirely to its burrowing
habit.

The fine chalky fill of the ditch (Fig. 3A, 162·5–215 cm) is representative of an unstable environment with a surface colonized here and
there by plants, but for the most part bare. This is amply reflected in
the fauna which is dominated by *Hygromia hispida* and *Vallonia
costata*, the latter present in its rupestral capacity. *Vallonia excentrica*,
characteristic of more stable grassland, shows a marked decline from
its abundance in the pre-barrow soil. The similarity of this fauna to
that of the Late-glacial period is remarkable. Even to the presence of
Cochlicopa spp. and of *Vitrina pellucida*, of *Punctum pygmaeum* as
opposed to *Discus rotundatus* and of *Retinella radiatula* as opposed to
Retinella pura do the faunas match closely. The marked increase of
Hygromia hispida is of interest and again reflects the broken nature of
the substratum and the ability of this species to colonize freshly
exposed chalk surfaces. Kerney (1963) notes its abundance in zone III
deposits, though here a response to increased humidity may also be
involved.

There is, however, another facies to this fauna reflecting the
presence of patches of leaf-litter or living vegetation. The two zonitids,
Vitrea contracta (Westerlund) and *Oxychilus cellarius* (Müller), are
hardly species one would associate with bare chalk slopes and their
presence in low numbers suggests at least some stability in the environment at this stage in the filling of the ditch.

The shade-loving species which are so often found attached to logs
or stones do not become present, however, until the A/C-horizon of the
buried soil (150–162·5 cm). This rupestral element comprises such
species as *Discus rotundatus*, *Acanthinula aculaeta*, *Ena* spp. and the
Clausiliidae. Through this horizon, this group together with the
Zonitidae, *Carychium tridentatum* and *Pomatias elegans* show an increase
while *Vitrina pellucida*, *Retinella radiatula*, *Punctum pygmaeum* and
the more strictly grassland species show a decline. These changes are
associated with the local shading-over of the environment which took
place towards the end of the Neolithic period.

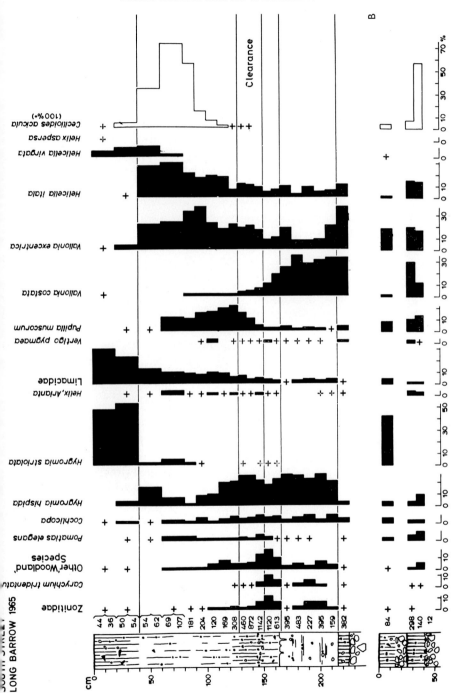

Fig. 3. South Street long barrow. A, Ditch; B, barrow mound.

Above 150 cm changes in the opposite direction take place and these, at any rate in the early stages, are associated with the clearance of the area by the Beaker folk (see below). *Vallonia excentrica* and *Helicella itala* regain their former abundance and *Pupilla muscorum* shows a steady increase through the depth of the soil profile. *Vallonia costata* on the other hand, shows a rather marked decline and by about 125 cm is virtually absent, indicating a certain impoverishment of the environment when compared with that in the pre-barrow soil.

In the ploughwash deposits above 130 cm, probably referable in the main to the Iron Age period, the fauna is an open-country one dominated by the xerophiles *Helicella itala*, *Vallonia excentrica* and *Pupilla muscorum* with the Limacidae becoming important for the first time. The numbers of Molluscs decrease towards the surface and the picture suggested is of an environment becoming gradually impoverished by agriculture.

At about 40 cm well-marked changes in the fauna occur. *Hygromia striolata* comes in strongly for the first time, almost completely replacing *H. hispida*. *Pupilla muscorum*, *Vallonia excentrica* and *Helicella itala* become virtually excluded, while the Limacidae show a sharp increase. These changes are in part due to the introduction of certain species by man, and their taking over the niches occupied by members of the indigenous fauna, and in part to the intensive methods of modern ploughing.

THE BEAKER PERIOD

At about 2000 B.C. occupation by the Beaker folk began and there is evidence that this was accompanied by a further opening-up of the environment brought on by changes in agricultural practice. At South Street (Fig. 3A), plough-marks scored on the weathering ramps of the ditches and transversing the ditch fill as drag lines of chalk rubble (150 cm) are associated with a faunal change towards the open-country aspect, and pottery evidence is in keeping with a Beaker date for this phase. *Arianta arbustorum*, *Helix hortensis* and *H. nemoralis* are still present but *Vallonia excentrica* shows an increase at the expense of *V. costata*. So too with the Windmill Hill Long Barrow (Connah and McMillan, 1964), a fauna from the upper levels of the ditch soil dated to the Beaker period shows a fairly open aspect with *V. excentrica* almost twice as abundant as *V. costata*, foreshadowing the expansion which this species makes in the Iron Age. Even at Knap Hill the upper levels of the ditch begin to show a more open aspect, perhaps referable to Beaker activity.

The fauna from the soil beneath the Hemp Knoll Round Barrow reflects the situation on downland and is still similar to that from the Windmill Hill, South Street and West Kennet Long Barrows (Figs 2 and 4c). The fauna from the ditch soil of this barrow—probably also referable to the Beaker period—shows a reversion to a more shaded facies (Fig. 4b) though with *Helicella itala* and the two species of *Vallonia* maintaining their abundance; *Arianta arbustorum*, *Helix hortensis* and *H. nemoralis* are present. The fauna from a recent fallow soil in the upper levels of the ditch (Fig. 4a) directly below the modern plough soil shows a completely different pattern. *Vallonia costata* has become virtually excluded at the expense of *V. excentrica*, and *Pupilla muscorum* has become an important member of the fauna.

The distribution of the various species of Mollusca in these faunas demonstrates nicely the effect of the environment on the pattern of molluscan populations. In the buried soil (c), representing a downland environment in the Beaker period, no species reaches more than 32% abundance, though there is dominance by the two *Vallonia* species. In the ditch soil (b) the maximum relative abundance value falls to 18% (*V. costata*) and there is a more even distribution of the various species present. In the fallow soil (a), probably representative of the present day rather severe conditions, *V. excentrica* assumes almost total dominance (58%) with *Pupilla muscorum* (23%) as the only other important member of the fauna.

Other faunas of Beaker date have been recorded from Beckhampton (Howard, *in* Young, 1950) where ". . . the most numerous shells were *Vallonia pulchella* and *Vallonia excentrica* which intergrade and cannot be separated", with small numbers of other species including *Helix nemoralis*, absent from the area today; from Beaker pits at West Kennet (Evans, *in* Smith, 1965b) where *Arianta arbustorum* and *Helix nemoralis* were recovered with *Carychium tridentatum*, *Discus rotundatus* and other shade-loving species; and from Fargo Plantation in South Wiltshire (Kennard, *in* Stone, 1938) dominated by *Hygromia hispida*, *Pupilla muscorum* and *Vallonia excentrica* with *V. costata*, *Helicella itala*, *Vertigo pygmaea* and a few shade-lovers in lesser abundance. *Arianta arbustorum*, *Helix hortensis* and *H. nemoralis* were recorded from the West Kennet Avenue (A. S. Kennard, unpublished), where now only *H. hortensis* occurs; and the two species of *Helix* were recorded from a Beaker barrow on Overton Down (D. D. A. Simpson, personal communication).

Thus through the Beaker period the shade-loving species maintain themselves; *Arianta* and the two species of *Helix* still occur in association; evidence from one or two sites suggests an increase in *Vallonia*

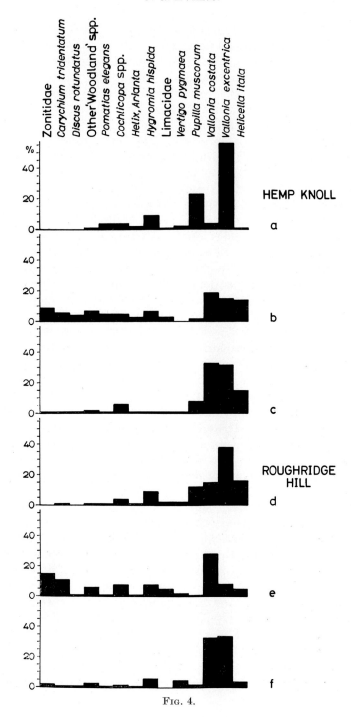

FIG. 4.

excentrica at the expense of *V. costata* but there is as yet little inkling of the enormous expansion which *Pupilla muscorum* is to make at a later stage.

Little information is available for the Bronze Age period but what there is suggests again an open environment of dry grassland, yet with refuges large enough to harbour species such as *Ena montana*.

At Roughridge Hill (Fig. 4d, e and f) a similar sequence to that at Hemp Knoll is preserved. The buried soil beneath a Bronze Age barrow (Figs 2 and 4f) is dominated by the two species of *Vallonia* in equal abundance. A fauna from a contemporary barrow about 50 yd away shows a similar pattern but with a greater abundance of *Pupilla muscorum*, *Vertigo pygmaea* and *Helicella itala*, approaching more the situation at Hemp Knoll (Fig. 4c). Such differences probably reflect lateral variation over the Bronze Age land surface and have no chronological implications. In the barrow ditch (Fig. 4e) there is a distinct increase in the shade-loving aspect though the paucity of certain species notably *Discus rotundatus* and *Pomatias elegans* suggests a degree of impoverishment and drying out of the landscape not present at Hemp Knoll. *Vallonia costata* maintains its abundance but the other open-country species fall to a minimum. *Arianta arbustorum*, *Helix hortensis* and *H. nemoralis* are present. In a recent plough soil close to the surface of the ditch (Fig. 4d), *Vallonia excentrica* increases notably, at the expense of *V. costata*, while *Pupilla muscorum* and *Helicella itala* become more prominent.

The fauna from a Bronze Age pit near Upavon (J. G. Evans, unpublished), and from a Bronze Age cremation pit near West Kennet (Evans, *in* Smith, 1965b) both show a distinctly open-country aspect dominated by *Pupilla muscorum*, *Vallonia excentrica* and *Helicella itala*, with *Vallonia costata*, *Vertigo pygmaea* and a few other species in small numbers. The soil in the ditch of the barrow from which this cremation pit derived contained a fauna showing a more shaded aspect, though with the notable absence of *Discus rotundatus*, a species well represented in Beaker pits on this site. At West Overton snails picked out during excavation from a Bronze Age pit included *Vallonia excentrica*,

---

FIG. 4. Molluscan assemblages from Hemp Knoll (Beaker) and Roughridge Hill (Bronze Age). a, Recent fallow soil in barrow ditch; b, ditch soil; c, buried soil beneath barrow mound; d, recent plough soil in barrow ditch; e, ditch soil; f, buried soil beneath mound.

Helicella itala, *Hygromia hispida*, *Discus rotundatus*, *Helix hortensis* and *Helix nemoralis* (Simpson, personal communication).

There is thus evidence that during the Bronze Age, and following the trend already set in the Beaker period, the fauna took on a more xerophilous aspect, associated with an impoverishment and drying-out of the environment.

In South Wiltshire too, a similar pattern seems to have obtained. Thus Christie (1964) of a site near Stonehenge, notes: "Examination of land Mollusca and the bones of small mammals adds yet further to the evidence gradually accumulating from other sites, that a very dry, open environment, with short-turfed grassland, prevailed." The fauna (Kerney, *in* Christie, 1964) shows an open aspect, dominated by *Vallonia excentrica* and *Pupilla muscorum* with *Vallonia costata*, *Vertigo pygmaea*, *Helicella itala* and the rare xerophile, *Truncatellina cylindrica* (Férussac), in lesser abundance.

<h2 style="text-align:center">LATER CLEARANCE</h2>

With the Late Bronze Age and Early Iron Age a change from an essentially pastoral economy in the Early and Middle Bronze Age to one of intensive arable farming appears to have taken place. Evidence from a number of sites (e.g. Sparks and Lewis, 1957; Kerney *et al.*, 1964; Evans, 1966b) suggests that it was at the beginning of this period that the main hillwash accumulations which now fill many of the dry valleys in the chalk country were initiated, and while lynchet formation and cross-ploughing can be traced as far back as the Neolithic period, it is not until the Late Bronze Age that the larger lynchets began to accumulate. The sequence through two of these demonstrates nicely the way in which molluscan populations have responded to agriculture during the Iron Age (Fig. 5).

At Fyfield Down, the lower levels of the deposits have already been discussed (Fig. 5; Fyfield Down I, Roothole fill and Ploughed pre-lynchet soil), and the fauna is probably representative of a shaded and relatively undisturbed environment. The fauna in the lynchet deposits is characteristic of an open environment. The abundance of *Vallonia costata* in the lowermost sample may reflect a slightly more stable and richer environment at this level, but above, the dominance by *V. excentrica* and *Helicella itala*, the latter as a major element for the first time, attests to the very dry nature of the environment in the later stages of the lynchet. Archaeological evidence suggests accumulation to have ceased by the end of the Romano-British period and in fact the latter is only represented by a thin layer of material near the surface, the majority of the accumulation being Early Iron Age in origin.

The sharp distinction between the fauna in the lynchet deposits and that in the modern soil tends to support this idea, the latter being dominated by *Pupilla muscorum* and to a lesser extent by *Vallonia excentrica*, all other species having fallen to a minimum.

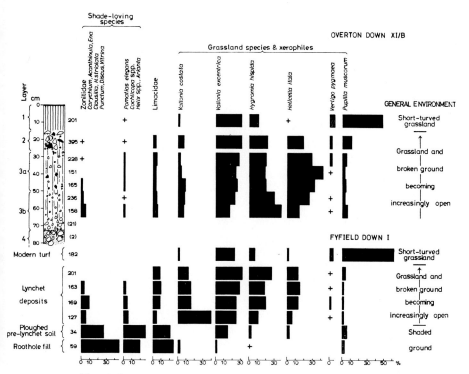

Fig. 5. Overton Down and Fyfield Down lynchets (Iron Age).

This series demonstrates nicely the effect of increasingly adverse conditions on the Mollusca, from the reasonably rich fauna before ploughing began where no one species is dominant, and none (apart from the composite group Limacidae) being represented by more than 14% abundance; through a phase of impoverishment in the lynchet where first *Vallonia costata* and the *V. excentrica* reach 30% abundance; to the highly restricted fauna in the modern turf of no more than six species, in which *Pupilla muscorum* reaches 56% abundance and reflects the rather severe conditions obtaining at the present day.

At Overton Down (Fig. 5; Overton Down XI/B) a similar situation is present, though due to the more intensive ploughing which took place in the initial stages of cultivation all traces of the original soil have been removed. No pre-lynchet fauna is present therefore, but otherwise the sequence compares closely with that from Fyfield Down. In Fig. 2, Overton Down Lynchet, the differences between the open-country fauna in the lynchet deposits and those from the Bronze Age and Beaker soils are clearly brought out.

During the Romano-British period an extension of the clearance process by man can be recognized and, as with the Beaker and Iron Age periods, this is not merely a continuation, but a distinct phase of agricultural activity perhaps accompanied by new methods of ploughing and land use. Thus at Overton and Fyfield Downs, the flint line (layer 2) yielded large numbers of Romano-British sherds but little or nothing else. The main ploughwash accumulation in the ditch of the round barrow by West Kennet (see above) is clearly of Romano-British or later origin; a fauna from this material was dominated by *Helicella itala* and *Hygromia hispida* with *Vallonia excentrica*, *V. costata*, *Pupilla muscorum* and one or two others in small numbers. Burchard (1966) has put forward evidence for Romano-British scrub clearance at Bayardo Farm, West Overton in the first century A.D.

At Waden Hill (Evans, 1966c), ploughwash deposits associated with, and overlying a Romano-British burial indicate two distinct phases of ploughing separated by a fallow stage. This site is in the Winterbourne valley on the very edge of the flood plain and the molluscan sequence was analysed in some detail (Fig. 6). Below 150 cm the deposits belong to the Late-glacial period and have been discussed above. Between 110 and 150 cm the deposits constitute a brown, chalky loam which is evidently a ploughwash deriving from the Romano-British period as shown by the masses of pottery therein. The fauna is characterized by large numbers of *Pupilla muscorum*, and it is suspected that these may derive in part at least from the Late-glacial deposits. Above this level (*ca* 80–110 cm) a feebly developed soil is present in which the fauna is dominated by *Hygromia hispida* with *Helicella itala* and the two species of *Vallonia* in equal abundance. This is the level from which the burial was made and it is evident therefore that this soil was lying fallow at the time. The fauna certainly supports this contention.

A further phase of ploughing is indicated to have taken place between 30 and *ca* 80 cm and here again pottery evidence suggests a Romano-British origin. The fauna shows a distinct increase in *Helicella itala*, *Vallonia excentrica* and the Limacidae, and a corresponding decrease in *Hygromia hispida* and *Vallonia costata*. As at Overton and Fyfield

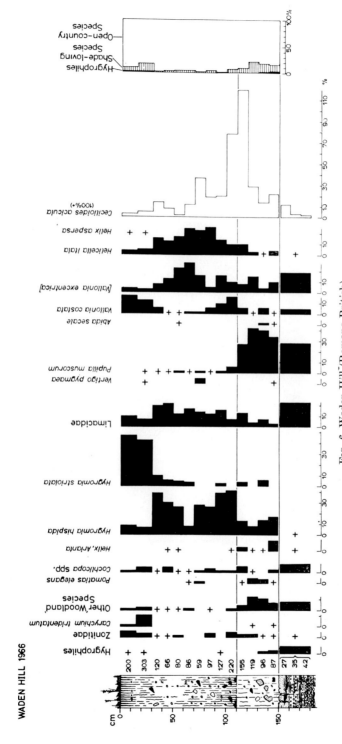

FIG. 6. Waden Hill (Romano-British).

Downs, *Helicella itala* is present in an arable environment. This is curious in the light of the present day ecology of the species which generally seems to confine it to rather remote areas of downland. Evidence from a site in Dorset (J. G. Evans, unpublished) and from the ditch at South Street (Fig. 3A) suggests, however, that in antiquity, at least from the Iron Age to the Romano-British period, *H. itala* was a characteristic member of the fauna of ploughed fields.

Above 30 cm changes in the reverse direction take place and a return to the more amenable conditions of stable grassland which were once present is indicated. The present day environment is one of tall grassland. It is interesting to note how *Hygromia striolata* persists and markedly increases almost entirely replacing *H. hispida*. A similar pattern from the bank of the West Kennet Long Barrow and from the ditch at South Street (Fig. 3A) was noted.

GENERAL CONCLUSIONS: THE MODERN FAUNA

Grundy (1939) demonstrates that since the Roman period the area has been free of woodland, and considering the extensive distribution of "Celtic" fields on the Marlborough Downs it is likely that complete clearance had taken place by the Iron Age period. In fact from the beginning of the Neolithic period, the evidence here presented shows that an impoverishment and drying-out of the environment, due to the activities of man, has been going on and this has had a pronounced effect on the molluscan population.

The beginnings of Neolithic agriculture saw the extinction of *Lauria anglica, Vertigo substriata* (Jeffreys) and *Acanthinula lamellata* at an early date. *Vertigo pusilla, Clausilia rolphi, Acicula fusca* and *Ena montana* persisted for longer, perhaps well into the Bronze Age period, and the latter three may still occur in nearby woodland. Grassland faunas during the Neolithic period were dominated by *Vallonia costata* and *V. excentrica* with *Pupilla muscorum* and *Helicella itala* in lesser abundance. There is ample evidence that the Neolithic people cultivated the soil and grew cereal crops (Smith, 1965a). At South Street, plough-marks scored on the subsoil surface beneath the barrow have been ascribed to the initial breaking-up of the soil and clearance of the area, and the soil itself shows evidence of at least two phases of cultivation. In view of this, the faunal composition of the soil is truly remarkable when compared with faunas from present day arable land. The latter are dominated by *Hygromia striolata*, introduced species of *Helicella* such as *H. gigaxi* (L. Pfeiffer) and slugs such as *Agriolimax reticulatus* (Müller) and are completely devoid of *Vallonia, Pupilla* and

Helicella itala. These differences may be associated with a delicate balance between the various species present in such severe environments, so that any change in favour of one species may cause an increase in that species perhaps at the expense of another. One can point to the climatic deterioration which has gone on since the Neolithic period, the more intensive and prolonged ploughing and the introduction of new species as possible agents in bringing about long-term faunal changes in what are apparently similar environments. In addition, it is possible that *Vallonia* is able to colonize open ground more successfully and quickly than are *Pupilla* and *Helicella* and this may have had some influence on the pattern of faunal change.

Through the Beaker and Bronze Age periods an increase in the abundance of *Pupilla muscorum* and *Helicella itala* took place while *Vallonia costata* showed a decline, but it was not until the Iron Age that the latter became excluded from areas of high downland and *Helicella itala* attained a position of dominance. *Helix hortensis* and *Arianta arbustorum* persisted likewise on the downs up to the Iron Age period. The present abundance of *Pupilla muscorum*, however, may well stem from a later date but evidence about this is inconclusive. A number of species which were once quite widespread are now rather local in their occurrence and often absent from apparently suitable habitats. Most notable in this respect are *Pomatias elegans*, *Carychium tridentatum*, *Vallonia costata*, *Arianta arbustorum*, *Helix hortensis*, *Helicella itala* and *Discus rotundatus*.

As Kerney (1965) pointed out, the effects of agriculture, may, by creating a diversity of habitats, be more beneficial than otherwise so that a once rather uniform environment has given way to one containing a variety of habitats each harbouring a distinct snail population. Hedgerows, ditches and river banks (e.g. Waden Hill, modern turf; Fig. 2) provide the best refuges for Mollusca while prehistoric mounds untouched by ploughing may have quite a rich fauna in comparison with the surrounding area. The Clausiliidae, *Ena obscura* (Müller), *Carychium tridentatum*, *Vallonia costata* and the Zonitidae may all abound in such habitats.

Faunas from more open places are characterized by being dominated by one or two species; *Pupilla muscorum* and *Vallonia excentrica* at Overton Down (Fig. 2; Overton Down, modern turf), Fyfield Down and Hemp Knoll; and *Hygromia hispida* and *Vallonia excentrica* at Knap Hill and Windmill Hill (Fig. 2; Windmill Hill, modern turf). *Vertigo pygmaea* and *Cochlicopa lubricella* (Porro) may comprise a part of these faunas but never in abundance. Ploughed fields constitute the most severe environment in the area, and at South Street this is reflected in a fauna (Fig. 2; South Street, modern plough soil) in which only two species of snail (apart from *Cecilioides acicula*), namely *Hygromia striolata* (47%) and *Helicella virgata* (da Costa) (7%), and the Limacidae are present. Analysis of the living fauna has shown the latter to be entirely *Agriolimax reticulatus*; *Arion hortensis* Férussac is present in small numbers. The one or two examples of other species are subfossil, incorporated from underlying deposits.

In such severe habitats, competition for food and shelter may well be a decisive factor in influencing the composition of the faunas. For instance, the virtual absence of *Vallonia costata* from high downland may relate to the same mechanism which has excluded *Helix hortensis* from these areas. The increase of *Hygromia striolata* since the Roman period may have brought about the exclusion of *Hygromia hispida* and *Helicella itala* from the more arable habitats and it is interesting to note that in the downland areas not frequented by *H. striolata*, *H. hispida* often occurs in enormous numbers. The spread of the introduced species of *Helicella* may well have enhanced this effect.

REFERENCES

Ant, H. (1963). Faunistische, ökologische und tiergeographische Untersuchungen zur Verbrietung der Landschnecken in Nordwest deutschland. *Abh. Landesmus. Naturk. Münster* **25**: 1–125.
Bowen, H. C. and Fowler, P. J. (1962). The archaeology of Fyfield and Overton Downs, Wilts (Interim Report). *Wilts. archaeol. nat. Hist. Mag.* **58**: 98–115.
Boycott, A. E. (1934). The habitats of land Mollusca in Britain. *J. Ecol.* **22**: 1–38.
Burchard, A. M. (1966). Ancient scrub clearance (?) at Bayardo Farm, West Overton. *Wilts. archaeol. nat. Hist. Mag.* **61**: 98.
Bury, H. and Kennard, A. S. (1940). Some Holocene deposits at Box (Wilts.). *Proc. Geol. Ass.* **51**: 225–229.
Cain, A. J. and Currey, J. D. (1963). Area effects in *Cepaea. Phil. Trans. R. Soc.* (B) **246**: 1–81.
Christie, P. M. (1964). A Bronze Age round barrow on Earl's Farm Down, Amesbury. *Wilts. archaeol. nat. Hist. Mag.* **59**: 30–45.
Connah, G. (1965). Excavations at Knap Hill, Alton Priors, 1961. *Wilts. archaeol. nat. Hist. Mag.* **60**: 1–23.
Connah, G. and McMillan, N. F. (1964). Snails and archaeology. *Antiquity* **38**: 62–64.
Cunnington, M. E. (1931). The "Sanctuary" on Overton Hill, near Avebury. *Wilts. archaeol. nat. Hist. Mag.* **45**: 300–335.
Curwen, E. Cecil, (1936). Excavations in Whitehawk Camp, Brighton: Third season, 1935. *Sussex archaeol. Collns* **77**: 60–92.
Drew, H. and Piggott, S. (1936). The excavation of long barrow 163a on Thickthorn Down, Dorset. *Proc. prehist. Soc.* **2**: 77–96.
Ellis, A. E. (1926). *British snails*. London: Oxford Univ. Press.
Elton, C. (1927). *Animal ecology*. London: Sidgwick and Johnson.

Evans, J. G. (1966a). Land Mollusca from the Neolithic enclosure on Windmill Hill. *Wilts. archaeol. nat. Hist. Mag.* **61**: 91–92.

Evans, J. G. (1966b). Late-glacial and Post-glacial deposits at Pitstone, Bucks. *Proc. Geol. Ass.* **77**: 347–364.

Evans, J. G. (1966c). A Romano-British interment in the bank of the Winterbourne, near Avebury. *Wilts. archaeol. nat. Hist. Mag.* **61**: 97–98.

Godwin, H. (1956). *The history of the British flora.* Cambridge: Univ. Press.

Godwin, H. and Tansley, A. G. (1941). Prehistoric charcoals as evidence of former vegetation, soil and climate. *J. Ecol.* **29**: 117–126.

Gray, H. St. G. (1935). The Avebury excavations, 1908–1922. *Archaeologia* **84**: 99–162.

Grundy, G. B. (1939). The ancient woodland of Wiltshire. *Wilts. archaeol. nat. Hist. Mag.* **48**: 530–598.

Kerney, M. P. (1963). Late-glacial deposits on the Chalk of South-East England. *Phil. Trans. R. Soc.* (B) **246**: 203–254.

Kerney, M. P. (1965). Snails and man in Britain. *J. Conch., Lond.* **26**: 3–14.

Kerney, M. P., Brown, E. H. and Chandler, T. J. (1964). The Late-glacial and Post-glacial history of the Chalk escarpment near Brook, Kent. *Phil. Trans. R. Soc.* (B) **248**: 135–204.

Smith, I. F. (1965a). *Windmill Hill and Avebury: Excavations by Alexander Keiller*, 1925–1939. London: Oxford Univ. Press.

Smith, I. F. (1965b). Excavation of a bell barrow, Avebury G.55. *Wilts. archaeol. nat. Hist. Mag.* **60**: 24–46.

Sparks, B. W. and Lewis, W. V. (1957). Escarpment dry valleys near Pegsdon, Hertfordshire. *Proc. Geol. Ass.* **68**: 26–38.

Stone, J. F. S. (1938). An early Bronze Age grave in Fargo Plantation. *Wilts. archaeol. nat. Hist. Mag.* **48**: 357.

Waldén, H. W. (1955). The land Gastropoda of the vicinity of Stockholm. *Ark. Zool.* (2) **7**: 391–448.

Warren, S. H. (1945). Some geological and prehistoric records on the north-west border of Essex. *Essex Nat.* **27**: 273–280.

Young, W. E. V. (1950). A Beaker interment at Beckhampton. *Wilts. archaeol. nat. Hist. Mag.* **53**: 311–327.

Symp. zool. Soc. Lond. (1968). No. 22, 319–346.

HABITAT DISTRIBUTION OF
SOLOMON ISLAND LAND MOLLUSCA

JOHN F. PEAKE

British Museum (Natural History), London, England

SYNOPSIS

In 1965 a Royal Society expedition visited the British Solomon Islands. A preliminary analysis of the distribution of Mollusca found in the forest habitat is presented here. Although the origin of the Solomon Islands probably dates from the Mesozoic, local topography was largely determined during the Pliocene-Pleistocene periods. Therefore the present spatial arrangement of molluscan species and habitats probably dates from that period. Climatically the islands can be included with extreme forms of the humid tropics. The forests constitute the dominant vegetation; they are divisible into four altitudinal zones; each division being characterized by its physiognomy and molluscan fauna. There is a decrease in faunal diversity with increasing altitude; this can be correlated with climatic factors and changes in the vegetation. There is a clearly defined vertical stratification within each altitudinal zone with a maximum of three strata in the lowland rain forest. The molluscan fauna confined to the ground stratum is dominated by small prosobranch species; their distribution being limited by microclimatic factors rather than local variations in geology. A further association with much wider taxonomic affinities occurs at this level; these are species with a vertical distribution extending into the arboreal strata. The arboreal region is divisable into two strata, each with two groups of niches, one on the tree trunks and branches and the other on the foliage. Prosobranch and pulmonate taxa are found in the lower arboreal but only pulmonate species with either a large body mass or slug-like form extend into the upper arboreal. The adaptations of arboreal fauna are varied, but of particular interest are the utilization of arboreal sites for oviposition and the development of conspicuous and cryptic colour patterns. The extension of the vertical distribution, into the upper canopy, of Mollusca exhibiting a slug-like form, as typified by *Helixarion*, is possibly associated with temperature depression resulting from water evaporation from an extensive body surface. The endemic genus *Cryptaegis* exhibits complete adaptation to the arboreal habit. Human disturbance produces marked modifications of both the habitat and the faunal composition. The fauna is divisible into four elements with varying biogeographic affinities and differing habitat distributions. The Recent introduced species are associated with human disturbance. *Partula*, a member of the Pacific Ocean element, is confined to the coastal zone. This suggests a more recent immigration to the Solomons than the other two elements, the Southern and the Palaeo-Oriental, which have a much wider altitudinal and habitat range.

INTRODUCTION

In 1965 the author spent six months, from July to December, in the British Solomon Islands Protectorate as a member of a Royal Society expedition. Prior to this expedition very few detailed investigations had been made on the fauna of these islands, although taxonomic studies on a few selected groups were available. In this context Guppy's

two books (1887a,b) are important, as they provide a wealth of background information that is unavailable elsewhere; more recent supple-
mentary data are contained in the reports of Lever (e.g. 1937a,b).
During the Royal Society expedition nine islands were visited (Fig. 1)

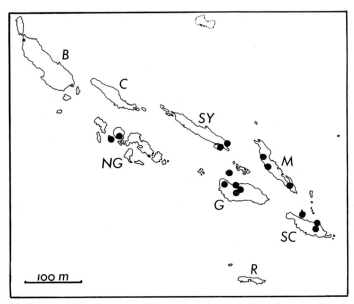

Fig. 1. Solomon Islands: areas visited by the author during the Royal Society
expedition are indicated by dots. B, Bougainville; C, Choiseul; NG, New Georgia group;
SY, Santa Ysabel; G, Guadalcanal; M, Malaita; SC, San Cristobal; R, Rennell.

and transects made from sea-level to the summits of three mountains.
Collections of land and freshwater Mollusca were obtained from a wide
variety of habitats, but the present paper is a preliminary analysis of
the distribution of only those species that occurred in the forests.
It is based entirely on field records; it is hoped, at a future date, to
implement these with further systematic data. The paucity of information on the ecology of land Mollusca in tropical rain forests emphasizes
the preliminary nature of this account. It has therefore been important
to consider the structure of the habitat in the Solomons, so that comparisons can be made with other tropical regions during future investigations.

GEOGRAPHICAL POSITION

The Solomon Islands occupy a central position in the Melanesian
Arc of islands that extends from New Guinea in the west to Fiji in the

east. Administratively they are divided into Australian and British sectors; the former includes the island of Bougainville, whilst the latter embraces the remainder of the central archipelago, the Santa Cruz group and outlying islands such as Rennell and Ontong Java. The central archipelago, lying between latitudes 5°S. and 12°S. and longitudes 152°E. and 163°E., constitutes a compact assemblage of seven large islands arranged in two parallel lines, orientated in an ESE. and WNW. direction, and separated by a narrow expanse of sea known as The Slot. Between these islands the sea gap is never greater than 50 miles. Included in the northern chain of islands, on the Pacific side, are Choiseul, Santa Ysabel, and Malaita, and in the inner series, on the Australian side are New Georgia, Guadalcanal and San Cristobal. Bougainville lies at the north-western end linking the two chains. For the purpose of this account the term Solomon Islands is restricted to the compact central archipelago as outlined above, and not to the administrative areas.

Characteristically the larger islands of the central archipelago have an attenuated form and a central backbone of mountains. These rise to over 9000 ft on Bougainville (where there is still an active volcano) and to 8000 ft on Guadalcanal. The New Georgia group provides an exception to this generalized pattern, for it includes a number of smaller irregularly shaped islands arranged in close proximity to one another; the highest being an isolated extinct volcano, Kolombangara, at 5540 ft. The mountainous terrain of all these islands is dissected by the many large rivers that display ample evidence of frequent flooding resulting from the torrential rainstorms so typical of this region. The dominant vegetation is the ubiquitous rain forest, covering the surface of the islands from the lowlands to the tops of the mountains, so that any minute areas of open habitat are extremely conspicuous.

Climatically the Solomon Islands constitute an extreme form of the humid tropics, exhibiting a complete absence of prolonged periods of dry conditions and a non-seasonal temperature regime, the average daily maximum being approximately 80°F. A recent paper by Fitzpatrick, Hart and Brookfield (1966) analyses the available information on the seasonality of the rainfall in the south-west Pacific. Utilizing the classification proposed in this paper, extensive areas of the Solomons fall into the regime type of continuous wet, with only limited tracts exhibiting moderate seasonal alteration. Local modifications are produced by the topography, for example, on the north side of Guadalcanal, in the region of Honiara, there is pronounced seasonal alternation with a long dry season. This feature is associated with exposure to the north and west and shelter by the high mountains of the central backbone to the south and east. In those regions of the Solomons with continuous wet or only moderate seasonality the rate of evaporation of water from the ground surface probably never or very rarely exceeds that derived from rainfall. In fact, there is usually an excess of water available as run-off; if dry periods do occur their duration must be short and their intensity low. Only in regions with marked seasonal patterns are the periods of drought prominent and of regular occurrence.

The intensity of rainfall has a pronounced effect on the stability of the ground stratum on the steeper slopes. At the commencement of the expedition this was convincingly demonstrated during a period of torrential rain; 110 in. in 10 days being recorded at one meteorological station. The effect was catastrophic. Complete hillsides were rapidly covered by sheets of flowing water; leaf litter, branches and logs were moved considerable distances and on the steeper slopes land slips were common.

GEOLOGY

The Solomon Islands, together with others in the Melanesian Arc, lie within the Andesite Line and are therefore classed, geologically, as continental. Evidence for former land bridges between the Solomons and surrounding archipelagos is absent, although on the basis of plant and animal distributions biologists have classified them as oceanic, subcontinental and even continental depending on which elements of the biota were being considered.

Detailed information concerning the geology of the Solomon Islands dates from the formation of the British Solomon Islands Geological Survey and expeditions from the University of Sydney (Grover et al., 1965). The only important investigation prior to these being the study of Guppy in the last century (1887a). Coleman (1963) has concluded that these islands may represent a youthful stage in the development of island arcs. Potassium-argon measurements (Richards, Cooper, Webb and Coleman, 1966), together with geomorphological studies (Coleman, 1966), have suggested that their origin need not be considered as earlier than the late Mesozoic. The Pleiocene-Pleistocene period was, however, the era of major geological change in the region; block faulting gave rise to the massive mountain systems, as occur on Guadalcanal, and intensive vulcanism was renewed. This vulcanism persisted until the recent past. The coastal fringes of raised reef masses and conglomerates are the result of the Pleistocene activity. Probably island complexes like the New Georgia group, that are assemblages of volcanic cones, originated during this Pleiocene–Pleistocene activity, certainly their history may not extend back into the Miocene.

The present habitat distribution and spatial arrangement of the flora and fauna in the Solomon Islands in, at least, the mountainous areas must have developed since the period of Pliocene–Pleistocene activity. Although dry land has probably existed continuously in this region for a much longer period and certain taxa may have a much older origin within the archipelago, having been distributed to the islands prior to the Pliocene period. The evidence from the New Georgia group suggests, however, that the development of local Rassenkreise of molluscan species and other faunal groups throughout the archipelago could be comparatively recent, that is post Pliocene.

BIOGEOGRAPHIC DISTRIBUTION PATTERNS

On the basis of the distribution of a wide variety of plants and animals numerous schemes have been proposed for the division of the Pacific basin into biogeographic regions (Thorne, 1963). However, no single scheme reflects the detailed distribution of all taxa; there is, for example, a contrast between those produced for vertebrate groups and those for many insect taxa. In Indonesia insects do not exhibit a region of faunal balance between the two divergent faunas of Asia and Australia as do many families of vertebrates. Instead there is an extension and gradual attenuation of the fauna from South East Asia eastwards, with peaks of radiation on the larger islands such as New Guinea (Gressitt, 1961). Many groups of terrestrial Mollusca probably follow similar patterns to those produced by the insects (Hedley, 1899; Solem, 1959), although Rensch (1936) has suggested otherwise.

An understanding of the distribution of terrestrial Mollusca in the south-west Pacific has benefited considerably from the earlier studies of Hedley (1899) and Pilsbry (1900a,b). Hedley demonstrated two migration routes into Melanesia; one from continental Asia, with New Guinea as an important area of diversification, the other a southern route including New Zealand and New Caledonia. Today these would not be considered migration routes, but rather areas demonstrating faunal affinities. Pilsbry's taxonomic revisions, of particular families of pulmonate snails from islands in the central Pacific, emphasized the isolated systematic position of these taxa, and established the presence of a distinctive Pacific molluscan fauna.

The molluscan fauna of Melanesia is divisible into four elements having affinities with adjacent geographical areas. These are characterized as follows; the terminology is adapted from that proposed by Solem (1958, 1959).

(1) *Palaeo-Oriental fauna.* This is the continental fauna that
extends eastwards from the Malayan peninsula through New Guinea to
Fiji and down the east coast of Australia, from Queensland to New
South wales (McMichael and Iredale, 1959). Key taxa are the families
Cyclophoridae, Camaenidae, Rathouisiidae and particular subfamilies
of the Ariophantacea. Other taxa have a more limited distribution.

(2) *Southern fauna.* The south of Australia, Tasmania, New Zealand
and New Caledonia is dominated by this fauna, but components extend
further north and west. Key taxa are the families Paryphantidae,
Bulimulidae, Athoracophoridae and selected groups of the superfamily
Endodontacea.

(3) *Pacific Ocean fauna.* This is the Polynesian fauna of many
authors (for example, Pilsbry) that dominates the high islands of the
central Pacific Ocean. Key taxa are the families Partulidae, Tornatel-
linidae, Microcystinae and particular subfamilies of the Endodontacea.
The Achatinellidae and Amastridae are confined to the Hawaiian
Islands.

(4) *Introduced species.* Included in this category are those species
frequently referred to as "tropical tramps", "coral atoll" taxa or
"tourist snails". While it is possible to recognize that certain species
have been recently introduced into Melanesia from other tropical areas
and the Palaearctic, it is difficult to draw a sharp division between
species that possess extremely efficient modes of dispersal, those transported by the indigenous human inhabitants and more recent intro-
ductions by Caucasians. Harry (1966) has suggested that all seventeen
species of land snails occurring on Ulithi Atoll, Caroline Islands, have
been recently introduced by man, as a surprisingly wide range of taxa is
included it suggests that a careful reappraisal must be made of island
faunas to determine those species that could have been "introduced".
Although not considered important zoogeographically, this element is
extremely significant ecologically, for some introduced species attain
such high densities as to dominate particular habitats.

These four divisions of the molluscan fauna intermingle in varying
proportions in the different archipelagos of the Melanesian Arc. New
Guinea, Bismarks and the Solomons are dominated by the Palaeo-
Oriental fauna, Fiji by the Pacific Ocean, New Caledonia and the
Loyalty Islands by the Southern, with the New Hebrides constituting
an area of faunal balance between the three groups (Solem, 1958).
This pattern has been determined by the geological history of the land
masses, the evolutionary history of the terrestrial Mollusca and the
ecological characteristics of both the taxa and the land masses.

THE FOREST HABITAT; ALTITUDINAL ZONATION

Within the forest areas visited by the expedition it has been possible to recognize four altitudinal divisions; three correspond in many respects to the vegetational formation types proposed by botanists for the classification of tropical rain forests (Richards, 1952; Grubb, Lloyd, Pennington and Whitmore, 1963). An overwhelming advantage of the botanical scheme is that the groups are based on the physiognomy of the forests and have universal application. While closely following the botanical divisions the units proposed below are arbitrary, in that they cannot be defined quantitatively but are the result of qualitative appraisal of field collecting. As the physiognomy is already well documented by botanists, emphasis is placed on attributes of the forest thought to be important to the distribution of the Mollusca. The four altitudinal divisions are as follows: (1) coastal forest; (2) lowland rain forest; (3) lower montane forest; (4) upper montane forest. Extensive modifications to this scheme are found in areas where the vegetation is grossly disturbed by human interference, it is then frequently impossible to produce an adequate synthesis of the habitats.

While these zones form an altitudinal series progressing from sea-level to 8000 ft, it is impracticable to define absolutely the height for the transition from one zone to the next. The series must be considered as a continuum, with local topography and climate being of paramount importance. A manifestation of the latter is the Massenerhebung effect, that is the lowering of the altitudinal limits of a particular zone on small isolated peaks. Grubb and Whitmore (1966) discuss this effect in relation to vegetational zonation.

Coastal forest

The unifying feature of the mosaic of habitats included in this division is their close proximity to the sea; therefore many of the features of the environment on large coral islands can be equated with those occurring in this zone. A similar classification of habitats could be used for these two formations (Usinger and La Rivers, 1953; Gressitt, 1954), with the important proviso that on islands with a large central land mass there is an extensive reservoir of potential plant and animal colonists abutting directly onto the coastal region. In many tropical regions it is frequently difficult to define the seaward extension of this zone, for a sharp distinction between the marine and terrestrial elements of the biota does not exist. It might be possible to select the upper strand line near the high tide mark as an arbitrary demarcation point, but even this is impossible in extensive mangrove swamps that are in close proximity to forest habitats.

The local topography of the coastal zone has been determined largely by land movements of submergence and emergence from the sea, a process that is still continuing today at varying rates throughout the Solomon Islands. Where submergence has occurred or is occurring coastal habitats are restricted to a very narrow belt and the lowland rain forest approaches close to the sea. Emergence results in a fringing strip of raised coral reef that can produce a form of habitat similar to that present on alluvial deposits.

Typically the vegetation consists of tall, well spaced trees with an open canopy and an understorey of shrubs about 12 ft high. The nature of the ground stratum depends on the substrate, slight variations in the elevation of the land surface and the height of the underlying water table. The latter will probably follow the same principles as the fresh-water lens on coral atolls, that is the Glyten-Herzberg law (Wiens, 1962), with the important difference that there will be continual replenishment of water loss from the rivers and drainage from steep hillsides. On the extreme seaward edge the ground layer is usually well drained, because of a slight rise in the land surface as well as a thinning of the freshwater lens. Behind this edge, however, the ground can be waterlogged over large areas and open stretches of water may occur during periods of heavy rainfall. The substrate varies from alluvium, broken coral fragments and pockets of leaf litter to thick mud in the wetter localities. Logs in all stages of decay are frequent.

Human disturbance is considerable, for in this zone the majority of villages and coconut plantations are established, and through it there must be access to inland habitations. In the Solomons this disturbance is increasing as there is continual migration of people from the interior of the islands to the coasts. Gardens are closely associated with every village, but typically they are established in areas of good soil development in lowland rain forest or on alluvial plains in the river valleys. Pigs occur around human habitation, but in the following zone their influence is more important.

Lowland rain forest

The lowland rain forest represents the richest development of vegetation in the Solomons, both in area covered and number of plant and animal species present. All the alluvial plains, lower slopes of the mountains and even the extremely steep sides of the valleys reaching angles of 40° plus are covered with this forest type.

Characteristically the canopy of the forest forms an upper closed stratum at a height between 80 and 120 ft, and above this are only the occasional emergent crowns of still taller trees. The understorey cannot readily be divided into clear definable strata as there is a continuous vertical distribution of climbers, shrubs and trees to the base of the canopy. The larger climbers flourish in the upper canopy and maintain connexions with the ground by means of thick woody stems. The majority of trees have a smooth bark, but numerous crevices are found on the bole of those trees produced by the encircling figs and bayans. The herb stratum is poorly developed. Impenetrable tangles of shrubs and herbaceous creepers are associated with riverine conditions and gaps in the upper canopy that permit light penetration to the lower levels.

Leaf litter is scattered as a thin layer over large areas of the ground surface. Rotten logs and branches are frequent, but occasionally these are supported in their original upright position by climbers linking the branches with those of neighbouring trees. Hillsides become completely awash during periods of extremely high rainfall, and the surface of every slope becomes a stream 2–3 in. deep. Rivers fail to cope with the increased load and overflow their banks. Leaves and debris, like twigs and branches, are moved considerable distances and thick layers of litter only accumulate around such obstructions as the large tree buttresses, so characteristic of this zone.

Extensive areas of forest are cleared around villages to provide garden land, this is then utilized for a few years before being abandoned, allowing recolonization by the vegetation and the fauna from the surrounding forest. However, the practice of keeping large herds of pigs has possibly had a more disastrous effect on the flora and fauna over large areas. The pigs must kill or eat many animals and young plants; while the continual rooting disturbs the litter and soil, permitting rapid desiccation and the creation of an unfavourable environment.

Lower montane forest

The difference between this zone and its neighbours is a matter of degree. Compared with the lowland rain forest the canopy is lower, more open and there is a significant increase in the numbers of herbaceous climbers and non-vascular epiphytes, usually mosses. Of particular interest is the abundance of vascular epiphytes on the trunks of large trees at a height of between 10 and 50 ft above the ground; these are usually large specimens of *Asplenium* spp.

Upper montane forest

In the Solomons a rise in altitude is usually synonymous with an increasing angle of slope, thus this upper forest zone exists on the steeper mountainside and dissected valleys, with extensions onto the flatter mountain tops. The open canopy of the trees is low, usually 10 and 50 ft and beneath this the understorey is either absent or poorly defined. There is a pronounced herb layer. An extreme development is exhibited by the "Elfin" type, where the canopy is even lower and the trees more stunted and twisted. The predominant leaf form is microphyll, in contrast to the larger, mesophyll, form of the lower altitudinal zones (Raunkiaer, 1934). Epiphytic mosses cover all suitable surfaces and no tree trunk or branch is visible under the thick enveloping layers that are up to 9 in. thick and usually saturated with water. The only vegetation to escape colonization by mosses are the long smooth stems of bamboo that grow in scattered clumps throughout this zone. The forest floor is extremely irregular with mosses extending from the tree trunks to form a continuous carpet covering all the protruding roots. Leaf litter collects in any slight depression.

STRATIFICATION OF THE MOLLUSCAN FAUNA IN LOWLAND RAIN FOREST

The majority of ecological studies on the terrestrial Mollusca of forests have been made in the northern temperate regions. There the fauna is confined to the lower strata, predominantly in the ground stratum with only a few species extending into the lower arboreal. This form of spatial arrangement contrasts sharply with that present in the rain forests of the Solomon Islands, where molluscs are found at all levels from the soil surface to the canopy. For the purpose of analysing this arrangement three vertical strata are recognized; these divisions cannot be equated with those proposed by botanists (Richards, 1952). Other schemes have been suggested by Allee (1926a,b) and Harrison (1962) for particular zoological investigations, but for the purpose of this study they are unnecessarily complex.

The vertical strata are: (1) ground stratum; (2) lower arboreal stratum; (3) upper arboreal stratum.

Ground stratum

Snails are found in the leaf litter, on the soil surface, under and on logs and small twigs, but are noticeably absent from the soil and rock surfaces. Waterlogged conditions and running water have probably contributed to their absence from these niches. The highest densities occur in the prevailing form of ground cover, the leaf litter, where the dominant species are prosobranchs, members of the subfamily Diplommatininae. These and other prosobranch taxa, e.g. *Nesopoma* and *Omphalotropis*, are confined to the litter. The second association found in this niche is that of species occurring in both the ground and higher strata, it includes pulmonate and prosobranch species that are predominantly larger in size than related groups confined to the litter. Characteristic taxa are *Eustomopsis eustoma* (Pfr.), *E. quercina* (Pfr.), *Placostylus founaki* (H. and J.), *P. guppyi* Smith and *Ouagapia* spp. and many species of Helicininae. *Placostylus founaki* exhibits an obvious behavioural distinction between different age groups; the adults occur in the ground and arboreal strata, while the young are found exclusively on the foliage in the lower arboreal stratum. It is impossible, at present, to provide data on the number of taxa in the two groups contributing to the litter fauna. Observations have demonstrated, however, the importance of those species that have the widest vertical distribution, for they contribute the greater biomass to the ground stratum. Individuals of pulmonate species characteristic of the upper arboreal stratum are found infrequently in the litter, their presence is possibly fortuitous and results from being accidentally dislodged from the foliage.

Variations in the density of snails can be correlated with local topography and corresponding changes in the distribution of litter. Dense aggregations of the smaller species are found in the deep litter that collects between the large tree buttresses, these probably result from the passive transportation of the animals with the leaf litter during periods of high rainfall. Animals deposited in unfavourable situations would incur a heavy mortality or fail to reproduce successfully. Distribution within the thick layers of litter depends on drainage, where it is poor and the site remains under water for long periods snails are rare or absent. When flooding does not occur the lower layers of deep litter are probably still always damp, as thin films of water persist on the leaf surfaces for long periods. This phenomenon accounts for the presence in this niche of such groups as polychaetes and ostracods (Greenslade, personal communication).

Comparisons between the litter on andesite and limestone in areas of high rainfall have failed to demonstrate any important differences in species composition between the two types. The variations in density can be correlated with the increased frequency of suitable niches in the well-drained soil and litter of the limestone regions. The short period spent collecting on ultra-basic rocks suggested that the molluscan fauna might be depleted when compared to that occurring on other geological formations, but it was impossible to evaluate the importance of changes in the structure of the vegetation and the ground stratum in relation to the underlying geology.

This description of the molluscan fauna in the ground stratum has emphasized the role of the leaf litter to the exclusion of other niches, particularly those associated with twigs and rotten logs. Unfortunately the numbers of individuals collected from these loci are small compared to those from the litter, but this is due undoubtedly to the lower density and not to variations in collecting techniques. Sufficient data are available to indicate that the faunal assemblage in or on logs is different from other niches in the ground stratum. Characteristic taxa are Pupininae, *Ouagapia* and certain species of Endodontacea, but when the water table is near the ground surface and the litter is saturated, the small prosobranch snails are found in greater abundance on small twigs or the underside of logs than in the leaf litter.

In disturbed localities, near villages, the patterns already described are completely modified, for during disturbance the litter becomes rapidly desiccated and molluscan species, not endemic to the Solomons, are frequently introduced. These include *Subulina octona* (Brug.) and *Lamellaxis gracilis* (Hutton). Members of the Veronicellidae also occur around human habitation and are probably also introduced; they are particularly common in the drier areas that are found in the rain shadow of the Guadalcanal plains.

Lower arboreal stratum

This stratum is inclusive of all the vegetation between the ground and the tree canopy, that is the herb layer and the forest understorey. Within this broad stratum the rich molluscan fauna, composed of both pulmonate and prosobranch taxa, is divisible into two categories: the first includes species found on branches, tree trunks and thick woody stems of creepers, while the other embraces all those living on the foliage. Characteristic taxa of the first group are *Trochomorpha*, *Pseudocyclotus*, *Placostylus* (other than those specified below) *Orpiella*, *Eustomopsis*, *Ouagapia*, *Syncera* and species of Helicininae. Typical taxa from the foliage are Papuininae, *Leptopoma*, *Nerita*, *Dendrotrochus*, *Cryptaegis*, *Crystallopsis*, *Helixarion*, *Placostylus miltocheilus* (Reeve) and *P. strangei* (Pfr.). While these taxa are considered typical of the respective categories within this stratum, they are not mutually exclusive and specimens do occur in niches other than the one in which they are most frequently found. The vertical distribution of many species included in the first group, i.e. snails found on branches, tree trunks and thick woody stems of creepers, extends down into the ground stratum. The importance of this behaviour has already been discussed.

The distribution of snails on the plant foliage is probably limited by structural features that are a function of leaf area and animal size. For example, in the undisturbed forests the majority of the larger molluscan species, e.g. *Placostylus miltocheilus* and members of the Papuininae, are found on broad flat leaves while the smaller species, e.g. *Dendrotrochus cineraceus* (H. and J.) and *Helixarion planospira* (Pfr.), occur on a much broader spectrum of leaf sizes, that includes smaller leaves. In disturbed areas of the forest there is typically an increase in broad leaved plants, e.g. gingers and bananas, these may grow in such profusion as to alter the local distribution of Mollusca. Undoubtedly collecting in these localities introduces a bias, as it is always easier to find snails on broad leaved plants where they are more conspicuous. An analysis of behavioural and structural modifications of molluscan taxa found on foliage is included in the Discussion (p. 338).

Upper arboreal stratum

The molluscan fauna of this stratum is closely associated with that of the lower arboreal, and many taxa are common to both levels. Distinguishing features are the complete absence of any prosobranch taxa and the impoverishment at the family level of the Pulmonata, only four families being represented by a few species that reach their peak of abundance in this stratum. As in the lower arboreal, the fauna is divisible into two categories, one inhabiting the foliage, the other the branches.

STRATIFICATION IN OTHER FOREST ZONES

Coastal

The complex mosaic of vegetation types and the high degree of human disturbance makes it extremely difficult to generalize on the distribution of land Mollusca within this zone. In close proximity to the sea, in the region of the upper strand line, the ground stratum is dominated by *Truncatella guerinii* Villa, *Pythia scarabeus* L. and *Melampus* spp. The latter are both members of the Ellobidae, a pulmonate family of world-wide distribution and marked preferences for habitats in or near the littoral zone. *Pythia scarabeus* is the only species that extends inland, distances up to 1 mile being recorded; its distribution is probably dependent on a high water table maintaining extremely moist conditions in the litter. Where the leaf litter is well drained the facies of the fauna approaches that of the lowland rain forest, but in the damper regions snails may exist only in low densities and then only on twigs and logs. The majority of species comprising the arboreal fauna of lowland forest are common to this zone, but typically the density in the upper arboreal stratum is low and in the lower arboreal a few species frequently become locally extremely abundant. This abundance is usually associated with disturbance and radical changes in the vegetational and faunal balance. *Partula* is the only arboreal taxon confined to the coastal zone.

Lower montane forest

In the Solomon Islands the lower montane forest is frequently characterized by a decrease in the species diversity of the molluscan fauna compared to the lowland form. Only on the massif surrounding Mt. Popomanasiu on Guadalcanal, however, is this zone clearly defined by distinctive species restricted to it. These are all arboreal and include *Placostylus sellersi* and undescribed species of *Helixarion* and *Trochomorpha*. In the ground stratum there is a reduction in the frequency of aggregations of small prosobranch snails. The vertical distribution of species belonging to the Diplommatininae extends into the damp but well-drained litter that collects 10–50 ft above the ground around the frond bases of *Asplenium* or in the leaf axils of *Pandanus*. In this niche the associated fauna consists of other elements from the ground stratum, e.g. earthworms.

Upper montane forest

As in the previous zone there is a decrease in the species diversity, but in this forest type it is of almost catastrophic proportions. The arboreal fauna is reduced to two or three species that exist at extremely low densities. On the forest floor the dense layers of leaf litter that collect in the hollows are saturated with water, while on the apex of any undulation the litter is thinner, scattered and comparatively well drained. The impoverished ground fauna is confined exclusively to these small accumulations of well-drained litter.

DISCUSSION

Observations made in the rain forests of the Solomon Islands have demonstrated geographical, altitudinal and vertical patterns in the distribution of the land Mollusca. There is evidence, however, to suggest these are not universal for all tropical forests. For example, in the

Solomons no differences could be demonstrated in the species composition between calcareous and non-calcareous localities. This contrasts sharply with the pattern found in Malaya where many small species of snails are restricted to calcareous habitats on limestone hills. Tweedie (1961) postulated that in Malaya, populations of the snails should exist in the intervening forest areas, as these would aid dispersal between the isolated limestone hills. There is no evidence to support these suggestions and it is probably unnecessary to imagine the existence of hypothetical populations or changes in the biology of the species to account for disjunct distributions. Many isolated islands in the central Pacific have been colonized by transoceanic dispersal of small species. Similar problems exist, on a reduced scale, with the occurrence of small prosobranch snails in the litter around arboreal epiphytes and in the ground stratum, but not the intervening regions. The variations in the distribution patterns of allied groups of snails, particularly species of Diplommatininae, in the Solomons and Malaya are probably associated with differences in the climate. In temperate regions increasing climatic severity for a particular species, as occurs at the periphery of its range, can result in compensating shifts in the calcium requirements of the animal (Boycott, 1934; Boycott and Oldham, 1936; Baker, 1958). Species, and groups of taxa, change from being indifferent calcicoles under favourable regimes to obligatory calcicoles when these are severe; the physiological basis for such a mechanism is not understood. In the tropics a factor that could influence such behaviour is the periodicity of the rainfall or rather the probability of prolonged dry periods. Droughts may be extremely infrequent and irregular in occurrence, but the facies of the fauna can still be altered by either the prevention of colonization or the extermination of existing populations of snails. The effect would be maintained if the rate of recolonization by the fauna was correspondingly slow. The meagre evidence from the Solomons suggests that in many regions prolonged dry periods must be extremely rare, if they occur at all (Fitzpatrick et al., 1966). This contrasts with the lowland regions of Malaya where the frequency and intensity of droughts is higher (Nieuwolt, 1965), confirming that an important factor influencing the distribution of terrestrial Mollusca in these two regions could be the frequency and intensity of the dry period. During the expedition the majority of collections were made in inland localities where this factor is unimportant, and the only collections made around Honiara, where there is a strongly seasonal climate, were unfortunately only in calcareous habitats. There is, however, a single record of two species of Diplommatina from a non-calcareous region in Malaya; this was from an isolated locality at an altitude of between

3000 and 5000 ft where the chance of prolonged periods of dry conditions would be very small. Together with similar records from Indonesia these data tend to confirm the importance of the rainfall pattern in determining the distribution of certain groups of land Mollusca in these areas. The influence of varying ecological conditions on the fragmentation of populations and speciation in the Diplommatininae awaits investigation.

Detailed information on the microclimates found in the rain forests of the Solomon Islands is not available. It is, however, possible to extrapolate from data collected in other tropical regions that the configuration of the daily cycles is probably similar, although there will be variations in the actual parameters. Within the lowland rain forest of the Solomons the daily patterns of temperature and humidity may be expected to follow those shown in Figs 2 and 3, which are based on data from Nigeria and Surinam (Evans, 1939; Schulz, 1960). They will probably conform closest to those produced during the wet season in these other regions, with short periods approaching those for the dry. An important feature demonstrated by these graphs is the climatic stability in the lower strata contrasting with the extreme variability in the canopy. Many measurements are available for light penetration and composition, but as there is no information on the responses of the Mollusca to these stimuli the data are meaningless. Modifications of the microclimate as found in montane regions can be associated with a decrease in temperature with increasing altitude and the insulation conferred by long periods of increased cloud cover. It is important, however, in these zones to distinguish periods with and without cloud cover (Grubb and Whitmore, 1966), for with complete cloud cover there is a reduction in the daily temperature range, a decreased saturation deficit in all strata and a decrease in the temperature difference between the canopy and the ground. In the absence of cloud the temperature pattern follows closely those recorded for the lowland rain forest but with lower mean maxima because of the rise in altitude. Therefore the net result is wider variation in the microclimates of the upper montane zone. It is difficult to generalize about the coastal zone, but the open structure and exposure to any air movements suggest that the microclimates of all strata are liable to wide fluctuations and high saturation deficits.

There is a trend of increasing shell size from the lower to the higher strata, i.e. from the ground to the upper arboreal. The range of shell height for the molluscan fauna of each stratum is given in Fig. 4, but there is no doubt this tendency would be further emphasized if data for the shell volume of each taxon were available. Size graduation can be

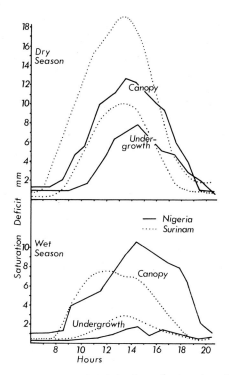

FIG. 2. Daily march of saturation deficit in the undergrowth and just above the main canopy of lowland rain forest in Nigeria and Surinam during wet and dry periods. (After Evans, 1939; Schulz, 1960.)

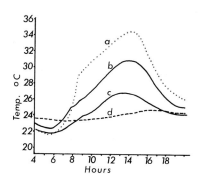

FIG. 3. Daily march of air and soil temperatures in different strata of lowland rain forest; data from Surinam. (After Schulz, 1960.)

Air temperatures: a, in clearing 1½ m above ground surface (daily patterns will be equivalent to those in the canopy); b, in rain forest 1½ m above ground surface; recorded during a hot period; c, as above, but recorded during a cool period. Soil temperature: d, at −5 cm in soil under rain forest.

correlated with variations in the structure of the habitat and the associated microclimates. The larger species equilibrate with the external temperature of the environment at a much slower rate than the smaller, and these dominate the higher strata where there is extreme variability of the microclimates while smaller taxa are dominant in the ground stratum, where the microclimates are comparatively stable. An exception to this generalization is provided by the group of molluscan species that move regularly between or exist in both the ground and higher strata. These exhibit a wide variety of shell sizes, but uniformity of size could not be expected in such a heterogeneous group occupying such an extensive range of niches.

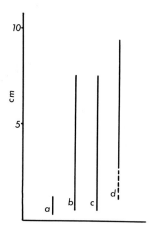

Fig. 4. Range of maximum and minimum shell sizes of molluscan taxa occurring in each of the three vertical strata of lowland rain forest. a, ground stratum; species confined to this stratum; b, as above, but species with greater vertical distribution, i.e. not confined to this stratum; c, lower arboreal stratum; d, upper arboreal stratum; lower dotted section indicates size range of *Helixarion* species.

The molluscan fauna of the exposed foliage in the upper arboreal stratum is dominated by two morphological types that also extend down into the lower arboreal stratum; the large shelled species of the Papuininae and the Bulimulidae, and a second group of the Helicarionidae and a single species of Papuininae that exhibit extensive expansion of the mantle over the surface of the shell. The latter are particularly interesting as they have been described as slug-like, yet there is a complete contrast in the niches occupied by these shelled forms and slugs. The slug form permits the animal to move without the impediment of a shell and to utilize small crevices or even progress through the soil. The slug-like form retains the shell and in the majority of tropical regions is arboreal. Species with this type of organization have a polyphyletic origin.

Hogben and Kirk (1944) described temperature regulation, in terrestrial Mollusca, by means of water evaporation. They concluded that a slug responds to changes in temperature and humidity in almost exactly the same manner as a wet bulb thermometer, yet the temperature of a snail, *Helix pomatia* L., withdrawn into the shell gradually approaches air temperature (Fig. 5). The rate to equilibrate will depend

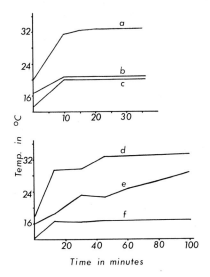

FIG. 5. Top temperature responses of a slug (*Arion ater* (L.)). a, air temperature; b, internal temperature of slug; c, wet bulb temperature. Below, temperature response of snail (*Helix pomatia* L.) withdrawn into shell. d, air temperature; e, internal temperature of snail; f, wet bulb temperature.

on the critical factor of the ratio, mass to surface area. The species of Papuininae can be equated with *Helix*, although the presence of a large shell with thick walls, shiny periostracum and the absence of dark colours will tend to prolong the period taken to equilibrate (an upper limit to shell size will possibly be determined by the surface area of the leaves in the canopy). The slug-like species of *Helixarion* are smaller than those of the Papuininae and could therefore be expected to equilibrate at a faster rate. It is suggested, however, that the slug-like species possess a mechanism that would interfere with this principle. When the mantle and body lobes are completely expanded and enveloping the shell the animal would react, like a slug, as a wet bulb thermometer and the internal temperature would be depressed compared to that of the environment. The lobes are under muscular control and can be partially or completely withdrawn into the shell, even when the animal is fully active; this behaviour could provide a regulatory mechanism. During periods of unfavourable environmental conditions or when the pressure of the hydrostatic skeleton is lowered the complete animal is able to retract into the shell and thereby reduce or eliminate evaporation from the body surface. Further, the presence of a shell reduces the problems of over-hydration that might occur in an extremely wet environment. It is possibly this hazard that proves to be a limiting factor to the shell-less slug. Machin (1964, 1966) has recently demonstrated in *Helix aspersa* Müller that inactive specimens are able to regulate the evaporative water loss from the mantle. This property is unique to mantle tissue, but there is unfortunately no information to determine whether such a mechanism operates in the expanded mantle lobes as found in *Helixarion* or is indeed universal in terrestrial Mollusca. If so, it could enhance the efficiency of temperature regulation in these slug-like forms. A limiting factor to temperature depression by water evaporation is the availability of water during a 24-h period. It is known that some species, e.g. *Limax tenellus*, can lose up to 80% of their body weight by loss of water. Replenishment is no problem in a rain forest, as typically it rains every day and at night the temperature depression is sufficient for dew to form on the foliage of the canopy. Mucus is extremely hydroscopic and would easily hydrate from conditions of 100% humidity (Hogben and Kirk, 1944; Machin, 1964).

There is a complete absence from the rain forests of the Solomon Islands of the slug form that has evolved in a number of diverse families of the Stylommatophora and is common to a wide range of habitats in temperate regions. The few records of slugs from this region all refer to species of the order Soleolifera; those of the family Veronicellidae are closely associated with human disturbance and are presumably introduced; it is impossible to comment on the only specimen of the Rathouisiidae that has been discovered. The absence of stylommatophoran slugs is possibly associated with the unfavourable climatic conditions of extremely high rainfall and humidity, as they are numerous in other tropical regions where there is a lower annual rainfall.

Behavioural and structural modifications or adaptations associated with an arboreal habit can be summarized as follows.

(1) *Oviviparity*. Many arboreal snails are oviparous, but at present there is no direct evidence regarding Solomon Island species. It is widespread in related taxa from adjacent archipelagos, e.g. members of the Papuininae and Partulidae.

(2) *Oviposition sites.* Various arboreal sites are utilized for oviposition; in *Placostylus miltocheilus* the detritus and soil that collects in the leaf axils of *Pandanus* spp. and around the large epiphytic *Asplenium* spp. are regularly employed. Other species use the same sites but it has been impossible to identify the eggs. The unusual site, on the surface of leaves, chosen by *Cryptaegis pilsbryi* is paralleled in a few other species not found in the Solomons. In *Cochlostyla* the eggs are laid between folded leaves (Sarasin and Sarasin, 1898) and *Amphidromus* lays large numbers of eggs in "nests" of folded leaves, up to 234 being recorded for *A. palaceus* var. *pura* (Laidlow and Solem, 1961). While the eggs of *Cochlostyla* and *Amphidromus* need protection from rapid desiccation, those of *Cryptaegis* are protected by an enveloping gelatinous coating (Figs 6 and 7). Undoubtedly the vertical distribution of certain arboreal Mollusca will be limited by their dependence on the ground stratum for suitable oviposition sites.

(3) *Radula modifications.* Pilsbry (1894) records the differences between the radular teeth of the arboreal Papuininae and those of closely related groups that occur in the ground stratum. The important features are the great breadth and bluntness of the radular cusps in the arboreal forms. These features are also combined with a herbivorous diet and probably this generalization cannot be applied to taxa other than the Camaenidae, for example, *Helixarion planospira* and *Dendrotrochus cineraceus* are both very common arboreal species that have completely different radular forms (Hoffmann, 1931; Rensch, 1934).

(4) *Mucus.* Field observations have suggested that many of the snails from the upper arboreal stratum have an extremely viscous mucus, when compared with those from lower strata and temperate regions. Probably it is this mucus that enables many arboreal snails to remain *in situ* during strong winds and even when a tree falls.

(5) *Colouration.* It is frequently presumed that snails with conspicuous polymorphic colour patterns are characteristic of arboreal habitats, but in the Solomons a high proportion of the species found in this habitat are cryptically coloured. Those occurring regularly on branches and tree trunks are dull coloured, usually brown, and not obviously polymorphic, e.g. members of the genera *Trochomorpha*, *Eustomopsis* and *Placostylus*. Cryptic species living on the foliage are green, and in the dim and greenish light of the forest they are extremely difficult to distinguish. This colouration is produced by pigments in the surface of the mantle, these are visible through the thin and translucent shell. The green is more pronounced in juvenile animals when the shell is

particularly thin; at maturity some of this initial transparency is lost, as a result of deposition of thick calcareous layers. Unfortunately the pigments responsible for the green coloration have not been identified. Species exhibiting this form of crypsis have a polyphyletic origin, three families of Pulmonata and one of Prosobranchia are represented. Species are as follows.

Pulmonata   Bulimulidae   *Placostylus miltocheilus* (Reeve)
                                 Placostylus sellersi (Cox)
              Camaenidae   *Crystallopsis*, many species.
              Helicarionidae *Helixarion* n.sp., in this species the mantle lobes
                               are green.
Prosobranchia Cyclophoridae *Leptopoma perlucidum* (Grateloup)

FIG. 6. *Cryptaegis pilsbryi*. Adult snail crawling on surface of leaf.

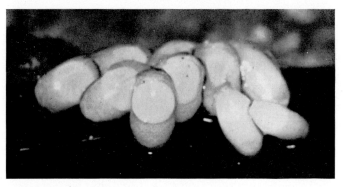

FIG. 7. Eggs of *C. pilsbryi* attached to surface of palm leaf. Each individual egg is enclosed in a gelatinous mass.

This type of pigmentation is not universal, for there are a few species of arboreal snails with a green shell, although none occurs in the Solomon Islands, an example is provided by *Papustyla pulcherimma* (Rensch) from Manus Island, near New Guinea. *Papuina vexillaris* (Pfr.) from the island of Kolombangara presents a novel form of obtaining a green colouration. The surface of a clean shell is pure white with a finely reticulate sculpture, but when the snail is found living in the rain forest the grooves of this sculpture contain a rich growth of green algae that also covers the intervening surfaces. The development of the algae must be assisted by the presence of the sculpture, the absence of a highly polished periostracum and the apparent failure of the animal to clean the surface of the shell. This feature is reminiscent of the weevils, found in New Guinea, with lichens growing on their carapace (Gressitt, 1966a,b).

Various associations of colour and arboreal habit have been suggested for species with conspicuous and polymorphic patterns. Clarke (1962) quotes Hartley's experimental demonstration in British snails "that climbing tends to reduce the risks from predators, at any rate from birds". Quoting further "at the same time it probably renders the snail more obvious though less attainable and may reduce the likelihood of effective crypsis. Such a habit may therefore favour apostatic polymorphism." In the Solomon Islands this category is represented by the polymorphic and brightly coloured species of Papuininae, for example *Solmopina boivini* (Petit). Many of these species live on the foliage in the upper arboreal stratum and occupy a conspicuous position. In the tropics it is probably this factor of position, together with the colour polymorphisms, that assists apostatic selection and not just the arboreal habit (for a discussion of apostatic selection, see Clarke, 1962). Not all species occurring on foliage are brightly coloured, for example *Dendrotrochus cineraceus* is brown and polymorphic for banding patterns, nor are all the dull coloured species polymorphic. Bright polymorphic colour forms can possibly be associated with occupation of a conspicuous niche and dull colouration, whether polymorphic or not, with niches in less conspicuous positions in, for example, the lower arboreal strata. At least a single species of the green cryptic forms is polymorphic; in *Crystallopsis aphrodite* (Pfr.) from San Cristobal the commonest morph is lime green, with blue and rarely blue-green individuals also being found.

An important difficulty in attempting to assess the selective advantage of colour patterns is the lack of information regarding predators. Shells showing signs of having been attacked by a bird or mammal were never found during the expedition. The only records of predation are the recovery of *Papuina ambrosia* from the stomachs of the introduced toad, *Bufo marinus*, and the unidentified specimens of snails recorded by Cain and Galbraith (1956) from the stomachs of four species of birds. There are, however, a number of potential predators amongst the avifauna and arboreal reptiles, amphibians and mammals, but there is an absence of information.

The arboreal species *Cryptaegis pilsbryi* Clapp warrants special attention; it is endemic to the island of San Cristobal possibly reaching the highest density at the eastern end. Originally there was doubt concerning the taxonomic position of this species, but Hoffmann (1931) on the basis of the internal anatomy included it with the Papuininae in the Camaenidae. The confusion is attributable to the presence of a non-retractile extension of the mantle completely enclosing the shell, such a morphological form is exceedingly rare in the Helicacea. Colour patterns vary, the young snails are brown, but there is a gradual development of pigmented cells with increasing age. In mature specimens the dominant colour of the mantle is green with a scattered distribution of darker granules and raised ridges of bright yellow tissue. Usually the dorsal surface of the body has fawn to orange bands interspaced with white unpigmented areas. *Cryptaegis* occupies an unique position amongst those taxa restricted to the arboreal habitat. In size it is equivalent to many species of Papuininae, being larger than *Helixarion*. The colouration of the animal at rest, that is retracted into the shell, is basically cryptic with only smaller areas of brighter pigmentation. Morphologically it is included with the slug-like species, as the shell is completely enclosed in a fused extension of the mantle. The complete adaptation of *Cryptaegis* to the arboreal environment is emphasized by the choice of oviposition site; eggs are encapsulated in individual gelatinous masses and attached to the surface of leaves in groups of five to sixteen.

The pattern of altitudinal distribution is one of decreasing species diversity with increasing altitude; undoubtedly climate is a major limiting factor. Changes in habitat structure will also influence distribution patterns. There is a decrease in suitable surfaces and a firm substrate over which snails can crawl in the montane zones. This is due to the gradual increase of epiphytic mosses that finally results in all tree trunks and branches being covered in thick layers of mosses saturated with water. The predominantly smaller leaf size in this zone will provide a further limiting factor on the size of snails existing on the foliage.

Information has been presented on the division of the molluscan fauna of the Solomon Islands into four elements having affinities with the surrounding regions. Within the islands these elements possess different distribution patterns. The Palaeo-Oriental and Southern faunas have a wide occurrence throughout all the major habitat divisions. There is little evidence of any isolated populations of molluscs on the higher mountains that are not derivable from taxa existing at lower levels and belonging to one of these faunal groups. The distribution of the Pacific Ocean element provides a marked contrast, for example, the genus *Partula* is restricted to the coastal zone and small isolated islands where it is found only in the lower arboreal stratum. Populations exhibit little or no polymorphic variation. The widest degree of species diversity and polymorphic variation of *Partula* is found in the Society Islands, here species have a wide habitat distribution and extend inland to the tops of mountains at altitudes up to 6000 ft (Crampton, 1916). This evidence suggests that the dispersal centre for at least this member of the Pacific Ocean element is in Polynesia, and the Solomon Island species are comparatively recent additions to the fauna of the archipelago. If *Partula* had been established in the Solomons for a longer period a greater range of habitat and species diversity would be expected, similar to that exhibited by the other faunal elements.

An examination of charts illustrating the geographical juxtaposition and climate of land masses and islands in South East Asia and the Pacific demonstrates many discontinuities. The importance of these, relative to the geological history of the region, in determining the distribution of the biota is at present unknown. This preliminary study of the terrestrial Mollusca has examined an element of the biota existing in regions with an extreme climatic regime and with a comparatively recent geological history. Investigations on islands exhibiting a wider variation of climate and location are required before any conclusions on the influence of these features on the present geographical distribution of Mollusca can be considered.

ACKNOWLEDGEMENTS

I am indebted to the Trustees of the British Museum (Natural History) and The Royal Society for enabling me to participate in the expedition. To fellow members of the expedition and to Dr. P. J. M. Greenslade, in particular, I am grateful for the many hours spent discussing biogeographic problems relating to the Solomon Islands and for assistance with field collecting.

REFERENCES

Papers marked * are not cited in the text but are relevant to the subject and were consulted during the preparation of this article.

Allee, W. C. (1926a). Measurement of environmental factors in the tropical rain forest of Panama. *Ecology* **7**: 273–302.
Allee, W. C. (1926b). Distribution of animals in a tropical rain forest with relation to environmental factors. *Ecology* **7**: 445–468.
Baker, H. B. (1958). Land snail dispersal. *Nautilus* **71**: 141–148.
*Bentham Jutting, W. S. S. van (1948). On the present state of the malacological research in the Malay archipelago. *Chronica Nat.* **104**: 129–138.
Boycott, A. E. (1934). The habitats of land mollusca in Britain. *J. Ecol.* **22**: 1–38.
Boycott, A. E. and Oldham, C. (1936). A conchological reconnaissance of the limestone in West Sutherland and Ross. *Scott. nat.* **1936**: 47–52, 65–71.
Cain, A. J. and Galbraith, I. C. J. (1956). Field notes on birds of the eastern Solomon Islands. *Ibis* **98**: 100–134, 262–295.
*Clapp, W. F. (1923). Some mollusca from the Solomon Islands. *Bull. Mus. comp. Zool. Harv.* **65**: 351–418.
Clarke, B. (1962). Balanced polymorphism and the diversity of sympatric species. *Publs Syst. Ass.* No. 4: 47–70.
*Clench, W. J. (1941). The land mollusca of the Solomon Islands. (Succineidae, Bulimulidae and Partulidae). *Am. Mus. Novit.* No. 1129: 1–21.
*Clench, W. J. (1958). The land and freshwater mollusca of Rennell Island, Solomon Islands. No. 27. In *The natural history of Rennell Island, British Solomon Islands* **2**: 155–202. Wolff, T. (ed.) Copenhagen: Danish Science Press.
Coleman, P. J. (1963). Tertiary foraminifera of the British Solomon Islands, southwest Pacific. *Micropaleontology* **9**: 1–38.
Coleman, P. J. (1966). The Solomon Islands as an island arc. *Nature, Lond.* **211**: 1249–1251.
*Coleman, P. J., Grover, J. C., Stanton, R. L. and Thompson, R. B. (1965). A first geological map of the British Solomon Islands. In *Br. Solomon Isl. Geol. Rec.* **2** (1959–62). Honiara, Guadalcanal, British Solomon Islands Protectorate.
Crampton, H. E. (1916). Studies on the variation, distribution and evolution of the genus *Partula*. The species inhabiting Tahiti. *Publs Carnegie Instn* No. 228: 1–311.
*Dainton, B. H. (1954). The activity of slugs. 1. The induction of activity by changing temperatures. *J. exp. Biol.* **31**: 165–187.
Evans, G. C. (1939). Ecological studies on the rain forest of Southern Nigeria. II. The atmospheric environmental conditions. *J. Ecol.* **27**: 436–482.
Fitzpatrick, E. A., Hart, D. and Brookfield, H. C. (1966). Rainfall seasonality in the tropical southwest Pacific. *Erdkunde* **20**: 181–194.
Gressitt, J. L. (1954). Insects of Micronesia. Introduction. *Insects Micronesia* **1**: 1–257.
Gressitt, J. L. (1961). Problems in the zoogeography of Pacific and Antarctic insects. *Pacif. Insects* **2**: 1–94.
Gressitt, J. L. (1966a). Epizoic symbiosis: the Papuan weevil genus *Gymnopholus* (Leptopiinae) symbiotic with cryptogamic plants, oribatid mites, rotifers and nematodes. *Pacif. Insects* **8**: 221–280.

Gressitt, J. L. (1966b). Epizoic symbiosis: cryptogamic plants growing on various weevils and on a colydid beetle in New Guinea. *Pacif. Insects* **8**: 294–297.

Grover, J. C., Thompson, R. B., Coleman, P. J., Stanton, R. L., Bell, J. D. *et al.* (1965). *Br. Solomon Isl. Geol. Rec.* **2**—(1959–62). Honiara, Guadalcanal, British Solomon Islands Protectorate.

Grubb, P. J., Lloyd, J. R., Pennington, I. D. and Whitmore, T. C. (1963). A comparison of montane and lowland rain forest in Ecuador. I. The forest structure, physiognomy and floristics. *J. Ecol.* **51**: 567–601.

Grubb, P. J. and Whitmore, T. C. (1966). A comparison of montane and lowland rain forest in Ecuador. II. The climate and its effects on the distribution and physiognomy of the forest. *J. Ecol.* **54**: 303–334.

Guppy, H. B. (1887a). *The Solomon Islands and their natives.* London: Swan Sonnenschein.

Guppy, H. B. (1887b). *The Solomon Islands: their geology, general features, and suitability for colonisation.* London: Swan Sonnenschein.

Harrison, J. L. (1962). The distribution of feeding habits among animals in a tropical rain forest. *J. Anim. Ecol.* **31**: 53–63.

Harry, H. W. (1966). Land snails of Ulithi Atoll, Caroline Islands: A study of snails accidentally distributed by man. *Pacif. Sci.* **20**: 212–223.

Hedley, C. (1899). A zoogeographic scheme for the mid Pacific. *Proc. Linn. Soc. N.S.W.* **1899**: 391–417.

Hoffmann, H. (1931). Uber zwei Halbnacktschnecken von den Salomoninseln. *Z. wiss. Zool.* **138**: 99–136.

Hogben, L. and Kirk, R. L. (1944). Studies on temperature regulation. 1. Pulmonates and oligochaeta. *Proc. R. Soc.* **B132**: 239–252.

*Ladd, H. S. (1960). Origin of the Pacific Island molluscan fauna. *Am. J. Sci.* **258A** (Bradley Vol.): 137–150.

Laidlow, F. F. and Solem, A. (1961). The land snail genus *Amphidromos.* A synoptic catalogue. *Fieldiana, Zool.* **41**: 505–677.

Lever, R. J. A. W. (1937a). The physical environment, fauna and agriculture of the British Solomon Islands. *Trop. Agr.* **14**: 281–285, 307–312.

Lever, R. J. A. W. (1937b). Economic insects and biological control in the British Solomon Islands. *Bull. ent. Res.* **28**: 325–331.

Machin, J. (1964). The evaporation of water from *Helix aspersa.* 1. The nature of the evaporating surface. *J. exp. Biol.* **41**: 759–769.

Machin, J. (1966). The evaporation of water from *Helix aspersa.* IV. Loss from the mantle of the inactive snail. *J. exp. Biol.* **45**: 269–278.

McMichael, D. F. and Iredale, T. (1959). The land and freshwater mollusca in Australia. *Monographiae biol.* **8**: 224–245.

*Mayr, E. (1944). Wallace's Line in the light of recent zoogeography studies. *Q. Rev. Biol.* **19**: 1–14.

Nieuwolt, S. (1965). Evaporation and water balance in Malaya. *J. trop. Geogr.* **20**: 34–53.

Pilsbry, H. A. (1894). Guide to the studies of *Helices. Man. Conch.* (2) **9**: 1–366.

Pilsbry, H. A. (1900a). On the zoological position of *Partula* and *Achatine la. Proc. Acad. nat. Sci. Philad.* **1900**: 561–567.

Pilsbry, H. A. (1900b). The genesis of mid-Pacific faunas. *Proc. Acad. nat. Sci. Philad.* **1900**: 568–581.

Raunkiaer, C. (1934). *The life forms of plants.* Oxford: Univ. Press.

*Rees, W. J. (1964). A review of breathing devices in land operculate snails. *Proc. malac. Soc. Lond.* **36**: 55–66.

Rensch, B. (1936). *Die Geschichte des Sundabogens*. Berlin.

Rensch, I. (1934). Studies on *Papuina* and *Dendrotrochus*, pulmonate mollusks from the Solomon Islands. *Am. Mus. Novit.* No. 763: 1–26.

*Rensch, I. and Rensch, B. (1935). Systematische und tiergeographische Studien über die Landschnecken der Salomonen. I. *Revue suisse Zool.* **42**: 51–86.

*Rensch, I. and Rensch, B. (1936). Systematische und tiergeographische Studien über die Landschnecken der Salomonen. II. *Revue suisse Zool.* **43**: 653–695.

Richards, J. R., Cooper, J. A., Webb, A. W. and Coleman, P. J. (1966). Potassium-argon measurements of the age of the basal schists in the British Solomon Islands. *Nature, Lond.* **211**: 1251–1252.

Richards, P. W. (1952). *The tropical rain forest*. Cambridge: University Press.

Sarasin, P. and Sarasin, F. (1898). *Die süsswasser-mollusken von Celebes*. Wiesbaden: C. W. Kreidel.

Solem, A. (1958). Biogeography of the New Hebrides. *Nature, Lond.* **181**: 1253–1255.

Solem, A. (1959). Systematics and zoogeography of the land and fresh-water Mollusca of the New Hebrides. *Fieldiana, Zool.* **43**: 1–359.

Schulz, J. P. (1960). Ecological studies on the rain forest of northern Suriname. *Verh. K. ned. Akad. Wet.* **53**: 1–267.

Thorne, R. F. (1963). Biotic distribution patterns in the tropical Pacific. In *Pacific Basin biogeography*: 311–350. Gressit, J. L. (ed.) Hawaii: Bishops Museum Press.

*Thornthwaite, C. W. (1952). The water balance in tropical climates. *Bull. Am. met. Soc.* **32**: 166–173.

Tweedie, M. W. F. (1961). On certain molluscs of the Malayan limestone hills. *Bull. Raffles Mus.* No. 26: 49–65.

Usinger, R. L. and La Rivers, I. (1953). The insect life of Arno. *Atoll Res. Bull.* **15**: 1–28.

*Wallace, A. R. (1876). *The geographical distribution of animals*. 2 vols. London: Macmillan.

*Wallace, A. R. (1880). *Island life*. London: Macmillan.

Wiens, H. J. (1962). *Atoll environment and ecology*. New Haven: Yale Univ. Press.

ADDENDUM

While this paper was in press the following abstract was published; it provides further information on the molluscan faunas of the limestone hills in Malaya.

Purchon, R. D. (1968). The distribution of terrestrial species of prosobranch and pulmonate snails on the limestone hills of Malaya. *Symposium on Mollusca*. Issued by *The Marine Biological Association of India*.

AUTHOR INDEX

SYSTEMATIC INDEX

SUBJECT INDEX

A

A-band area, 200, 201, 207, 208
Aberrant forms, 136, 148
Abundance, 293–298, 301, 302, 304, 306, 307, 309–311, 314
Accessory sex organs, 214, 225, 232
Acetic acid, 22
Acetylcholine, 8, 13, 15, 33, 34, 36, 37, 38, 41, 43, 44, 46, 48–54, 56–66, 68–73
Acetylthiocholine, 15
Acini, 213–216, 218, 222, 223, 225, 226, 230, 233
Actinomyosin, 207
Action potential, 8, 20, 69
Acyl group, 36
Adaptability, 136
Adaptation, 122, 135, 139, 142, 144, 147, 152, 168, 184, 191, 258, 280, 297, 319, 338, 342
Adaptive radiation, 109, 128–130
Adduction, 167, 169, 170–174, 177, 178, 180, 183, 184
Adductor, 109, 113, 116–121, 123, 130, 173–175, 180–184, 193
  anterior, 168
  muscle scar, 110, 129
  posterior, 168
  smooth, 193, 194, 196, 197
  striated, 193, 194, 207–209
  unstriated, 207
Adhesiveness, 144, 147
Adrenaline, 22
Adrenergic ending, 15
Adsorption, 96, 98–100
Adult, 111, 113, 144, 152, 154, 156, 187, 213, 214, 218–220, 223, 233, 329, 340
Aeolid, 144
Age, 219, 223, 329
Agonists, 33–36, 39, 71
  cholinergic, 71
  muscarinic, 33, 44, 54, 70, 72
  nicotinic, 33, 54, 70, 72
Agriculture, 293, 303, 306, 310, 312, 314, 315

Algae, 109, 122, 128–131, 341
Aliesterase, 71
Alimentary system, 117, 125, 127, 130
Allantoicase, 190
Allantoinase, 190
Allerød Interstadial, 275
Alluvium, 326
Alpine, 275
Altitude, 299, 319, 325, 328, 332–334, 342, 343
America, 109, 111, 115, 151, 259
Amines, 1, 8, 11, 14, 15, 29
Amoebocytes, 83–85
Amoeboid movement, 215
Amphibia, 204
Anaesthetization, 194
Ancestors, 110, 113, 127
Andesite, 329
Andesite line, 322
Annual
  cycle, 213
  ring, 106
Antagonists, 33–40, 63, 69, 71
  acetylcholine, 70
  cholinergic, 37, 69, 71
  competitive, 69, 70
  dopamine, 58
  histamine, 58
  mecholine, 44
  nicotinic, 44, 70
  vertebrate, 72
Anthropophiles, 287
Anthropophobes, 276, 278, 285, 287, 300
Anticholinesterase, 54
Anus, 116, 117, 121, 124–126, 142, 152, 158
Aorta, 2–4, 158, 160, 161
Aragonite, 75, 81, 90
Arboreal
  sites, 319, 339
  strata, 319, 329
Archaeogastropods, 190
Arctic circle, 281, 284, 286
Arecoline, 33, 35, 44, 49, 51, 70, 72
Argentine land snail, 20

357